# Tensor Categories and Hopf Algebras

CONTEMPORARY MATHEMATICS

728

# Tensor Categories and Hopf Algebras

Scientific Session of the
Mathematical Congress of the Americas
Hopf Algebras and Tensor Categories
July 27–28, 2017
Montreal, Canada

Nicolás Andruskiewitsch
Dmitri Nikshych
Editors

AMERICAN MATHEMATICAL SOCIETY

2010 *Mathematics Subject Classification*. Primary 16D90, 16E65, 16T05, 16T20, 16W50, 17B37, 17D05, 18D10, 20G05.

---

**Library of Congress Cataloging-in-Publication Data**

Names: Mathematical Congress of the Americas on Hopf Algebras and Tensor Categories (2017 : Montreal, Canada.) Scientific session, author. | Andruskiewitsch, Nicolás, 1958- editor. | Nikshych, Dmitri, 1973- editor.

Title: Tensor categories and Hopf algebras : Scientific session of the Mathematical Congress of the Americas on Hopf Algebras and Tensor Categories, July 27–28, 2017, Montreal, Canada / Nicolás Andruskiewitsch, Dmitri Nikshych, editors.

Description: Providence, Rhode Island : American Mathematical Society, [2019] | Series: Contemporary mathematics ; Volume 728 | Includes bibliographical references.

Identifiers: LCCN 2018042384 | ISBN 9781470443214 (alk. paper)

Subjects: LCSH: Hopf algebras – Congresses. | Tensor fields–Congresses.

Classification: LCC QA613.8 .M38 2019 | DDC 512/.55–dc23

LC record available at https://lccn.loc.gov/2018042384

Contemporary Mathematics ISSN: 0271-4132 (print); ISSN: 1098-3627 (online)

DOI: https://doi.org/10.1090/conm/728

---





∞ The paper used in this book is acid-free and falls within the guidelines
established to ensure permanence and durability.
Visit the AMS home page at `https://www.ams.org/`

10 9 8 7 6 5 4 3 2 1    24 23 22 21 20 19

# Contents


Preface vii

On finite GK-dimensional Nichols algebras of diagonal type
NICOLÁS ANDRUSKIEWITSCH, IVÁN ANGIONO, and ISTVÁN HECKENBERGER 1

On nonassociative graded-simple algebras over the field of real numbers
YURI BAHTURIN and MIKHAIL KOCHETOV 25

Nonsemisimple Hopf algebras of dimension $8p$ with the Chevalley property
MARGARET BEATTIE, GASTÓN A GARCÍA, SIU-HUNG NG, and JOLIE ROAT 49

Pointed braided tensor categories
COSTEL-GABRIEL BONTEA and DMITRI NIKSHYCH 67

Eigenvalues of the squared antipode in finite dimensional weak Hopf algebras (with an appendix by Gregor Schaumann)
PAVEL ETINGOF 95

Cohen-Macaulay invariant subalgebras of Hopf dense Galois extensions
JI-WEI HE and YINHUO ZHANG 119

Bialgebra coverings and transfer of structure
AARON LAUVE and MITJA MASTNAK 137

Classifying braidings on fusion categories
DMITRI NIKSHYCH 155

Remarks on global dimensions of fusion categories (with an appendix by Pavel Etingof and Victor Ostrik)
VICTOR OSTRIK 169

On Hopf algebras with triangular decomposition
CRISTIAN VAY 181

# Preface

The scientific session "Hopf algebras and tensor categories" was held at the McGill University in Montreal (Canada) on July 27-28, 2017 as a part of the Mathematical Congress of the Americas. The session consisted of 15 talks by researchers from six different countries. The list of speakers and abstracts of the talks can be found online at `https://mca2017.org/prog/session/hat`.

Many important advances in the theory of tensor categories are motivated by the theory of Hopf algebras and, conversely, categorical methods play an increasingly important role in Hopf algebra studies. The recent achievements in the classification of pointed Hopf algebras and fusion categories further stimulate a cross-fertilization between these areas. The main idea of the session was to bring together researchers in Hopf algebras, quantum groups, and tensor categories to exchange the most current ideas and techniques. The organizers were Nicolás Andruskiewitsch (Universidad Nacional de Córdoba, Argentina) and Dmitri Nikshych (University of New Hampshire, USA). We would like to thank all participants and speakers at the session for making it such an interesting and fruitful event.

The present volume contains ten articles contributed by the invited speakers of the session. These articles cover a wide variety of topics in the theory of tensor categories and Hopf algebras and contain a number of original results:

- Andruskiewitsch, Angiono, and Heckenberger study the finiteness of the Gelfand-Kirillov dimension of Nichols algebras of diagonal type.
- Bahturin and Kochetov develop the theory of non-associative graded simple algebras over real numbers.
- Beattie, Garcia, Ng, and Roat advance the classification of non-semisimple Hopf algebras with the Chevalley property.
- Bontea and Nikshych classify finite pointed braided tensor categories.
- Etingof presents a categorical interpretation of eigenvalues of the squared antipode in a weak Hopf algebras.
- He and Zhang prove the Cohen-Macaulay property of invariant subalgebras of Hopf dense Galois extensions.
- Lauve and Mastnak introduce and study the bicategory of bialgebra coverings.
- Nikshych classifies braidings on fusion categories in geometric terms.
- Ostrik establishes a non-trivial lower bound for the global dimension of a spherical fusion category.
- Vay gives a survey of Hopf algebras with a triangular decomposition.

We thank the editorial staff of *Contemporary Mathematics* and especially Christine Thivierge for the help with the preparation of this volume and the American

Mathematical Society for supporting this project. We are also grateful to the individual contributors and referees who invested quite a lot of their valuable time in writing and refereeing the articles.

Contemporary Mathematics
Volume **728**, 2019
https://doi.org/10.1090/conm/728/14653

# On finite GK-dimensional Nichols algebras of diagonal type

Nicolás Andruskiewitsch, Iván Angiono, and István Heckenberger

ABSTRACT. We conjectured in a previous paper that a Nichols algebra of diagonal type with finite Gelfand-Kirillov dimension has finite (generalized) root system. We prove the conjecture assuming that the rank is 2. We also show that a Nichols algebra of affine Cartan type has infinite Gelfand-Kirillov dimension.

## 1. Introduction

In this paper we contribute to the classification of Hopf algebras with finite Gelfand-Kirillov dimension, GKdim for short. Specifically we propose:

CONJECTURE 1.1. [**AAH**, Conjecture 1.5] *If $V$ is a braided vector space of diagonal type and dimension $\theta \in \mathbb{N}$ such that the* GKdim *of its Nichols algebra $\mathcal{B}(V)$ is finite, then its (generalized) root system is finite.*

Since the classification of the Nichols algebras of diagonal type with finite generalized root system is known [**H2**], a positive answer to the Conjecture 1.1 would imply the classification of the Nichols algebras of diagonal type with finite GKdim. Indeed, the converse in the Conjecture is clearly true. The defining relations of these Nichols algebras are also known [**An1**]. The GKdim of these Nichols algebras can be computed (2.6). Our main result provides partial answers to this Conjecture:

THEOREM 1.2. *Conjecture* 1.1 *holds in the following cases:*
(a) *$V$ is of affine Cartan type.*
(b) $\theta = 2$.

Theorem 1.2 (a) for generic parameter agrees with the well-known fact that $\operatorname{GKdim} U(\mathfrak{g}) = \dim \mathfrak{g}$ for any Lie algebra $\mathfrak{g}$. The reader should not confuse the growth of a graded Lie algebra $\mathfrak{g}$ with the growth of its enveloping algebra $U(\mathfrak{g})$, see [**Ma**].

We collect some necessary definitions and concepts in Section 2; Section 3 is devoted to general results that might be of interest elsewhere. Part (a) is proved in Proposition 3.1. Part (b), which is Theorem 4.1, is proved in §4.2, after some preparatory Lemmas in §4.1.

---

2010 *Mathematics Subject Classification.* Primary 16W30.
The work of N. A. and I. A. was partially supported by CONICET, Secyt (UNC), the MathAmSud project GR2HOPF. The work of I. A. was partially supported by ANPCyT (Foncyt). The work of N. A., respectively I. A., was partially done during a visit to the University of Marburg, respectively the MPI (Bonn), both visits supported by the Alexander von Humboldt Foundation.

## 2. Preliminaries

**2.1. Conventions.** Let $\Bbbk$ be an algebraically closed field of characteristic zero. All the vector spaces, algebras and tensor products are over $\Bbbk$. Let $\mathbb{N} = \{1, 2, \dots\}$, $\mathbb{N}_0 = \mathbb{N} \cup 0$. Given $\theta \in \mathbb{N}$, we set $\mathbb{I}_\theta = \{1, 2, \dots, \theta\}$, or simply $\mathbb{I}$ if $\theta$ is clear from the context. In the polynomial ring $\mathbb{Z}[\mathsf{q}]$, we denote

$$(n)_{\mathsf{q}} = \sum_{j=0}^{n-1} \mathsf{q}^j, \qquad (n)^!_{\mathsf{q}} = \prod_{j=1}^{n} (j)_{\mathsf{q}}, \qquad n \in \mathbb{N}_0.$$

If $q \in \Bbbk$, then $(n)_q$, $(n)^!_q$ are the evaluations at $q$.

We denote by $\widehat{\Gamma}$ the group of multiplicative characters (one-dimensional representations) of a group $\Gamma$. Let $\mathbb{G}_N$ be the group of roots of unity of order $N$, $\mathbb{G}'_N$ the subset of primitive roots of order $N$, and $\mathbb{G}_\infty = \bigcup_{N \in \mathbb{N}} \mathbb{G}_N$.

Let $H$ be a Hopf algebra (always with bijective antipode). A braided Hopf algebra means a Hopf algebra in the category ${}^H_H\mathcal{YD}$ of Yetter-Drinfeld modules over $H$. If $R$ is a Hopf algebra in ${}^H_H\mathcal{YD}$, then $R\#H$ is the bosonization of $R$ by $H$. Let ad be the adjoint action of $R\#H$ and $\operatorname{ad}_c$ the braided adjoint action of $R$. Then $\operatorname{ad}_c x \otimes \operatorname{id} = \operatorname{ad}(x\#1)$ for $x \in R$. If $x \in R$ is primitive, then $\operatorname{ad}_c x(y) = xy - \text{multiplication} \circ c(x \otimes y)$ for all $y \in R$.

If $(V, c)$ is a braided vector space, then $\mathcal{B}(V) = T(V)/\mathcal{J}(V)$ is the Nichols algebra of $V$, see [**AS1, A, AA2**] for surveys on this notion.

A *braided vector space of diagonal type* is a pair $(V, c)$, where $V$ is a vector space of dimension $\theta$ with a basis $(x_i)_{i \in \mathbb{I}_\theta}$, and the braiding $c \in GL(V \otimes V)$ is given by $c(x_i \otimes x_j) = q_{ij} x_j \otimes x_i$ for all $i, j \in \mathbb{I}$; here $\mathbf{q} = (q_{ij})_{i,j \in \mathbb{I}_\theta} \in (\Bbbk^\times)^{\theta \times \theta}$, $q_{ii} \neq 1$. Let

$$\widetilde{q}_{ij} := q_{ij} q_{ji}, \qquad i \neq j \in \mathbb{I}_\theta.$$

The generalized Dynkin diagram of $\mathbf{q}$ is a graph with set of points $\mathbb{I}_\theta$ with the vertex $i$ decorated with $q_{ii}$; and for $i \neq j \in \mathbb{I}_\theta$, no edge between $i$ and $j$ when $\widetilde{q}_{ij} = 1$, otherwise there is an edge decorated with $\widetilde{q}_{ij}$.

**2.2. The Gelfand-Kirillov dimension.** A comprehensive exposition is [**KL**]. For further use, we recall statements from [**AAH**, §2.3.2] inspired by [**R**, Lemma 19].

LEMMA 2.1. *Let $\mathcal{B} = \oplus_{n \geq 0} \mathcal{B}^n$ be a finitely generated graded algebra with $\mathcal{B}^0 = \Bbbk$. Let $(y_k)_{k \geq 0}$ be a family of homogeneous elements of $\mathcal{B}$ such that*

$$(y_{i_1} \dots y_{i_l} : i_j \in \mathbb{N}, i_1 < \dots < i_l) \tag{2.1}$$

*is a family of linearly independent elements. If there exist $m, p \in \mathbb{N}$ such that $\deg y_i \leq mi + p$, for all $i \in \mathbb{N}$, then* $\operatorname{GKdim} \mathcal{B} = \infty$. □

The following Lemma is due to Rosso [**R**, Lemma 14, Corollary 18]. Let $(U, c)$ be a braided vector space of diagonal type, with respect to a basis $x_1, x_2$ and a matrix $(q_{ij})_{i,j \in \mathbb{I}_2}$. We set

$$\mu_k = \prod_{i=0}^{k-1} (1 - q_{11}^i \widetilde{q}_{12}), \qquad y_k = (\operatorname{ad}_c x_1)^k x_2. \tag{2.2}$$

LEMMA 2.2. (a) *If* $k \in \mathbb{N}$, *then* $y_k = 0$ *iff* $\mu_k x_1^k = 0$, *iff* $(k)_{q_{11}}!\mu_k = 0$.
(b) *If* $y_k \neq 0$ *for all* $k \in \mathbb{N}$, *then the set* (2.1) *is linearly independent.*
(c) *If* $(k)_{q_{11}}!\mu_k \neq 0$ *for every* $k \in \mathbb{N}$, *then* $\operatorname{GKdim} \mathcal{B}(U) = \infty$.
(d) *If* $q_{11} = 1$ *and* $\widetilde{q}_{12} \neq 1$, *then* $\operatorname{GKdim} \mathcal{B}(U) = \infty$.

**2.3. Nichols algebras of diagonal type.** We fix a braided vector space of diagonal type $(V, c)$ with notation as in §2.1. *We assume that the generalized Dynkin diagram is connected.*

Let $(\alpha_i)_{i\in\mathbb{I}}$ be the canonical basis of $\mathbb{Z}^\theta$. Let $\mathbf{q}$ be the bicharacter on $\mathbb{Z}^\theta$ such that $\mathbf{q}(\alpha_i, \alpha_j) = q_{ij}$ for all $i, j \in \mathbb{I}$; we set $q_{\alpha\beta} = \mathbf{q}(\alpha, \beta)$ for $\alpha, \beta \in \mathbb{Z}^\theta$.

Then $T(V)$ and $\mathcal{B}(V)$ are $\mathbb{Z}^\theta$-graded with $\deg x_i = \alpha_i$ for all $i \in \mathbb{I}$. For each $\alpha \in \mathbb{Z}^\theta$, $\mathcal{B}^\alpha(V)$ denotes the homogeneous component of degree $\alpha$.

Let $\Gamma = \mathbb{Z}^\theta$. We realize $V$ in ${}^{\Bbbk\Gamma}_{\Bbbk\Gamma}\mathcal{YD}$ by choosing the family $(\chi_i)_{i\in\mathbb{I}}$ in $\widehat{\Gamma}$ such that $\chi_j(\alpha_i) = q_{ij}$ for all $i, j \in \mathbb{I}$. Then $\mathcal{B}(V)$ becomes an $\mathbb{Z}^\theta$-graded object in ${}^{\Bbbk\Gamma}_{\Bbbk\Gamma}\mathcal{YD}$. There are skew-derivations $\partial_i$, $i \in \mathbb{I}$ of $\mathcal{B}(V)$, such that $\partial_i(x_j) = \delta_{ij}$ and

$$\partial_i(xy) = x\partial_i(y) + \partial_i(x)(\alpha_i \cdot y), \qquad x, y \in \mathcal{B}(V).$$

Let $\mu_k$, $y_k = (\operatorname{ad}_c x_1)^k x_2$ as in (2.2). We notice that

$$\partial_1(y_k) = 0, \qquad \partial_2(y_k) = \mu_k x_1^k, \qquad \text{for all } k \in \mathbb{N}_0. \tag{2.3}$$

Also, it is well-known, and easy to check by a recursive argument, that

$$\Delta(y_k) = y_k \otimes 1 + \sum_{i=0}^{k} \binom{k}{i}_{q_{11}} \frac{\mu_k}{\mu_i} x_1^{k-i} \otimes y_i. \tag{2.4}$$

By [**Kh**], there is a totally ordered subset $L \subset \mathcal{B}(V)$ consisting of $\mathbb{Z}^\theta$-homogeneous elements such that

$$\{\ell_1^{m_1} \cdots \ell_k^{m_k} \mid k \in \mathbb{N}_0, \ell_1 > \cdots > \ell_k \in L, 0 < m_i < N_{\ell_i} \text{ for all } i \in \mathbb{I}_k\} \tag{2.5}$$

is a linear basis of $\mathcal{B}(V)$ (a so called restricted PBW basis); here

$$N_\ell = \min\{n \in \mathbb{N} : (n)_{q_{\deg \ell, \deg \ell}} = 0\} \in \mathbb{N} \cup \infty$$

is called the height of $\ell$. Notice that 1 belongs to (2.5), by taking $k = 0$.

LEMMA 2.3. *If* $\mathcal{B}(V)$ *has a restricted homogeneous PBW basis with infinitely many PBW generators of infinite height, then* $\operatorname{GKdim} \mathcal{B}(V) = \infty$.

PROOF. By assumption there exists $L$ as above and $I \subseteq L$ infinite such that $N_\ell = \infty$ for all $\ell \in I$. Let $d \in \mathbb{N}$, $F_d \subset I$ with $|F_d| = d$ and $V_d$ the subspace generated by 1, the $x_i$'s and all $\ell \in F_d$. Then $(V_d)^{n+1}$ contains the ordered monomials in $F_d$ of degree $\leq n+1$, hence $\dim(V_d)^{n+1} \geq \binom{n+d}{d}$. Hence $\limsup \log_n \dim(V_d)^{n+1} \geq d$. Since $d$ is arbitrary, $\operatorname{GKdim} \mathcal{B}(V) = \infty$. □

Assume that $L$ is finite (the Conjecture 1.1 says that this is the case when $\operatorname{GKdim} \mathcal{B}(V) < \infty$). By [**KL**, Theorem 12.6.2], we conclude that

$$\operatorname{GKdim} \mathcal{B}(V) = |\{\ell \in L : N_\ell = \infty\}|. \tag{2.6}$$

Let $\mathbf{\Delta}_+^V = \mathbf{\Delta}_+ = (\deg \ell)_{\ell\in L}$ be the family of positive roots of $\mathcal{B}(V)$ (with multiplicities). By [**HS**, Lemma 4.7], it is uniquely determined, i.e. it does not depend on $L$.

We say that *we can reflect* $V$ *at* $i \in \mathbb{I}$ if, for all $j \neq i$, there exists $n \in \mathbb{N}_0$ such that $(n+1)_{q_{ii}}(1-q_{ii}^n q_{ij} q_{ji}) = 0$. In such case, following [**H1**] we define a generalized Cartan matrix $(c_{ij})$ by $c_{ii} = 2$ and

$$c_{ij} := -\min\{n \in \mathbb{N}_0 : (n+1)_{q_{ii}}(1 - q_{ii}^n q_{ij} q_{ji}) = 0\}, \qquad j \neq i. \tag{2.7}$$

Let $s_i \in GL(\mathbb{Z}^\theta)$ be given by

$$s_i(\alpha_j) = \alpha_j - c_{ij}\alpha_i, \qquad j \in \mathbb{I}. \tag{2.8}$$

The reflection at the vertex $i$ of $\mathbf{q}$ is the matrix $\mathcal{R}^i(\mathbf{q}) = (t_{jk})_{j,k\in\mathbb{I}}$, where

$$t_{jk} := q_{s_i(\alpha_j), s_i(\alpha_k)} = q_{jk} q_{ik}^{-c_{ij}} q_{ji}^{-c_{ik}} q_{ii}^{c_{ij}c_{ik}}, \qquad j, k \in \mathbb{I}. \tag{2.9}$$

Let $\mathcal{R}^i(V)$ be the braided vector space of diagonal type with matrix $\mathcal{R}^i(\mathbf{q})$.

THEOREM 2.4. [**H1, AA1**] $\operatorname{GKdim} \mathcal{B}(\mathcal{R}^i(V)) = \operatorname{GKdim} \mathcal{B}(V)$. □

We say that $V$ *admits all reflections* if we can reflect $V$ at every $i_1 \in \mathbb{I}$, then we can reflect $\mathcal{R}^{i_1}(V)$ at every $i_2 \in \mathbb{I}$ and so on, we can reflect $\mathcal{R}^{i_k} \dots \mathcal{R}^{i_1}(V)$ at every $i_{k+1} \in \mathbb{I}$ for all $k$.

If $V$ admits all reflections, then we denote by $\mathcal{X}$ the collection of all braided vector spaces of diagonal type obtained from $V$ by a finite number of successive reflections at various vertices. Here any two braided vector spaces with the same braiding matrix are identified. The collection $(\mathbf{\Delta}_+^U)_{U\in\mathcal{X}}$ is the *generalized root system* of $V$.

REMARK 2.5. If $\operatorname{GKdim} \mathcal{B}(V) < \infty$, then we can reflect $V$ at every $i \in \mathbb{I}$ by Lemma 2.2 (c). Hence $V$ admits all reflections by Theorem 2.4.

## 3. General results

Recall that $(V, c)$ is of *Cartan type* if there exist $a_{ij} \in \mathbb{Z}_{\leq 0}$ such that

$$q_{ij} q_{ji} = q_{ii}^{a_{ij}}, \qquad i \neq j \in \mathbb{I}. \tag{3.1}$$

Set $a_{ii} = 2$, $i \in \mathbb{I}$. If $q_{ii} \in \mathbb{G}_\infty$, then we choose $a_{ij} \in (-\operatorname{ord} q_{ii}, 0]$, when $i \neq j$; otherwise it is uniquely determined. In any case, $\mathbf{a} = (a_{ij})_{i,j\in\mathbb{I}_\theta}$ is an indecomposable symmetrizable generalized Cartan matrix [**K**]. These matrices are of three types: finite, affine or indefinite. If $(V, c)$ is of Cartan type and $\operatorname{GKdim} \mathcal{B}(V) < \infty$, then Conjecture 1.1 predicts that $\mathbf{a}$ is of finite type. Here is the confirmation for the affine type.

PROPOSITION 3.1. *If* $\mathbf{a}$ *is of affine type, then* $\operatorname{GKdim} \mathcal{B}(V) = \infty$.

PROOF. Let $\mathbf{\Delta}^{\mathrm{re}}$ denote the set of real roots corresponding to $\mathbf{a}$. There exists a positive imaginary root $\delta$ such that $\mathbf{\Delta}^{\mathrm{re}} + \delta = \mathbf{\Delta}^{\mathrm{re}}$ [**K**, Proposition 6.3 d)]. Let $m$ be the height of $\delta$ and let $\alpha$ be a simple root. Choose a homogeneous restricted PBW basis of $\mathcal{B}(V)$. Then for all $k \geq 0$ there exists a PBW generator $y_k$ of degree $\alpha + k\delta$, hence $\deg y_k = mk + 1$. Therefore $\operatorname{GKdim} \mathcal{B}(V) = \infty$ by Lemma 2.1. □

An indecomposable generalized Cartan matrix is *compactly hyperbolic* if it is of indefinite type and every proper minor is of finite type. If $\mathbf{a} \in \mathbb{Z}^{\theta\times\theta}$ is compactly hyperbolic, then $\theta \leq 5$. In fact, the classification of compactly hyperbolic generalized Cartan matrices is known [**C+**]; there are the matrices $\begin{pmatrix} 2 & a \\ b & 2 \end{pmatrix}$ with $ab > 4$; 31 matrices in $\mathbb{Z}^{3\times 3}$; 3 matrices in $\mathbb{Z}^{4\times 4}$; 1 matrix in $\mathbb{Z}^{5\times 5}$. To prove Conjecture

1.1 in the Cartan case, it would be enough to verify it for compactly hyperbolic generalized Cartan matrices with $3 \leq \theta \leq 5$, as the case $\theta = 2$ is taken care by Theorem 4.1.

Back to the general diagonal type, we distinguish three classes of Nichols algebras. Given $\mathbf{q}$ as above, we say that

- $\circ$ $\mathbf{q}$ is of *torsion class* if $q_{ii}, q_{ij}q_{ji} \in \mathbb{G}_\infty$ for all $i \neq j \in \mathbb{I}$;
- $\circ$ $\mathbf{q}$ is *generic*, if $q_{ii} \notin \mathbb{G}_\infty$, and $q_{ij}q_{ji} = 1$ or $q_{ij}q_{ji} \notin \mathbb{G}_\infty$, for all $i \neq j \in \mathbb{I}$.
- $\circ$ $\mathbf{q}$ is *semigeneric* if it is neither generic nor of torsion class.

REMARK 3.2.

(1) If $\mathbf{q}$ is of torsion class, then Conjecture 1.1 says that $\operatorname{GKdim} \mathcal{B}(V) < \infty$ implies $\operatorname{GKdim} \mathcal{B}(V) = 0$. Indeed all roots are real by [**CH**], and they would have finite non-trivial order by Lemma 2.2 (d), hence (2.6) applies.

(2) If $\mathbf{q}$ is of torsion class, then the set $\mathcal{X}$ defined after Theorem 2.4 is finite. Indeed, there are finitely many matrices with the shape (2.9).

(3) [**R**, **AA1**] If $\mathbf{q}$ is generic, then $\operatorname{GKdim} \mathcal{B}(V) < \infty$ if and only if there exists a Cartan matrix of finite type $\mathbf{a} = (a_{ij})$, with symmetrizing diagonal matrix $(d_i)$, and $q \notin \mathbb{G}_\infty$ such that $q_{ii} = q^{2d_i}$ and $q_{ij}q_{ji} = q^{2d_i a_{ij}}$ for all $i \neq j \in \mathbb{I}$. Thus Conjecture 1.1 holds in this case.

(4) A semigeneric matrix with finite generalized root system is either of super type or else one of two exceptions of ranks 2 and 4:

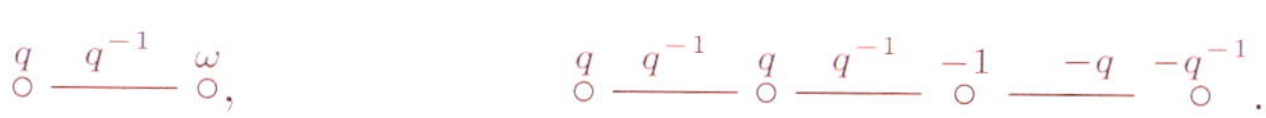


Here $\omega \in \mathbb{G}'_3$ and $q \notin \mathbb{G}_\infty$; the first corresponds to Yamane's exotic quantum groups [**Y**] while the second is the row 14 in [**H2**, Table 3].

**3.1. Semigeneric diagonal type.** Let us fix $\mathbf{q} = (q_{ij})_{i,j \in \mathbb{I}}$ semigeneric with $\operatorname{GKdim} \mathcal{B}(V) < \infty$. Let $\mathbb{J} = \{i \in \mathbb{I} : q_{ii} \notin \mathbb{G}_\infty\}$ be the set of generic points of $\mathbf{q}$ and let $\mathbb{J}_1, \dots \mathbb{J}_t$ be the connected components of the generalized Dynkin diagram $\mathbf{q} = (q_{ij})_{i,j \in \mathbb{J}}$.

LEMMA 3.3. *If $i \in \mathbb{I}$ and $j \in \mathbb{J}$, then there exists $h \in \mathbb{N}_0$ such that $q_{jj}^{-h} = q_{ij}q_{ji}$. In particular, either $q_{ij}q_{ji} = 1$ or $\notin \mathbb{G}_\infty$.*

PROOF. By Lemma 2.2. □

LEMMA 3.4. *If $i \notin \mathbb{J}$, $j \in \mathbb{J}$ and $q_{ij}q_{ji} \neq 1$, then either* $\operatorname{ord} q_{ii} = 2$ *and $q_{ij}q_{ji} = q_{jj}^{-h}$ with $h \in \mathbb{I}_2$; or else* $\operatorname{ord} q_{ii} = 3$ *and $q_{ij}q_{ji} = q_{jj}^{-1}$.*

PROOF. First, there exists $h \in \mathbb{N}$ such that $q_{jj}^{-h} = q_{ij}q_{ji}$ by Lemma 3.3. Let $N = \operatorname{ord} q_{ii}$. We apply the reflection at $i$:

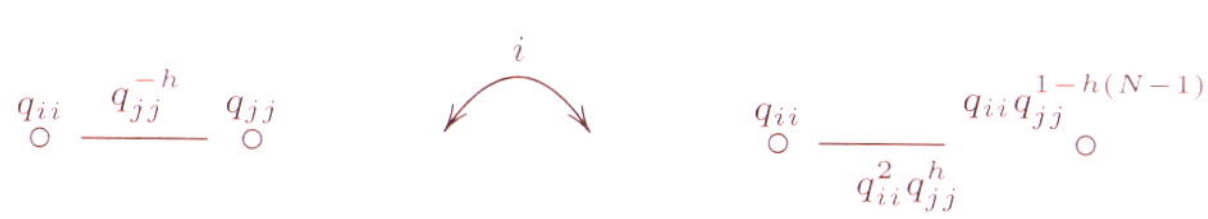


Then either $1 = h(N-1)$ that gives $h = 1$, $N = 2$; or else there exists $t \in \mathbb{N}$ such that $(q_{ii}q_{jj}^{1-h(N-1)})^{-t} = q_{ii}^2 q_{jj}^h$ by Lemma 3.3. A straightforward analysis yields the claim. □

As a consequence we derive the corresponding version of Theorem 4.1 for semigeneric braidings. It will be useful for the proof of the general case.

COROLLARY 3.5. *Let $V$ be a braided vector space of* semigeneric *diagonal type and dimension* 2 *such that the* GKdim *of its Nichols algebra $\mathcal{B}(V)$ is finite. Then its generalized root system is finite.*

PROOF. We may assume that $q_{11} \in \mathbb{G}_\infty$, $q_{22} \notin \mathbb{G}_\infty$ up to reflection. Indeed, if neither $q_{11}$ nor $q_{22}$ belong to $\mathbb{G}_\infty$, then $V$ is generic by Lemma 2.2 (c). So, either $q_{11} \in \mathbb{G}_\infty$ or else $q_{22} \in \mathbb{G}_\infty$. If both belong to $\mathbb{G}_\infty$, then $q_{ij}q_{ji} \notin \mathbb{G}_\infty$. Applying reflection at 1, we have that the new $q_{22} \notin \mathbb{G}_\infty$.

By Lemma 3.4, the Dynkin diagram of $V$ is one of the following:

$$\overset{-1}{\circ} \overset{q^{-1}}{\rule{40pt}{0.4pt}} \overset{q}{\circ}, \qquad \overset{-1}{\circ} \overset{q^{-2}}{\rule{40pt}{0.4pt}} \overset{q}{\circ}, \qquad \overset{\zeta}{\circ} \overset{q^{-1}}{\rule{40pt}{0.4pt}} \overset{q}{\circ}, \qquad q \notin \mathbb{G}_\infty,\ \zeta \in \mathbb{G}'_3.$$

All of them appear in [**H2**, Table 1], so $V$ has a finite root system. □

**3.2. Nichols algebras of indefinite Cartan type.** Let $A$ be an indecomposable generalized Cartan matrix. Let $W$ be the corresponding Weyl group, see [**K**].

LEMMA 3.6. *Assume that $A$ is of indefinite type. Let $Q$ be the root lattice corresponding to $A$ and let $Q_+ \subset Q$ be the submonoid generated by the simple roots. Then $W\gamma \cap Q_+$ is infinite for all $\gamma \in Q_+ - 0$.*

PROOF. Let $\pi$ denote the set of simple roots. Let $\gamma \in Q_+ - 0$. Assume that $W\gamma \cap Q_+$ is finite. Let $\beta = \sum_{\alpha\in\pi} c_\alpha \alpha \in W\gamma \cap Q_+$, where $c_\alpha \geq 0$ for all $\alpha \in \pi$, such that $w\gamma - \beta \notin Q_+ - 0$ for all $w \in W$. Then for all $\alpha \in \pi$ there exists $m_\alpha \geq 0$ such that $s_\alpha \beta = \beta - m_\alpha \alpha$. On the other hand, $s_\alpha\beta = \beta - \sum_{\alpha'\in\pi} c_{\alpha'} a_{\alpha\alpha'}\alpha$, that is, $A(c_{\alpha'})_{\alpha'\in\pi} \geq 0$. Since $A$ is of indefinite type and $(c_{\alpha'})_{\alpha'\in\pi}$, $A(c_{\alpha'})_{\alpha'\in\pi}$ have only non-negative entries, we have a contradiction. □

Let $V$ be a braided vector space of Cartan type with Cartan matrix $A$.

LEMMA 3.7. *If there exists a root $\gamma \in \boldsymbol{\Delta}_+^V$ of $\mathcal{B}(V)$ such that $q_{\gamma,\gamma} = 1$, then* $\operatorname{GKdim} \mathcal{B}(V) = \infty$.

PROOF. The Cartan matrix $A$ is not of finite type, since otherwise $q_{ii} \neq 1$ for all $i$ and hence $q_{\gamma,\gamma} \neq 1$ for all roots $\gamma$. If $A$ is of affine type, then $\operatorname{GKdim} \mathcal{B}(V) = \infty$ by Proposition 3.1. We assume then that $A$ is of indefinite type. By [**H1**], $s_\alpha(\boldsymbol{\Delta}_+^V - \{\alpha\}) = \boldsymbol{\Delta}_+^V - \{\alpha\}$ for all simple roots $\alpha$. Since $q_{w\gamma,w\gamma} = q_{\gamma,\gamma} = 1$ for all $w \in W$, all root vectors of degree $w\gamma$ with $w \in W$ have infinite height. By Lemma 3.6, $W\gamma$ is infinite. Thus $\mathcal{B}(V)$ has a restricted homogeneous PBW basis containing infinitely many PBW generators having infinite height. The claim follows by Lemma 2.3. □

**3.3. Braided coideal subalgebras.** Just in this Subsection, the field $\Bbbk$ is arbitrary. Let $H$ be a Hopf algebra with bijective antipode.

PROPOSITION 3.8. [**GH**, Prop. 2.1]. *Let $B$ be a bialgebra in ${}^H_H\mathcal{YD}$, let $K$ be a subalgebra of $B$ and let $I$ be a subobject of $K$ in ${}^H_H\mathcal{YD}$, such that it is a coideal of $B$, an ideal of $K$ and*

$$\Delta(K) \subseteq K \otimes K + I \otimes B. \tag{3.2}$$

*Then $K/I$ inherits a structure of bialgebra in ${}^H_H\mathcal{YD}$ from $B$.*

By (3.2), $K$ is a right coideal subalgebra of $B$.

PROOF. The existence of $\Delta$ is verified by usual chasing in the following commutative diagram:

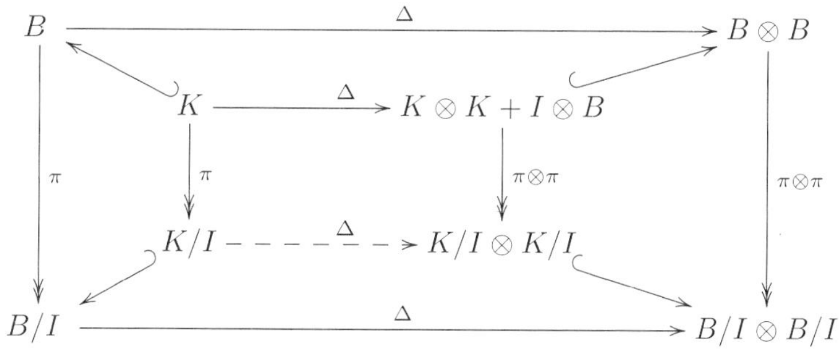

The associativity and compatibilities follow at once from those of $B$. □

We now apply Proposition 3.8 in the following context. Recall that $\alpha_1, \alpha_2$ is the canonical basis of $\mathbb{Z}^2$. Let $V = V_1 \oplus V_2$ be a direct sum in ${}^H_H\mathcal{YD}$. Then $\mathcal{B}(V)$ has a unique $\mathbb{N}_0^2$-grading (as a Hopf algebra in ${}^H_H\mathcal{YD}$)

$$\mathcal{B}(V) = \bigoplus_{\alpha \in \mathbb{N}_0^2} \mathcal{B}^\alpha(V)$$

such that $\deg V_1 = \alpha_1$ and $\deg V_2 = \alpha_2$. Let $r \in \mathbb{Q}_{\geq 0}$. We set

$$B_{\geq r} = \bigoplus_{\substack{\alpha = a_1\alpha_1 + a_2\alpha_2 \in \mathbb{N}_0^2: \\ a_1 \geq r a_2}} \mathcal{B}^\alpha(V), \qquad B_{>r} = \bigoplus_{\substack{\alpha = a_1\alpha_1 + a_2\alpha_2 \in \mathbb{N}_0^2: \\ a_1 > r a_2}} \mathcal{B}^\alpha(V),$$

$$K_{\geq r} = \{x \in \mathcal{B}(V) \mid \Delta(x) \in B_{\geq r} \otimes \mathcal{B}(V)\}, \qquad K_{>r} = K_{\geq r} \cap B_{>r}.$$

PROPOSITION 3.9. *Let $r \in \mathbb{Q}$ with $r \geq 0$. Then the braided bialgebra structure of $\mathcal{B}(V)$ induces a braided Hopf algebra structure on $K_{\geq r}/K_{>r}$.*

PROOF. We claim that:

(i) $K_{\geq r} \subseteq B_{\geq r}$,
(ii) $K_{\geq r}$ is a subalgebra of $\mathcal{B}(V)$ in ${}^H_H\mathcal{YD}$,
(iii) $\Delta(K_{\geq r}) \subseteq K_{\geq r} \otimes K_{\geq r} + K_{>r} \otimes \mathcal{B}(V)$,
(iv) $K_{>r}$ is an ideal of $K_{\geq r}$ and a coideal of $\mathcal{B}(V)$ in ${}^H_H\mathcal{YD}$.

For (i), apply $(\mathrm{id} \otimes \varepsilon)$ to the inclusion defining $K_{\geq r}$. Now (ii) follows since the multiplication and the Yetter-Drinfeld structure of $\mathcal{B}(V)$ are $\mathbb{N}_0^2$-graded.

For (iii), note that $K_{\geq r}$ is a right coideal since $\Delta$ is coassociative. Using this fact and that $\Delta$ is $\mathbb{N}_0^2$-graded,

$$\Delta(K_{\geq r}) \subseteq K_{\geq r} \otimes B_{\geq r} + K_{>r} \otimes B(V).$$

Indeed, let $x \in K_{\geq r}$ of degree $a_1\alpha_1 + a_2\alpha_2$. We write $\Delta(x) = \sum_i y_i \otimes z_i$ with $y_i, z_i$ $\mathbb{N}_0^2$-homogeneous. If $y_i$, $z_i$ have degree $b_1\alpha_1 + b_2\alpha_2$, $c_1\alpha_1 + c_2\alpha_2$, then either $b_1 > rb_2$ so $y_i \otimes z_i \in K_{>r} \otimes B(V)$, or else $b_1 = rb_2$, in which case $c_1 = a_1 - b_1 \geq ra_2 - rb_2 = rc_2$ and $y_i \otimes z_i \in K_{\geq r} \otimes B_{\geq r}$. Now

$$\begin{aligned}(\Delta \otimes \mathrm{id})\Delta(K_{\geq r}) &\subseteq (\Delta \otimes \mathrm{id})(K_{\geq r} \otimes B_{\geq r} + K_{>r} \otimes \mathcal{B}(V)) \\ &\subseteq K_{\geq r} \otimes K_{\geq r} \otimes B_{\geq r} + K_{\geq r} \otimes K_{>r} \otimes \mathcal{B}(V) + K_{>r} \otimes \mathcal{B}(V) \otimes \mathcal{B}(V).\end{aligned}$$

We apply $(\mathrm{id} \otimes \mathrm{id} \otimes \varepsilon)$ to the previous inclusion and get (iii).

Finally we prove (iv): Let $x \in K_{>r}$, $y \in K_{\geq r}$, we may assume they are homogeneous of degrees $\alpha = a_1\alpha_1 + a_2\alpha_2$, $\beta = b_1\alpha_1 + b_2\alpha_2$, so $a_1 > ra_2$, $b_1 \geq rb_2$. Hence $xy \in K_{\geq r}$ by (ii), and $xy$ has degree $\alpha + \beta = (a_1+b_1)\alpha_1 + (a_2+b_2)\alpha_2$ with $a_1+b_1 > r(a_2+b_2)$, so $xy \in K_{>r}$. Analogously, $yx \in K_{>r}$. Thus $K_{>r}$ is an ideal of $K_{\geq r}$. Now $\Delta(x) \in K_{\geq r} \otimes K_{\geq r} + K_{>r} \otimes \mathcal{B}(V)$ by (iii). For each $u \otimes v \in K_{\geq r} \otimes K_{\geq r}$ appearing in $\Delta(x)$, $u$, $v$ homogeneous of degrees $c_1\alpha_1 + c_2\alpha_2$, $d_1\alpha_1 + d_2\alpha_2$, we have that $c_1 \geq rc_2$, $d_1 \geq rd_2$. As $\Delta$ is $\mathbb{N}_0^2$-graded, $c_1 + d_1 = a_1$, $c_2 + d_2 = a_2$, so either $c_1 > rc_2$ or else $d_1 > rd_2$; in other words, either $u \in K_{>r}$ or else $v \in K_{>r}$, so $\Delta(x) \in K_{\geq r} \otimes K_{>r} + K_{>r} \otimes \mathcal{B}(V)$. Thus $K_{>r}$ is a coideal of $\mathcal{B}(V)$ in ${}^H_H\mathcal{YD}$.

Hence we may apply Proposition 3.8 and $K_{\geq r}/K_{>r}$ is a bialgebra in ${}^H_H\mathcal{YD}$. Since $\mathcal{B}(V)$ is $\mathbb{N}_0$-graded and connected and $\Delta$ is $\mathbb{N}_0$-graded, $K_{\geq r}$ and $K_{>r}$ are also $\mathbb{N}_0$-graded. Therefore $K_{\geq r}/K_{>r}$ is $\mathbb{N}_0$-graded and connected, so it is a Hopf algebra in ${}^H_H\mathcal{YD}$ by [**Mo**, 5.2.10]. □

## 4. Rank 2

This Section contains a proof of the following result:

THEOREM 4.1. *Let $V$ be a braided vector space of diagonal type and dimension 2 such that* $\operatorname{GKdim} \mathcal{B}(V) < \infty$. *Then its generalized root system is finite.*

Let $Z$ be a vector space and $(z_i)_{i\in I}$ a family of vectors in $Z$. Then $\langle z_i : i \in I\rangle$ denotes the subspace of $Z$ generated by the $z_i$'s.

**4.1. Lemmas for $\theta = 2$.** We establish some properties needed in the proof of Theorem 4.1. We keep the notation in §2.3; we assume that $\theta = 2$ and set as above $y_k = (\operatorname{ad} x_1)^k x_2 \in \mathcal{B}(V)$. We also set:

$$\beta_m = m\alpha_1 + \alpha_2, \qquad p_m = q_{\beta_m\beta_m} = q_{11}^{m^2} q_{12}^m q_{21}^m q_{22}, \qquad m \in \mathbb{N}_0. \tag{4.1}$$

4.1.1. *On the powers of the $y_n$'s.*

LEMMA 4.2. *Let $r, s, t \in \mathbb{N}_0$, $r \leq s$. Then*

$$\partial_1^t \partial_2(y_r y_s) = \begin{cases} 0, & s < t; \\ \mu_s (s)_{q_{11}} (s-1)_{q_{11}} \dots (s-t+1)_{q_{11}}\, y_r x_1^{s-t}, & r < t \leq s. \end{cases}$$

PROOF. By (2.3) we have that

$$\partial_2(y_r y_s) = \mu_r q_{21}^s q_{22} x_1^r y_s + \mu_s y_r x_1^s \qquad \text{for all } r, s \in \mathbb{N}_0. \tag{4.2}$$

Using (2.3) again, if $t > s$, then $\partial_1^t \partial_2(y_r y_s) = 0$, and if $r < t \leq s$,

$$\partial_1^t \partial_2(y_r y_s) = \partial_1^t(\mu_s y_r x_1^s) = \mu_s (s)_{q_{11}} (s-1)_{q_{11}} \dots (s-t+1)_{q_{11}}\, y_r x_1^{s-t}$$

by induction on $t$. □

LEMMA 4.3. *Let $l \in \mathbb{N}_0$. Assume that*

$$x_1^{2l+1} \neq 0, \qquad x_2^2 \neq 0, \qquad y_{l+1} \neq 0, \qquad y_{l+1} y_l \in \langle y_r y_s : 0 \leq r \leq s\rangle.$$

*Then $q_{11}^{l(l+1)} (\widetilde{q}_{12})^{l+1} q_{22} = 1$ and*

$$y_{l+1} y_l = q_{11}^{l(l+1)} q_{12}^{l+1} q_{21}^l q_{22}\, y_l y_{l+1}. \tag{4.3}$$

PROOF. The last assumption of the lemma implies that

$$y_{l+1}y_l \in \langle y_r y_s : 0 \le r \le s, \ r+s = 2l+1\rangle.$$

Using Lemma 4.2 we have that

$$\partial_1^{l+2}\partial_2(y_{l+1}y_l) = 0, \qquad \partial_1^{l+2}\partial_2(y_r y_s) \in \Bbbk y_r x_1^{s-l-2} - 0,$$

for all $r, s \in \mathbb{N}_0$ with $r+s = 2l+1$, $s \ge l+2$ such that $y_s \ne 0$. Thus there exists $\lambda \in \Bbbk$ such that $y_{l+1}y_l = \lambda y_l y_{l+1}$. Now we apply $\partial_1^{l+1}\partial_2$ to this equation. Using that $\partial_1^{l+1}(x_1^l) = 0$, $\partial_1^{l+1}(x_1^{l+1}) = (l+1)^!_{q_{11}} \ne 0$ we obtain

$$\mu_{l+1}q_{21}^l q_{22}(q_{11}^l q_{12})^{l+1} y_l = \lambda \mu_{l+1} y_l.$$

Thus $\lambda = q_{11}^{l(l+1)} q_{12}^{l+1} q_{21}^l q_{22}$, since $y_{l+1} \ne 0$. Finally we apply $\partial_2^2$ to (4.3). Since $\mu_{l+1} \ne 0$ and $x_1^{2l+1} \ne 0$, we obtain that

$$q_{21}^l q_{22} + q_{21}^l = q_{11}^{l(l+1)} q_{12}^{l+1} q_{21}^l q_{22}(q_{21}^{l+1}q_{22} + q_{21}^{l+1}).$$

As $x_2^2 \ne 0$, we have that $q_{22} \ne -1$. Hence $q_{11}^{l(l+1)}(\widetilde{q}_{12})^{l+1}q_{22} = 1$. □

LEMMA 4.4. *Let $l \in \mathbb{N}$. Assume that $y_l^2 \in \langle y_r y_s : 0 \le r < s\rangle$ and that $y_l \ne 0$. Then $y_{l+1} = 0$ and $q_{11}^{l^2}(\widetilde{q}_{12})^l q_{22} = -1$.*

PROOF. As $\mathcal{B}(V)$ is $\mathbb{N}_0$-graded, the last assumption says that

$$y_l^2 \in \langle y_r y_s : 0 \le r < s, \ r+s = 2l\rangle.$$

Using Lemma 4.2 we have that

$$\partial_1^{l+1}\partial_2(y_l^2) = 0, \qquad \partial_1^{l+1}\partial_2(y_r y_s) \in \Bbbk y_r x_1^{s-l-1} - 0,$$

for all $r, s \in \mathbb{N}_0$ with $r < l$, $r+s = 2l$, $y_s \ne 0$. Thus $y_l^2 = 0$. Since $l > 0$,

$$0 = \partial_1^{l-1}\partial_2(y_l^2) = a_1 x_1 y_l + a_2 y_l x_1$$

for some $a_1, a_2 \in \Bbbk - 0$. Since $x_1 y_l = y_{l+1} + q_{11}^l q_{12} y_l x_1$ and since $y_{l+1}$ and $y_l x_1$ are linearly independent whenever $y_{l+1} \ne 0$, the latter equation implies that $y_{l+1} = 0$. Therefore $x_1 y_l = q_{11}^l q_{12} y_l x_1$ and we have that

$$0 = \partial_2(y_l^2) = \mu_r(q_{21}^l q_{22}(q_{11}^l q_{12})^l + 1) y_l x_1^l.$$

This implies the last claim. □

We fix $n > 0$ and set $q = p_n$, cf. (4.1). Assume that $q \in \mathbb{G}'_N$ for some $N \ge 2$. In the next few Lemmas we prepare a condition for $y_n^N$ being a root vector. We start with some computations with $q$-numbers. Recall that

$$(t_1 + t_2)_q = (t_1)_q + q^{t_1}(t_2)_q, \qquad \text{for all } t_1, t_2 \in \mathbb{N}_0.$$

In particular, $(N)_q = 0$ and $(N-t)_q = -q^{-t}(t)_q$ for all $0 \le t \le N$.

LEMMA 4.5. *Let $r \in \{0, 1, \dots, N-2\}$ and $t \in \mathbb{N}_0$. Then*

$$\sum_{l=0}^{t} q^l (l+1)_q (l+2)_q \cdots (l+r)_q = \frac{(t+1)_q (t+2)_q \cdots (t+r+1)_q}{(r+1)_q}.$$

PROOF. Note that $(r+1)_q \neq 0$ since $1 \leq r+1 < N$. We proceed by induction on $t$. For $t = 0$ the claim is trivial. For any $t \geq 0$, for which the claim holds, we obtain that

$$\begin{aligned}
&\sum_{l=0}^{t+1} q^l (l+1)_q (l+2)_q \cdots (l+r)_q \\
&= \sum_{l=0}^{t} q^l (l+1)_q (l+2)_q \cdots (l+r)_q + q^{t+1}(t+2)_q \cdots (t+r+1)_q \\
&= \frac{(t+1)_q (t+2)_q \cdots (t+r+1)_q}{(r+1)_q} + q^{t+1}(t+2)_q (t+3)_q \cdots (t+r+1)_q \\
&= \frac{(t+1)_q + q^{t+1}(r+1)_q}{(r+1)_q} (t+2)_q (t+3)_q \cdots (t+r+1)_q \\
&= \frac{(t+2)_q (t+3)_q \cdots (t+r+1)_q (t+r+2)_q}{(r+1)_q}.
\end{aligned}$$

This proves the claim. □

For any $t \in \{0, 1, \dots, N-2\}$ let

$$Y(t) = \sum_{j=0}^{t} (q_{11}^n q_{12})^{-j} \frac{(N-t-1+j)_q^!}{(j)_q^!} y_n^{t-j} y_{n+1} y_n^j. \tag{4.4}$$

In particular, $Y(0) = (N-1)_q^! y_{n+1}$.

LEMMA 4.6. $\partial_1^{n-1} \partial_2(y_n^N) = -\mu_n (n)_{q_{11}}^! (q_{11}^n q_{12})^{-1} Y(N-2)$.

PROOF. First we obtain that

$$\partial_2(y_n^N) = \sum_{l=0}^{N-1} y_n^{N-1-l} \partial_2(y_n) \alpha_2 \cdot y_n^l = \mu_n \sum_{l=0}^{N-1} (q_{21}^n q_{22})^l y_n^{N-1-l} x_1^n y_n^l.$$

Since $\partial_1(y_n) = 0$, we have that

$$\partial_1^{n-1} \partial_2(y_n^N) = \mu_n (n)_{q_{11}}^! \sum_{l=0}^{N-1} (q_{11}^{n(n-1)} q_{12}^{n-1} q_{21}^n q_{22})^l y_n^{N-1-l} x_1 y_n^l.$$

The equation $y_{n+1} = x_1 y_n - q_{11}^n q_{12} y_n x_1$ implies that

$$x_1 y_n^l = q_{11}^{nl} q_{12}^l y_n^l x_1 + \sum_{j=0}^{l-1} (q_{11}^n q_{12})^{l-1-j} y_n^{l-1-j} y_{n+1} y_n^j. \tag{4.5}$$

Therefore

$$\begin{aligned}\partial_1^{n-1}\partial_2(y_n^N) =&\mu_n(n)^!_{q_{11}}\sum_{l=0}^{N-1}(q_{11}^{n(n-1)}q_{12}^{n-1}q_{21}^n q_{22})^l y_n^{N-1-l}\cdot\\ &\Big(q_{11}^{nl}q_{12}^l y_n^l x_1+\sum_{j=0}^{l-1}(q_{11}^n q_{12})^{l-1-j}y_n^{l-1-j}y_{n+1}y_n^j\Big)\\ =&\mu_n(n)^!_{q_{11}}\Big(\sum_{l=0}^{N-1}q^l y_n^{N-1}x_1\\ &+(q_{11}^n q_{12})^{-1}\sum_{j=0}^{N-2}\Big(\sum_{l=j+1}^{N-1}q^l\Big)(q_{11}^n q_{12})^{-j}y_n^{N-2-j}y_{n+1}y_n^j\Big).\end{aligned}$$

Hence the Lemma follows since $\sum_{l=0}^{N-1}q^l=0$ and $\sum_{l=j+1}^{N-1}q^l=-(j+1)_q$. □

For any $1\le t\le N-1$ let

$$d_t=1-q^{t+1}q_{11}^{2n}\widetilde{q}_{12}+\frac{q^t(1-q_{11}^n\widetilde{q}_{12})(n+1)_{q_{11}}}{(t)_q}. \tag{4.6}$$

Observe that $d_t$ depends on $n$.

LEMMA 4.7. *Let $t\in\{1,2,\dots,N-1\}$. Then*

$$\partial_1^n\partial_2(Y(t))=\mu_n(n)^!_{q_{11}}(q_{11}^n q_{12})^{-1}d_tY(t-1).$$

PROOF. Similarly to the calculation in Lemma 4.6 we have that

$$\begin{aligned}\partial_1^n\partial_2(Y(t))=&\sum_{l=0}^{t}(q_{11}^n q_{12})^{-l}\Big(\prod_{i=1}^{N-t-1}(l+i)_q\Big)\cdot\\ &\Big(\partial_1^n\partial_2(y_n)(t-l)_q q^{l+1}q_{11}^n q_{21}y_n^{t-1-l}y_{n+1}y_n^l\\ &+\partial_1^n\partial_2(y_n)(1-q_{11}^n\widetilde{q}_{12})(n+1)_{q_{11}}q^l y_n^{t-l}x_1y_n^l\\ &+\partial_1^n\partial_2(y_n)(l)_q y_n^{t-l}y_{n+1}y_n^{l-1}\Big).\end{aligned}$$

Using (4.5) this implies that

$$\begin{aligned}\partial_1^n\partial_2(Y(t))=&\ \partial_1^n\partial_2(y_n)\sum_{j=0}^{t-1}(q_{11}^n q_{12})^{-j}\Big(\prod_{i=1}^{N-t-1}(j+i)_q\Big)\cdot\\ &(-q^{t-j})(N-t+j)_q q^{j+1}q_{11}^n q_{21}y_n^{t-1-j}y_{n+1}y_n^j\\ &+\partial_1^n\partial_2(y_n)(1-q_{11}^n\widetilde{q}_{12})(n+1)_{q_{11}}\sum_{l=0}^{t}(q_{11}^n q_{12})^{-l}\Big(\prod_{i=1}^{N-t-1}(l+i)_q\Big)q^l\cdot\\ &\Big(q_{11}^{nl}q_{12}^l y_n^t x_1+\sum_{j=0}^{l-1}(q_{11}^n q_{12})^{l-1-j}y_n^{t-1-j}y_{n+1}y_n^j\Big)\\ &+\partial_1^n\partial_2(y_n)\sum_{j=0}^{t-1}(q_{11}^n q_{12})^{-j-1}\Big(\prod_{i=1}^{N-t-1}(j+1+i)_q\Big)(j+1)_q y_n^{t-1-j}y_{n+1}y_n^j.\end{aligned}$$

Lemma 4.5 tells that

$$\sum_{l=0}^{t} q^l \prod_{i=1}^{N-t-1} (l+i)_q = \prod_{i=1}^{N-t} (t+i)_q/(N-t)_q = 0$$

since $1 < t < N$. Therefore the terms $y_n^t x_1$ disappear in the above expression for $\partial_1^n \partial_2(Y(t))$. Moreover,

$$\begin{aligned}
&\sum_{l=0}^{t}\sum_{j=0}^{l-1} \Big( \prod_{i=1}^{N-t-1} (l+i)_q \Big) q^l (q_{11}^n q_{12})^{-1-j} \\
&= \sum_{j=0}^{t-1} (q_{11}^n q_{12})^{-1-j} \sum_{l=j+1}^{t} q^l \Big( \prod_{i=1}^{N-t-1} (l+i)_q \Big) \\
&= \sum_{j=0}^{t-1} (q_{11}^n q_{12})^{-1-j} \Big( \sum_{l=0}^{t} q^l \Big( \prod_{i=1}^{N-t-1} (l+i)_q \Big) - \sum_{l=0}^{j} q^l \Big( \prod_{i=1}^{N-t-1} (l+i)_q \Big) \Big) \\
&= \sum_{j=0}^{t-1} (q_{11}^n q_{12})^{-1-j} \Big( \prod_{i=1}^{N-t} (t+i)_q - \prod_{i=1}^{N-t} (j+i)_q \Big)/(N-t)_q \\
&= -\sum_{j=0}^{t-1} (q_{11}^n q_{12})^{-1-j} \Big( \prod_{i=1}^{N-t} (j+i)_q \Big)/(N-t)_q.
\end{aligned}$$

Therefore,

$$\begin{aligned}
\partial_1^n \partial_2(Y(t)) =& \partial_1^n \partial_2(y_n)(q_{11}^n q_{12})^{-1} \sum_{j=0}^{t-1} (q_{11}^n q_{12})^{-j} \Big( \prod_{i=1}^{N-t} (j+i)_q \Big) \cdot \\
& \Big( -q^{t+1} q_{11}^{2n} \widetilde{q}_{12} - \frac{(1-q_{11}^n \widetilde{q}_{12})(n+1)_{q_{11}}}{(N-t)_q} + 1 \Big) y_n^{t-1-j} y_{n+1} y_n^j.
\end{aligned}$$

Thus the claim follows from this equality and $(N-t)_q = -q^{-t}(t)_q$. □

PROPOSITION 4.8. *Assume that $y_n^N = 0$ and $y_{n+1} \neq 0$. Then $d_t = 0$ for some $1 \le t \le N-2$, see (4.6).*

PROOF. Lemma 4.6 implies that $Y(N-2) = 0$. Thus

$$(\partial_1^n \partial_2)^{N-2}(Y(N-2)) = 0.$$

By Lemma 4.7 either $d_{N-2} = 0$ or else $Y(N-3) = 0$. Recursively, if $Y(t) = 0$, then either $d_t = 0$ or else $Y(t-1) = 0$, for each $1 \le t \le N-2$. Since $Y(0) = (N-1)_q^! y_{n+1} \neq 0$, necessarily $d_t = 0$ for some $1 \le t \le N-2$. □

LEMMA 4.9. *If $y_n^N \neq 0$ then $N\beta_n$ is a root of $V$.*

PROOF. By the definition of roots of $\mathcal{B}(V)$, either $\deg y_n^N = N(n\alpha_1 + \alpha_2)$ is a root of $V$ or $y_n^N$ can be expressed as a linear combination of products $\ell_1^{m_1} \cdots \ell_k^{m_k}$ as in (2.5), where each $\ell_i$ corresponds either to a Lyndon word $l_i$ greater than $x_1^n x_2$, or else to a power of this kind of letters.

Assume the last case holds. Then each $l_i$ starts with $1^k 2$, where $k \le n$, and ends with 2. Since $l_i$ is a Lyndon word, any end of $l_i$ is larger than $v$, and hence it contains no subword $1^l 2$ with $l > k$. Therefore $l_i = x_1^{k_1} x_2 \cdots x_1^{k_r} x_2$ with $k_1, \dots, k_r \le n$

and $k_1 + \cdots + k_r < rn$. This implies that $\deg \ell_1 \cdots \ell_m \neq \deg y_n^N$ so $y_n^N = 0$, a contradiction. Thus $y_n$ has infinite height and hence $N\beta_n$ is a root of $\mathcal{B}(V)$. □

LEMMA 4.10. *Let $m \in \mathbb{N}_0$ with $y_m \neq 0$. Then $y_m^2 = 0$ if and only if $p_m = -1$ and $y_{m+1} = 0$.*

PROOF. If $y_m^2 = 0$, then Lemma 4.4 says that $y_{m+1} = 0$ and $p_m = -1$. Conversely, assume that $y_{m+1} = 0$, $y_m \neq 0$, and $p_m = -1$. Then

$$\begin{aligned}\partial_2(y_m^2) &= \mu_m(y_m x_1^m + x_1^m q_{21}^m q_{22} y_m)\\ &= \mu_m(1 + (q_{11}^m q_{12})^m q_{21}^m q_{22}) y_m x_1^m = 0.\end{aligned}$$

Since $\partial_1(y_m) = 0$, we conclude that $y_m^2 = 0$. □

LEMMA 4.11. *Let $m \in \mathbb{N}_0$. Assume that $p_m = -1$ and $y_{m+1} \neq 0$. Then $2\beta_m$ is a root of $V$ and $q_{2\beta_m\, 2\beta_m} = 1$.*

PROOF. As $y_{m+1} \neq 0$, Lemma 4.10 says that $y_m^2 \neq 0$. Now Lemma 4.9 applies and $2\beta_m$ is a root. Finally, $q_{2\beta_m\, 2\beta_m} = p_m^4 = 1$. □

4.1.2. *On the $w_m$'s.* We consider the following elements of $\mathcal{B}(V)$:

$$w_m = y_{m+2} y_m - q_{\beta_{m+2}, \beta_m} y_m y_{m+2}, \qquad m \in \mathbb{N}_0. \tag{4.7}$$

Notice that $w_m$ is $\mathbb{N}_0^2$-homogeneous of degree $2\beta_{m+1}$.

Let $m \in \mathbb{N}_0$. Assume that $y_{m+2} \neq 0$ and $q_{\beta_{m+1}\, \beta_{m+1}} \neq -1$. Let

$$\widetilde{w}_m = w_m - \frac{q_{\beta_{m+1}\, \beta_m} (m+2)_{q_{11}} (1 - q_{11}^{m+1} \widetilde{q}_{12})}{1 + q_{\beta_{m+1}\, \beta_{m+1}}} y_{m+1}^2. \tag{4.8}$$

Our next goal is to determine when $\widetilde{w}_m = 0$.

LEMMA 4.12. *Let $m \in \mathbb{N}_0$. Assume that*

$$y_{m+2} \neq 0, \qquad p_{m+1} \neq -1, \qquad w_k \in \Bbbk y_{k+1}^2, \quad 0 \le k < m.$$

*Then the following are equivalent:*

(1) $w_m \in \Bbbk y_{m+1}^2$.
(2) $\widetilde{w}_m = 0$.
(3) $\partial_1^m \partial_2(\widetilde{w}_m) = 0$.
(4) *The following equation holds:*

$$0 = \Big(1 - \frac{p_{m+1}}{q_{11}}\Big)\Big(1 + \frac{p_{m+1}}{q_{11}} + \frac{p_m (m+2)_{q_{11}}^! (1 - q_{11}^m \widetilde{q}_{12})(1 - q_{11}^{m+1}\widetilde{q}_{12})}{(m)_{q_{11}}^! (1 + q_{11})(1 + p_{m+1})}\Big).$$

PROOF. First we compute

$$\begin{aligned}\partial_1^m \partial_2(w_m) &= \partial_1^m \Big(\mu_{m+2} x_1^{m+2} q_{21}^m q_{22} y_m + y_{m+2} \mu_m x_1^m\Big)\\ &\quad - q_{\beta_{m+2}\, \beta_m} \Big(\mu_m x_1^m q_{21}^{m+2} q_{22} y_{m+2} + y_m \mu_{m+2} x_1^{m+2}\Big)\\ &= \mu_{m+2} q_{\beta_m\, \beta_m} \frac{(m+2)_{q_{11}}^!}{1 + q_{11}} (x_1^2 y_m - q_{11}^{2m} q_{12}^2 y_m x_1^2)\\ &\quad + \mu_m (m)_{q_{11}}^! (1 - q_{\beta_{m+2}\, \beta_m} q_{\beta_m\, \beta_{m+2}}) y_{m+2}.\end{aligned}$$

Similarly, we compute

$$\partial_1^m \partial_2(y_{m+1}^2) = \mu_{m+1}(m+1)^!_{q_{11}}\big(q_{\beta_m\,\beta_{m+1}} y_{m+2} + (1+p_{m+1}) y_{m+1} x_1\big),$$

see also the calculation in the proof of Lemma 4.10.

(1) $\Longrightarrow$ (2). Using the previous formulas one obtains quickly that

$$\partial_1^{m+1}\partial_2(w_m) = \mu_{m+2} q_{\beta_{m+1}\,\beta_m}(m+2)^!_{q_{11}} y_{m+1},$$
$$\partial_1^{m+1}\partial_2(y_{m+1}^2) = \mu_{m+1}(m+1)^!_{q_{11}}(1+p_{m+1}) y_{m+1}.$$

Since $y_{m+1}\neq 0$, this implies the claim.

(2) $\Longrightarrow$ (3). Trivial.

(3) $\Longleftrightarrow$ (4). We notice that

$$x_1^2 y_m - q_{11}^{2m} q_{12}^2 y_m x_1^2 = y_{m+2} + (1+q_{11}) q_{11}^m q_{12} y_{m+1} x_1.$$

Since $y_{m+2}$ and $y_{m+1}x_1$ are linearly independent in $\mathcal{B}(V)$, the formulas at the beginning of the proof imply that (3) is equivalent to

$$\frac{p_m(m+1)_{q_{11}}(m+2)_{q_{11}}(1-q_{11}^m\widetilde{q}_{12})(1-q_{11}^{m+1}\widetilde{q}_{12})}{1+q_{11}} + 1 - q_{11}^{-2}p_{m+1}^2$$
$$- \frac{q_{11}^{-1}p_m p_{m+1}(m+1)_{q_{11}}(m+2)_{q_{11}}(1-q_{11}^m\widetilde{q}_{12})(1-q_{11}^{m+1}\widetilde{q}_{12})}{1+p_{m+1}} = 0.$$

This gives the equivalence between (3) and (4).

(3) $\Longrightarrow$ (1). We prove by induction on $k$ that

$$\partial_1^{m-k}\partial_2(\widetilde{w}_m) = 0 \qquad \text{for all } 0\le k\le m. \tag{4.9}$$

If so, then $\partial_i(\widetilde{w}_m)=0$ for $i=1,2$ and then $\widetilde{w}_m=0$.

For $k=0$, (4.9) holds by assumption. Now let $k>0$ and assume the statement holds for $j<k$. Notice that

$$\ker\partial_1 \cap \mathcal{B}^{(m+k+2)\alpha_1+\alpha_2} = \Bbbk y_{m+k+2}$$
$$\Longrightarrow \partial_1^{m-k}\partial_2(\widetilde{w}_m) = b\, y_{m+k+2} \text{ for some } b\in\Bbbk.$$

We may assume that $y_{m+k+2}\neq 0$, otherwise the induction step holds.

Let $\iota: V\to V^*$ be the linear isomorphism with $x_i\mapsto\partial_i$ for $i\in\mathbb{I}$. Then $\iota$ is an isomorphism of braided vector spaces and hence induces an isomorphism between the Hopf algebras $\mathcal{B}(V)$ and $\mathcal{B}(V^*)$ in ${}^H_H\mathcal{YD}$. Notice that

$$\iota(y_{m-j})(\widetilde{w}_m) = \partial_1^{m-j}\partial_2(\widetilde{w}_m) = 0 \qquad \text{for all } 0\le j<k, \tag{4.10}$$

since $y_j = x_1^j x_2 +$ terms ending in $x_1$ and $\partial_1(\widetilde{w}_m)=0$. Hence

$$\begin{aligned}\iota(w_{m-k})(\widetilde{w}_m) &= \iota(y_{m-k+2}y_{m-k} - q_{\beta_{m-k+2},\beta_{m-k}} y_{m-k} y_{m-k+2})(\widetilde{w}_m)\\ &= \iota(y_{m-k+2})\iota(y_{m-k})(\widetilde{w}_m) = b\iota(y_{m-k+2})(y_{m+k+2})\\ &= b\partial_1^{m-k+2}\partial_2(y_{m+k+2}) = b\mu_{m+k+2}\frac{(m+k+2)^!_{q_{11}}}{(m-k+2)^!_{q_{11}}}.\end{aligned}$$

On the other hand, using that $w_{m-k} = a_{m-k} y_{m-k+1}^2$ for some $a_{m-k}\in\Bbbk$ and the inductive hypothesis,

$$\iota(w_{m-k})(\widetilde{w}_m) = a_{m-k}\iota(y_{m-k+1}^2)(\widetilde{w}_m) = a_{m-k}\iota(y_{m-k+1})\partial_1^{m-k+1}\partial_2(\widetilde{w}_m) = 0.$$

Hence $b=0$, which completes the inductive step. $\square$

REMARK 4.13. In the following cases, the right hand side of the equation in Lemma 4.12 (4) is equal to the following:

(i) $m=0$: $(1+q_{22})(1-\widetilde{q}_{12}q_{22})(1+q_{11}\widetilde{q}_{12}^2q_{22})(1+q_{11}\widetilde{q}_{12}q_{22})^{-1}$.

(ii) $m=1$, $q_{22}=-1$:
$$(1+q_{11}^3\widetilde{q}_{12}^2)(1-q_{11}^3\widetilde{q}_{12})(3)_{-q_{11}\widetilde{q}_{12}}(1+q_{11}^2\widetilde{q}_{12})^{-1}.$$

(iii) $m=1$, $\widetilde{q}_{12}q_{22}=1$:
$$(1-q_{11}^3\widetilde{q}_{12})(4)_{q_{11}}(3)_{-q_{11}^2\widetilde{q}_{12}}(1+q_{11}^4\widetilde{q}_{12})^{-1}.$$

(iv) $m=1$, $q_{11}\widetilde{q}_{12}^2q_{22}=-1$:
$$(1+q_{11}^2)(1-\widetilde{q}_{12}^{-1})(1-q_{11}^3\widetilde{q}_{12})(1-q_{11})^{-1}.$$

(v) $m=2$, $q_{22}=-1$:
$$(1+q_{11}^8\widetilde{q}_{12}^3)(1-q_{11}^4\widetilde{q}_{12})(3)_{q_{11}^3\widetilde{q}_{12}}^{-1}\cdot$$
$$\left(q_{11}^{10}\widetilde{q}_{12}^4+(q_{11}^7+q_{11}^6)\widetilde{q}_{12}^3-(3)_{q_{11}}q_{11}^4\widetilde{q}_{12}^2+(q_{11}^4+q_{11}^3)\widetilde{q}_{12}+1\right).$$

(vi) $m=2$, $\widetilde{q}_{12}q_{22}=1$, $q_{11}^2=-1$: $1-\widetilde{q}_{12}^4$.

(vii) $m=2$, $\widetilde{q}_{12}q_{22}=1$, $(3)_{-q_{11}^2\widetilde{q}_{12}}=0$:
$$(1+q_{11}^4\widetilde{q}_{12})(1-q_{11}^4\widetilde{q}_{12})(1-q_{11}^5\widetilde{q}_{12})(5)_{q_{11}}(3)_{-q_{11}}(1+q_{11}^9\widetilde{q}_{12}^2)^{-1}.$$

Recall the definition of $\widetilde{w}_m$, $m\in\mathbb{N}_0$, given in (4.8). We study in the next Lemmas when $\widetilde{w}_m\neq 0$ for small values of $m$.

LEMMA 4.14. *Assume that $y_2\neq 0$, $p_1\neq -1$. Then $\widetilde{w}_0=0$ if and only if*
$$(\widetilde{q}_{12}q_{22}-1)(q_{22}+1)(q_{11}\widetilde{q}_{12}^2q_{22}+1)=0.$$

PROOF. The claim follows by Lemma 4.12 and Remark 4.13 (i). □

Next we give conditions on the matrix $\mathbf{q}$ which are equivalent to the equation $\widetilde{w}_1=0$.

LEMMA 4.15. *Assume that $y_3\neq 0$, $p_2\neq -1$.*

(a) *If $q_{22}=-1$, then $\widetilde{w}_1=0$ if and only if*
$$(1-q_{11}^3\widetilde{q}_{12})(q_{11}^3\widetilde{q}_{12}^2+1)(3)_{-q_{11}\widetilde{q}_{12}}=0.$$

(b) *If $\widetilde{q}_{12}q_{22}=1$, then $\widetilde{w}_1=0$ if and only if*
$$(1-q_{11}^3\widetilde{q}_{12})(q_{11}^2+1)(3)_{-q_{11}^2\widetilde{q}_{12}}=0.$$

(c) *If $q_{11}\widetilde{q}_{12}^2q_{22}=-1$, then $\widetilde{w}_1=0$ if and only if $(1-q_{11}^3\widetilde{q}_{12})(q_{11}^2+1)=0$.*

PROOF. As we assume $y_3\neq 0$, we have that $(3)_{q_{11}}^!\mu_3\neq 0$. That is,
$$q_{11}\notin\mathbb{G}_2\cup\mathbb{G}_3, \qquad q_{11}^k\widetilde{q}_{12}\neq 1, \quad k=0,1,2.$$

Assume that $q_{22}=-1$, so we have that $p_1=-q_{11}\widetilde{q}_{12}\neq -1$. Hence $\widetilde{w}_0=0$ by Lemma 4.14. Now (a) follows by Lemma 4.12 and Remark 4.13 (ii).

To prove (b), we assume that $\widetilde{q}_{12}q_{22}=1$. Again we have that $p_1=q_{11}\neq -1$, so $\widetilde{w}_0=0$ by Lemma 4.14. Now we can apply Lemma 4.12 and Remark 4.13 (iii) and the claim follows.

Finally, if $q_{11}\widetilde{q}_{12}^2q_{22}=-1$, then $p_1=-\widetilde{q}_{12}^{-1}\neq -1$, so $\widetilde{w}_0=0$ by Lemma 4.14. Hence (c) follows by Lemma 4.12 and Remark 4.13 (iv). □

**4.2. Proof of Theorem 4.1.** First we extend Lemma 3.7 to any braided vector space of diagonal type and dimension two.

PROPOSITION 4.16. *Assume that $V$ is of dimension two and $\widetilde{q}_{12} \neq 1$. If there is a root $\gamma$ of $V$ such that $q_{\gamma\gamma} = 1$, then* $\operatorname{GKdim} \mathcal{B}(V) = \infty$.

PROOF. If $V$ does not admit all reflections, then $\operatorname{GKdim} \mathcal{B}(V) = \infty$ by Remark 2.5.

Now we assume that $V$ admits all reflections. If $V$ is generic, then $V$ is of Cartan type. As $\gamma$ cannot be a real root since $q_{\gamma\gamma} = 1$, $V$ is not of finite type. Then $\operatorname{GKdim} \mathcal{B}(V) = \infty$ by Remark 3.2 (3).

If $V$ is semigeneric, then $\operatorname{GKdim} \mathcal{B}(V) = \infty$. Indeed if we suppose that $\operatorname{GKdim} \mathcal{B}(V) < \infty$, then all roots $\beta$ satisfy $q_{\beta\beta} \neq 1$ since the root system is finite by Corollary 3.5, and this gives a contradiction.

Finally, if $V$ is of torsion class, then the set $\mathcal{X}$ is finite by Remark 3.2 (2). If the orbit of $\gamma$ is infinite, then $\operatorname{GKdim} \mathcal{B}(V) = \infty$ since there are infinitely many roots $\delta$ of $V$ with $q_{\delta,\delta} = 1$. Now assume that $\gamma$ has finite orbit. Let $s_1, s_2$ be the simple reflections corresponding to $V$, cf. (2.8). Then there exists $k > 0$ such that $(s_1s_2)^k(\gamma) = \gamma$ and $(\mathcal{R}^1\mathcal{R}^2)^k(V) = V$. Thus 1 is an eigenvalue of $(s_1s_2)^k \in \operatorname{Aut}(\mathbb{Z}^2)$. Since $\det(s_1s_2) = 1$, the other eigenvalue is also 1. Thus either $(s_1s_2)^k = \mathrm{id}$, or else $(s_1s_2)^k$ is a shear mapping. The first case implies that the set of real roots is finite, hence the Weyl groupoid is finite by [**CH**], and consequently all roots $\delta$ of $V$ are real [**CH**], a contradiction.

In the second case there exist $c_i \in \mathbb{Z}$ such that $(s_1s_2)^k(\alpha_i) = \alpha_i + c_i\gamma$ for $i = 1, 2$. Hence, each $\beta_n := (s_1s_2)^{nk}(\alpha_i) = \alpha_i + c_in\gamma$ is a real root for $n \in \mathbb{N}$ and $\operatorname{GKdim} \mathcal{B}(V) = \infty$ by Lemma 2.1. □

As a consequence of Proposition 4.16 we have:

COROLLARY 4.17. *Let $p \in \Bbbk^\times$ such that $p^4 \neq 1$. Assume that $q_{11} = p$ and $\widetilde{q}_{12} = q_{22} = p^4$. Then* $\operatorname{GKdim} \mathcal{B}(V) = \infty$.

PROOF. If $p$ is a root of 1 and $p^4 \neq 1$, then $V$ is of Cartan type, whose Cartan matrix we denote $A = (a_{ij})_{i,j\in\mathbb{I}_2}$. Recall that $m\alpha_1 + \alpha_2$ is a root if and only if $0 \leq m \leq -a_{12}$. We study three cases according with the order of $p$:

- If $p \in \mathbb{G}'_{4N}$, $N \geq 2$, then $-a_{12} = 4N - 4$. Thus $\gamma = \beta_{2N-2}$ is a root. As
$$q_{\gamma\gamma} = q_{11}^{(2N-2)^2}\widetilde{q}_{12}^{\,2N-2}q_{22} = p^{4N^2} = 1,$$
$\operatorname{GKdim} \mathcal{B}(V) = \infty$ by Proposition 4.16.
- If $p \in \mathbb{G}'_{4N+2}$, $N \geq 1$, then $-a_{12} = 4N - 2$. In particular, $\gamma = \beta_{2N-1}$ is a root. Now $2\beta_{2N-1}$ is a root by Lemma 4.11, since $y_{2N} \neq 0$ and
$$p_{2N-1} = q_{\gamma\gamma} = q_{11}^{(2N-1)^2}\widetilde{q}_{12}^{\,2N-1}q_{22} = p^{(2N+1)^2} = -1.$$
Thus $\operatorname{GKdim} \mathcal{B}(V) = \infty$ by Proposition 4.16.
- If $p \in \mathbb{G}'_{2N+1}$, $N \geq 1$, then $-a_{21} = 2N$. Thus $\gamma = N\alpha_2 + \alpha_1$ is a root, and
$$q_{\gamma\gamma} = q_{22}^{N^2}\widetilde{q}_{12}^{\,N}q_{11} = p^{(2N+1)^2} = 1,$$
$\operatorname{GKdim} \mathcal{B}(V) = \infty$, again by Proposition 4.16.

Finally, if $p$ is not a root of 1, then $\mathcal{B}(V)$ does not admit all reflections and hence $\operatorname{GKdim} \mathcal{B}(V) = \infty$. □

We apply next Corollary 4.17 in the braided Hopf algebra $K_{\geq m+1}/K_{>m+1}$. Since $x_1^k \in K_{>m}$ for all $k, m \in \mathbb{N}$, (2.4) implies that $y_m \in K_{\geq m}$ is a primitive element of $\mathbb{N}_0^2$-degree $\beta_m$ in $K_{\geq m}/K_{>m}$ for all $m \in \mathbb{N}$. Then (2.4) leads to the following shape of the coproduct of $w_m$ and $y_m^2$:

$$\begin{aligned}\Delta(w_m) &= w_m \otimes 1 + 1 \otimes w_m \\ &\quad + (m+2)_{q_{11}}(1 - q_{11}^{m+1}\widetilde{q}_{12})q_{\beta_{m+1}\,\beta_m}\, y_{m+1} \otimes y_{m+1} \\ &\quad + \text{ terms } x \otimes y,\ \deg x = k\alpha_1 + l\alpha_2,\ k \geq l(m+1)+1;\end{aligned} \tag{4.11}$$

$$\Delta(y_m^2) \in y_m^2 \otimes 1 + (1+p_m)y_m \otimes y_m + 1 \otimes y_m^2 + B_{>m} \otimes \mathcal{B}(V). \tag{4.12}$$

Assume that $p_{m+1} \neq -1$. By (4.11) and (4.12), $\widetilde{w}_m$ is a primitive element of $\mathbb{N}_0^2$-degree $2\beta_{m+1}$ in $K_{\geq m+1}/K_{>m+1}$.

LEMMA 4.18. *Let $m \in \mathbb{N}_0$ be such that $p_{m+1} \neq -1$. If $\widetilde{w}_m \neq 0$, then* $\operatorname{GKdim} \mathcal{B}(V) = \infty$.

PROOF. The subalgebra of $K_{\geq m+1}/K_{>m+1}$ generated by $y_{m+1}$ and $\widetilde{w}_m$ is a pre-Nichols algebra of diagonal type. Let $W$ be the $\Bbbk$-span of $y_{m+1}$ and $\widetilde{w}_m$, $p = q_{\beta_{m+1}\,\beta_{m+1}}$. The braiding matrix of $W$ is

$$\begin{pmatrix} p & p^2 \\ p^2 & p^4 \end{pmatrix}$$

If $p^4 = 1$ then $2\beta_{m+1}$ is a root of $V$ of infinite height. Thus $\operatorname{GKdim} \mathcal{B}(V) = \infty$ by Proposition 4.16.

If $p^4 \neq 1$, then $W$ satisfies the assumptions of Corollary 4.17 and hence $\operatorname{GKdim} \mathcal{B}(W) = \infty$. Since $\mathcal{B}(W)$ is a subquotient of $K_{\geq m+1}/K_{>m+1}$, and this is a subquotient of $\mathcal{B}(V)$, we have that $\operatorname{GKdim} \mathcal{B}(V) = \infty$. □

We now apply Lemma 4.18 combined with the Lemmas in §4.1.

LEMMA 4.19. *Assume that*

$$\widetilde{q}_{12} \neq 1, \quad (2)_{q_{11}}(1 - q_{11}\widetilde{q}_{12}) \neq 0, \quad (2)_{q_{22}}(1 - \widetilde{q}_{12}q_{22}) \neq 0, \quad q_{11}\widetilde{q}_{12}^2 q_{22} \neq -1.$$

*Then* $\operatorname{GKdim} \mathcal{B}(V) = \infty$.

PROOF. By Lemma 2.2, $y_2 \neq 0$. If $p_1 = -1$, then $2\beta_1$ is a root of $V$ by Lemma 4.11, so $\operatorname{GKdim} \mathcal{B}(V) = \infty$ by Proposition 4.16.

Assume now that $p_1 \neq -1$. Then

$$\Delta(y_1^2) = y_1^2 \otimes 1 + 1 \otimes y_1^2 + (1+p_1)y_1 \otimes y_1$$

in $K_{\geq 1}/K_{>1}$ and hence $\widetilde{w}_0$ is primitive in $K_{\geq 1}/K_{>1}$. By Lemma 4.14 $\widetilde{w}_0$ is non-zero. Now we apply Lemma 4.18. □

Without loss of generality, we assume that $|a_{12}^V| \geq |a_{21}^V| > 0$. We find a bound for $a_{12}^V$, $a_{21}^V$ to reduce the possibilities.

LEMMA 4.20. *Let $V$ be a braided vector space of diagonal type and dimension two such that $a_{12}, a_{21} \leq -3$. Then* $\operatorname{GKdim} \mathcal{B}(V) = \infty$.

PROOF. Suppose that $\operatorname{GKdim} \mathcal{B}(V) < \infty$. Then $\widetilde{w}_0 = 0$, so by Lemma 4.14 we have that $q_{11}\widetilde{q}_{12}^2 q_{22} = -1$. As $y_3 \neq 0$, Lemma 4.15(c) implies that $(1 - q_{11}^3\widetilde{q}_{12})(1 + q_{11}^2) = 0$, so $a_{12} = -3$ and analogously $a_{21} = -3$. We exclude the case $q_{11}^2 = q_{22}^2 = -1$: if this happens, then $1 = q_{11}^2\widetilde{q}_{12}^4 q_{22}^2 = \widetilde{q}_{12}^4$, so $\widetilde{q}_{12} \in \mathbb{G}_4$, but this gives a contradiction since $a^V_{12} = a^V_{21} = -3$. Hence we may assume $q_{22}^3\widetilde{q}_{12} = 1$. Let $r = q_{22}$, so $\widetilde{q}_{12} = r^{-3}$. As $-1 = q_{11}\widetilde{q}_{12}^2 q_{22} = r^{-5}q_{11}$, we have that $q_{11} = -r^5$.

First we assume that $q_{11}^3\widetilde{q}_{12} = 1$. Hence $q_{11}^3 = r^3$ so either $q_{11} = r \in \mathbb{G}'_8$ or $q_{11} = r^{17} \in \mathbb{G}'_{24}$. We compute $d_t$ as in (4.6) when $n = 1$, for each case:

- If $q_{11} = r \in \mathbb{G}'_8$, then $p_1 = r^7$, so $N = 8$ and

$$d_t = 1 - r^{7t+6} + \frac{r^{7t}(1 - r^{-2})(1 + r)}{(t)_{r^7}}$$
$$= 1 - r^{7t+6} + \frac{r^{7t-3}(r^2 - 1)^2}{1 - r^{7t}} = 1 - r^{7t+6} + \frac{2r^{7t+3}}{1 - r^{7t}}.$$

Hence $d_t = 0 \iff (1 - r^{7t+6})(1 - r^{7t}) = 2r^{7t+7}$. Now we check the validity of this equation for $1 \le t \le N - 2 = 6$:

$$\begin{aligned}
t = 1: &\qquad (1 - r^5)(1 - r^7) \neq 2r^6;\\
t = 2: &\qquad (1 - r^4)(1 - r^6) = 2(1 - r^6) \neq 2r^5;\\
t = 3: &\qquad (1 - r^3)(1 - r^5) \neq 2r^4;\\
t = 4: &\qquad (1 - r^2)(1 - r^4) = 2(1 - r^2) \neq 2r^3;\\
t = 5: &\qquad (1 - r)(1 - r^3) \neq 2r^2;\\
t = 6: &\qquad (1 - 1)(1 - r^2) = 0 \neq 2r.
\end{aligned}$$

Thus $d_t \neq 0$ for all $1 \le t \le 6$ so $y_1^8 \neq 0$ by Proposition 4.8, and hence $8\beta_1$ is a root by Lemma 4.9. This implies that $\operatorname{GKdim}\mathcal{B}(V) = \infty$ by Proposition 4.16.

- If $q_{11} = r^{17} \in \mathbb{G}'_{24}$, then $p_1 = r^{15}$, so $N = 8$ and

$$d_t = 1 - r^{15t-2} + \frac{r^{15t}(1 + r^2)(1 + r^{17})(1 + r^3)}{1 - r^{15t}}.$$

Hence $d_t = 0 \iff (1 - r^{15t-2})(1 - r^{15t}) = -r^{15t}(1 + r^2)(1 + r^{17})(1 + r^3)$. Now we check the validity of this equation for $1 \le t \le N - 2 = 6$:

$$\begin{aligned}
t = 1: &\qquad (1 - r^{13})(1 - r^{15}) \neq r^3(1 + r^2)(1 + r^{17})(1 + r^3);\\
t = 2: &\qquad (1 - r^4)(1 - r^6) \neq r^{18}(1 + r^2)(1 + r^{17})(1 + r^3);\\
t = 3: &\qquad (1 - r^{19})(1 - r^{21}) \neq r^9(1 + r^2)(1 + r^{17})(1 + r^3);\\
t = 4: &\qquad 2(1 - r^{10}) \neq (1 + r^2)(1 + r^{17})(1 + r^3);\\
t = 5: &\qquad (1 - r)(1 - r^3) \neq r^{15}(1 + r^2)(1 + r^{17})(1 + r^3);\\
t = 6: &\qquad (1 - r^{16})(1 - r^{18}) \neq r^6(1 + r^2)(1 + r^{17})(1 + r^3).
\end{aligned}$$

Thus $d_t \neq 0$ for all $1 \le t \le 6$ so $y_1^8 \neq 0$ by Proposition 4.8, and hence $8\beta_1$ is a root by Lemma 4.9. This implies that $\operatorname{GKdim}\mathcal{B}(V) = \infty$ by Proposition 4.16.

The last case is $q_{11} \in \mathbb{G}'_4$. Thus $r^5 = -q_{11}$. As $a^V_{12} = -3$ we have that $r \in \mathbb{G}'_{20}$. Again we compute $d_t$ for $n = 1$. Here, $p_1 = r^{13}$, so $N = 20$ and

$$d_t = 1 - r^{13t} + \frac{r^{13t}(1 + r^2)(1 + r^{15})(1 + r^3)}{1 - r^{13t}}.$$

Hence $d_t = 0 \iff (1-r^{13t})^2 = r^{13t+5}(1+r^2)(1+r^5)(1+r^3)$. Now we check the validity of this equation for $1 \le t \le N-2 = 18$:

$$\begin{aligned}
t=1: &\quad (1+r^3)^2 \neq r^{18}(1+r^2)(1+r^5)(1+r^3);\\
t=2: &\quad (1-r^6)^2 \neq r^{11}(1+r^2)(1+r^5)(1+r^3);\\
t=3: &\quad (1+r^9)^2 \neq r^{4}(1+r^2)(1+r^5)(1+r^3);\\
t=4: &\quad (1+r^2)^2 \neq r^{17}(1+r^2)(1+r^5)(1+r^3);\\
t=5: &\quad 2r^{15} \neq r^{10}(1+r^2)(1+r^5)(1+r^3);\\
t=6: &\quad (1+r^8)^2 \neq r^{3}(1+r^2)(1+r^5)(1+r^3);\\
t=7: &\quad (1+r)^2 \neq r^{16}(1+r^2)(1+r^5)(1+r^3);\\
t=8: &\quad (1-r^4)^2 \neq r^{9}(1+r^2)(1+r^5)(1+r^3);\\
t=9: &\quad (1+r^7)^2 \neq r^{2}(1+r^2)(1+r^5)(1+r^3);\\
t=10: &\quad 4 \neq r^{15}(1+r^2)(1+r^5)(1+r^3);\\
t=11: &\quad (1-r^3)^2 \neq r^{8}(1+r^2)(1+r^5)(1+r^3);\\
t=12: &\quad (1+r^6)^2 \neq r(1+r^2)(1+r^5)(1+r^3);\\
t=13: &\quad (1-r^9)^2 \neq r^{14}(1+r^2)(1+r^5)(1+r^3);\\
t=14: &\quad (1-r^2)^2 \neq r^{7}(1+r^2)(1+r^5)(1+r^3);\\
t=15: &\quad 2r^5 \neq (1+r^2)(1+r^5)(1+r^3);\\
t=16: &\quad (1-r^8)^2 \neq r^{13}(1+r^2)(1+r^5)(1+r^3);\\
t=17: &\quad (1-r)^2 \neq r^{6}(1+r^2)(1+r^5)(1+r^3);\\
t=18: &\quad (1+r^4)^2 \neq r^{19}(1+r^2)(1+r^5)(1+r^3).
\end{aligned}$$

Thus $d_t \neq 0$ for all $1 \le t \le 18$ so $y_1^{20} \neq 0$ by Proposition 4.8, and hence $20\beta_1$ is a root by Lemma 4.9. Again, $\operatorname{GKdim} \mathcal{B}(V) = \infty$ by Proposition 4.16. □

We finally assume that $\mathcal{B}(V)$ has finite GK-dimension. By Remark 2.5, $V$ admits all reflections. By Lemma 4.20, we may assume that $a_{21} \in \{-1,-2\}$. We consider below all possible cases under these assumptions and conclude in each case that the root system is finite–i. e. the Dynkin diagram appears in [**H2**, Table 1].

4.2.1. $a_{12}^V = a_{21}^V = -1$. We have that

$$(q_{11}\widetilde{q}_{12}-1)(2)_{q_{11}} = 0, \qquad (q_{22}\widetilde{q}_{12}-1)(2)_{q_{22}} = 0.$$

The four possible diagrams appear in [**H2**, Table 1, Rows 1 & 2].

4.2.2. $a_{12}^V = -2$, $a_{21}^V = -1$. We have that

$$(q_{11}^2\widetilde{q}_{12}-1)(3)_{q_{11}} = 0, \qquad (q_{22}\widetilde{q}_{12}-1)(2)_{q_{22}} = 0.$$

If $q_{22}\widetilde{q}_{12} = 1$, then we have [**H2**, Table 1, Rows 3 & 5].

Now we assume that $q_{22} = -1$. If $q_{11}^2\widetilde{q}_{12} = 1$, then we get [**H2**, Table 1, Row 4]. Let $q_{11} \in \mathbb{G}'_3$. For simplicity we set $q = q_{11}$, $r = \widetilde{q}_{12}$. Let $(t_{ij})_{i,j\in\mathbb{I}}$ be the braiding matrix of $\mathcal{R}^2(V)$: its Dynkin diagram is $\overset{-qr}{\circ}\xrightarrow{\ r^{-1}\ }\overset{-1}{\circ}$. We study the possible values of $a := a_{12}^{\mathcal{R}^2(V)}$. Notice that $a \le -2$.

- $a = -2$. If $1 = t_{11}^2\widetilde{t}_{12} = q^2r$, then $V$ appears in [**H2**, Table 1, Row 4]. Otherwise $1 = t_{11}^3 = -r^3$, so $r = -q^{\pm 1}$. As $1 \neq t_{11} = -qr$, we have that $r = -q$ and $V$ appears in [**H2**, Table 1, Row 6].
- $a = -3$. Either $1 = t_{11}^3\widetilde{t}_{12} = -r^2$, in which case $r \in \mathbb{G}_4'$ and $V$ appears in [**H2**, Table 1, Row 8], or else $1 = t_{11}^4 = qr^4$, in which case $r \in \mathbb{G}_{12}'$ and $V$ appears in [**H2**, Table 1, Row 7].
- $a \leq -4$. Notice that $\widetilde{w}_0 = 0$ by Lemma 4.14, and $\widetilde{w}_1 = 0$ by Lemma 4.15 (a) since $-t_{11}\widetilde{t}_{12} = q \in \mathbb{G}_3'$. Hence we may apply Lemma 4.12: as $\widetilde{w}_2 = 0$, the scalar in Remark 4.13 (v) is zero. That is,

$$0 = (1 + q^2r^5)(1 - qr^3)(1 + q^2r^2)(1 + q^2r^4).$$

  If $q^2r^5 = -1$, then $-r \in \mathbb{G}_{15}'$ and $V$ belongs to [**H2**, Table 1, Row 15]. If $qr^3 = 1$, then $r \in \mathbb{G}_9'$ and $V$ is in [**H2**, Table 1, Row 9]. If $q^2r^2 = -1$, then $r \in \mathbb{G}_{12}'$, $q = -r^2$ and $V$ belongs to [**H2**, Table 1, Row 7]. Otherwise $q^2r^4 = -1$, in which case $r \in \mathbb{G}_{24}'$ with $q = -r^4$, and $V$ is in [**H2**, Table 1, Row 12].

4.2.3. $a_{12}^V = -3$, $a_{21}^V = -1$ *with* $q_{22} = -1$. First we assume $q_{11}^3\widetilde{q}_{12} = 1$. Set $q = q_{11}$, so $\widetilde{q}_{12} = q^{-3}$. Let $(t_{ij})_{i,j\in\mathbb{I}}$ be the braiding matrix of $\mathcal{R}^2(V)$: its Dynkin diagram is $\overset{-q^{-2}}{\circ}\overset{q^3}{\text{———}}\overset{-1}{\circ}$.

- If $a_{12}^{\mathcal{R}^2(V)} = -2$, $\mathcal{R}^2(V)$ appears in §4.2.2, so $\mathcal{R}^2(V)$ has finite root system and then $V$ too.
- If $a_{12}^{\mathcal{R}^2(V)} = -3$, then then either $t_{11} \in \mathbb{G}_4'$, in which case $q \in \mathbb{G}_8'$ and $V$ belongs to [**H2**, Table 1, Row 11]; or else $t_{11}^3\widetilde{t}_{12} = 1$, in which case $q \in \mathbb{G}_6'$ and $V$ is of Cartan type $G_2$ [**H2**, Table 1, Row 10].
- If $a_{12}^{\mathcal{R}^2(V)} \leq -4$, then Remark 4.13 (v) says that $(1 + q^7)(1 - q^5)(5)_{-q^{-2}} = 0$. If $q^7 = -1$, then $V$ belongs to [**H2**, Table 1, Row 16]. If $q^5 = 1$, then $V$ belongs to [**H2**, Table 1, Row 13]. If $-q^2 \in \mathbb{G}_5'$, then $V$ belongs to [**H2**, Table 1, Row 14].

Finally, if $q_{11}^3\widetilde{q}_{12} \neq 1$, then $q_{11} = \eta \in \mathbb{G}_4'$. Set $q = \widetilde{q}_{12}$. By Lemma 4.15(a),

- either $\eta^3q^2 = -1$, so $q^2 = -\eta$ and $V$ belongs to [**H2**, Table 1, Row 11];
- or else $-\eta q \in \mathbb{G}_3'$, in which case $a_{12}^{\mathcal{R}^2(V)} = -2$, $a_{21}^{\mathcal{R}^2(V)} = -1$ since the diagram of $\mathcal{R}^2(V)$ is $\overset{-\eta q}{\circ}\overset{q^{-1}}{\text{———}}\overset{-1}{\circ}$. Hence $\mathcal{R}^2(V)$ appears in §4.2.2 and thus $\mathcal{R}^2(V)$ has finite root system.

4.2.4. $a_{12}^V = -3$, $a_{21}^V = -1$ *with* $q_{22}\widetilde{q}_{12} = 1$. If $q_{11}^3\widetilde{q}_{12} = 1$, then $V$ is of Cartan type $G_2$ and the root system is finite [**H2**, Table 1, Row 10]. Otherwise, $q_{11} = \eta \in \mathbb{G}_4'$. For simplicity we call $q = q_{22}$ so $\widetilde{q}_{12} = q^{-1}$. Let $(t_{ij})_{i,j\in\mathbb{I}}$ be the braiding matrix of $\mathcal{R}^1(V)$: its Dynkin diagram is $\overset{\eta}{\circ}\overset{-q}{\text{———}}\overset{q^{-2}\eta}{\circ}$. By Lemma 4.20, $a_{21}^{\mathcal{R}^1(V)} \geq -2$.

- If $a_{21}^{\mathcal{R}^1(V)} = -1$, then either $t_{22} = -1$, in which case $q^2 = -\eta$ and $V$ belongs to [**H2**, Table 1, Row 11]; or else $t_{22}\widetilde{t}_{12} = 1$, in which case $q = -\eta$ and $V$ is of Cartan type $G_2$.
- If $a_{21}^{\mathcal{R}^1(V)} = -2$, then either $1 = t_{22}^2\widetilde{t}_{12}$, in which case $q^3 = 1$, or else $1 = t_{22}^3$, which implies $q^6 = -\eta$. For the first case, we compute $p_1 = q^{-1}$, which has order $N = 3$, and by (4.6), $d_1 = -\eta q \neq 0$. Hence $y_1^3 \neq 0$ by Proposition 4.8. By Lemma 4.9, $3\beta_1$ is a root of $\mathcal{R}^1(V)$, thus $\operatorname{GKdim}\mathcal{B}(\mathcal{R}^1(V)) = \infty$ by Proposition 4.16. For the second case, $q \in \mathbb{G}_{24}'$ and $V$ belongs to [**H2**, Table 1, Row 12].

4.2.5. $a_{12}^V \le -4$, $a_{21}^V = -1$ *with* $q_{22} = -1$. For simplicity we set $q = q_{11}$, $r = \widetilde{q}_{12}$. Let $(t_{ij})_{i,j \in \mathbb{I}}$ be the braiding matrix of $\mathcal{R}^2(V)$: its Dynkin diagram is $\overset{-qr}{\circ} \overset{r^{-1}}{\text{------}} \overset{-1}{\circ}$. As $q^3 r \neq 1$, Lemma 4.15 (a) says that either $q^3 r^2 = -1$ or else $-qr \in \mathbb{G}'_3$.

- If $q^3r^2 = -1$, then $t_{11}^3 \widetilde{t}_{12} = -q^3 r^2 = 1$ so $a_{12}^{\mathcal{R}^2(V)} = -3$. Hence $\mathcal{R}^2(V)$ has a finite root system by §4.2.3, and $V$ too.
- If $-qr \in \mathbb{G}'_3$, then $a_{12}^{\mathcal{R}^2(V)} = -2$. Hence $\mathcal{R}^2(V)$ has a finite root system by §4.2.2, and $V$ too.

4.2.6. $a_{12}^V \le -4$, $a_{21}^V = -1$ *with* $q_{22}\widetilde{q}_{12} = 1$. For simplicity set $q = q_{11}$, $r = q_{22}$, so $\widetilde{q}_{12} = r^{-1}$. By Lemma 4.15 (b), $t := -q^2 r^{-1} \in \mathbb{G}'_3$ since $a_{12}^V \le -4$. As $y_4 \neq 0$ we may apply Lemma 4.12 and Remark 4.13 (vii):

$$0 = (1 - q^4 r^{-1})(1 + q^4 r^{-1})(1 - q^5 r^{-1})(3)_{-q}(5)_q.$$

- If $q^4 r^{-1} = 1$, then $V$ is of affine Cartan type and $\operatorname{GKdim} \mathcal{B}(V) = \infty$ by Proposition 3.1, a contradiction.
- If $q^4 r^{-1} = -1$, then $q^2 = (-q^4 r^{-1})(-q^2 r^{-1})^{-1} = t^2$, so $q = \pm t$, $r = -t$. If $q = t$, then $a_{12}^V \ge -2$; otherwise $q = -t$ and $q_{11}\widetilde{q}_{12} = qr^{-1} = 1$ so $a_{12}^V = -1$. In any case we obtain a contradiction with $a_{12}^V \le -4$.
- If $q^5 r^{-1} = 1$, then $q^3 = -t^{-1} \in \mathbb{G}'_6$. Hence $-q \in \mathbb{G}'_9$ and $p_3 = q^9 r^{-3} r = q^{-1} \in \mathbb{G}'_{18}$. Now we compute $d_l$ as in (4.6) for $n = 3$:
$$d_l = 1 - q^{-l-1} q^6 r^{-1} + \frac{q^{-l}(1 - q^3 r^{-1})(4)_q}{(l)_{q^{-1}}} = 1 - q^{-l} + \frac{q^{-l}(1 - q^{-2})(4)_q}{(l)_{q^{-1}}}$$
Thus $d_l = 0$ if and only if $q^{l+3}(1 - q^{-l})^2 = (q^2 - 1)(q^4 - 1)$, but this equality does not hold for $1 \le l \le 16$. Hence $18\beta_3$ is a root of $V$ by Proposition 4.8 and Lemma 4.9, so $\operatorname{GKdim} \mathcal{B}(V) = \infty$ by Proposition 4.16.
- If $-q \in \mathbb{G}'_3$, then either $t = -q$ or else $t = -q^{-1}$. Both are not possible since $a_{12}^V \le -4$.
- If $q \in \mathbb{G}'_5$, then $r = -t^{-1} q^2 \in \mathbb{G}'_{30}$. This is the case in [**H2**, Table 1, Row 15].

4.2.7. $a_{12}^V = a_{21}^V = -2$. We have that $(q_{11}\widetilde{q}_{12} - 1)(2)_{q_{11}}(q_{22}\widetilde{q}_{12} - 1)(2)_{q_{22}} \neq 0$,

$$(q_{11}^2 \widetilde{q}_{12} - 1)(3)_{q_{11}} = 0, \qquad (q_{22}^2 \widetilde{q}_{12} - 1)(3)_{q_{22}} = 0.$$

If $q_{11}^2 \widetilde{q}_{12} = 1 = q_{22}^2 \widetilde{q}_{12}$, then $V$ is of affine Cartan type, a contradiction with Proposition 3.1. Hence we may assume that $q_{11} \in \mathbb{G}'_3$. By Lemma 4.14, $q_{11}\widetilde{q}_{12}^2 q_{22} = -1$. Let $(t_{ij})_{i,j \in \mathbb{I}}$ be the braiding matrix of $\mathcal{R}^1(V)$:

$$t_{22} = q_{22}\widetilde{q}_{12}^2 q_{11}^4 = q_{22}\widetilde{q}_{12}^2 q_{11} = -1,$$

so $\mathcal{R}^1(V)$ appears in §4.2.2 since $\operatorname{GKdim} \mathcal{B}(\mathcal{R}^1(V)) < \infty$. Thus $\mathcal{R}^1(V)$ has a finite root system, and $V$ too.

4.2.8. $a_{12}^V \le -3$, $a_{21}^V = -2$. As $\widetilde{w}_0 = 0$, $q_{11}\widetilde{q}_{12}^2 q_{22} = -1$ by Lemma 4.14. As $\widetilde{w}_1 = 0$, either $q_{11}^3 \widetilde{q}_{12} = 1$ or else $q_{11}^2 = -1$ by Lemma 4.15 (c). Hence $a_{12} = -3$. We analyze the possible 4 cases.

If $q_{11}^3 \widetilde{q}_{12} = 1 = q_{22}^2 \widetilde{q}_{12}$, then $V$ is of indefinite Cartan type and

$$-q_{11}^2 q_{22} = (q_{11}\widetilde{q}_{12}^2 q_{22}) q_{11}^2 q_{22} = q_{11}^3 \widetilde{q}_{12} q_{22}^2 \widetilde{q}_{12} = 1,$$

thus $q_{22} = -q_{11}^{-2}$, and $1 = q_{22}^2 \widetilde{q}_{12} = q_{11}^{-7}$, so $q_{11} \in \mathbb{G}'_7$. Let $r := q_{22} \in \mathbb{G}'_{14}$, so $q_{11} = r^{10}$, $\widetilde{q}_{12} = r^{12}$. We compute $d_t$ as in (4.6) for $n = 1$. Here, $p_1 = r^9$, so

$N = 14$ and

$$d_t = 1 - r^{9t+13} + \frac{r^{9t}(1+r)(1-r^3)(1+r^2)}{1-r^{9t}}.$$

Hence $d_t = 0 \iff (1 - r^{9t+13})(1 - r^{9t}) = r^{9t}(1+r)(r^3-1)(1+r^2)$. Now we check the validity of this equation for $1 \le t \le N - 2 = 12$:

$$\begin{aligned}
t=1: &\quad (1+r)(1+r^2) \neq r^9(1+r)(r^3-1)(1+r^2);\\
t=2: &\quad (1-r^3)(1-r^4) \neq r^4(1+r)(r^3-1)(1+r^2);\\
t=3: &\quad (1+r^5)(1+r^6) \neq r^{13}(1+r)(r^3-1)(1+r^2);\\
t=4: &\quad 2(1+r) \neq r^8(1+r)(r^3-1)(1+r^2);\\
t=5: &\quad (1-r^2)(1-r^3) \neq r^3(1+r)(r^3-1)(1+r^2);\\
t=6: &\quad (1+r^4)(1+r^6) \neq r^{12}(1+r)(r^3-1)(1+r^2);\\
t=7: &\quad 2(1-r^6) \neq -(1+r)(r^3-1)(1+r^2);\\
t=8: &\quad (1-r)(1-r^2) \neq r^2(1+r)(r^3-1)(1+r^2);\\
t=9: &\quad (1+r^3)(1+r^4) \neq r^{11}(1+r)(r^3-1)(1+r^2);\\
t=10: &\quad (1-r^5)(1-r^6) \neq r^6(1+r)(r^3-1)(1+r^2);\\
t=11: &\quad 0 \neq r(1+r)(r^3-1)(1+r^2);\\
t=12: &\quad (1+r^2)(1+r^3) \neq r^{10}(1+r)(r^3-1)(1+r^2).
\end{aligned}$$

Thus $d_t \neq 0$ for all $1 \le t \le 12$ so $y_1^{14} \neq 0$ by Proposition 4.8, and hence $14\beta_1$ is a root by Lemma 4.9. Therefore $\operatorname{GKdim} \mathcal{B}(V) = \infty$ by Proposition 4.16.

Assume that $q_{11} = \eta \in \mathbb{G}'_4$, $q_{22}^2 \widetilde{q}_{12} = 1$: to simplify the notation, set $q = q_{22}$, so $\widetilde{q}_{12} = q^{-2}$. Then $-1 = q_{11}\widetilde{q}_{12}^2 q_{22} = \eta q^{-3}$, so $q^3 = -\eta$. The matrix $(t_{ij})_{i,j\in\mathbb{I}}$ of $\mathcal{R}^1(V)$ has diagram $\overset{\eta}{\circ} \xrightarrow{\;-q^2\;} \overset{-q^{-2}}{\circ}$. Thus $a_{21}^{\mathcal{R}^1(V)} = -1$ and then $\mathcal{R}^1(V)$ has finite root system, since $\operatorname{GKdim} \mathcal{B}(\mathcal{R}^1(V)) = \operatorname{GKdim} \mathcal{B}(V) < \infty$.

Assume now $q_{11}^3 \widetilde{q}_{12} = 1$, $q_{22} = \zeta \in \mathbb{G}'_3$. Let $q = q_{11}$ so $\widetilde{q}_{12} = q^{-3}$. The matrix $(t_{ij})_{i,j\in\mathbb{I}}$ of $\mathcal{R}^2(V)$ has diagram $\overset{-1}{\circ} \xrightarrow{\;\zeta^2 q^3\;} \overset{\zeta}{\circ}$. Thus $a_{12}^{\mathcal{R}^1(V)} = -1$ and then $\mathcal{R}^2(V)$ has finite root system, since $\operatorname{GKdim} \mathcal{B}(\mathcal{R}^2(V)) = \operatorname{GKdim} \mathcal{B}(V) < \infty$.

If $q_{11} = \eta \in \mathbb{G}'_4$, $q_{22} = \zeta \in \mathbb{G}'_3$, then $-1 = q_{11}\widetilde{q}_{12}^2 q_{22} = \eta q^2 \zeta$, where $q = \widetilde{q}_{12}$. Hence the matrix $(t_{ij})_{i,j\in\mathbb{I}}$ of $\mathcal{R}^2(V)$ has diagram $\overset{-1}{\circ} \xrightarrow{\;q^{-1}\zeta^2\;} \overset{\zeta}{\circ}$. Thus $\mathcal{R}^2(V)$ has finite root system, and $V$ too.

## References


[A] Nicolás Andruskiewitsch, *An introduction to Nichols algebras*, Quantization, geometry and noncommutative structures in mathematics and physics, Math. Phys. Stud., Springer, Cham, 2017, pp. 135–195. MR3751453

[AA1] Nicolás Andruskiewitsch and Iván Ezequiel Angiono, *On Nichols algebras with generic braiding*, Modules and comodules, Trends Math., Birkhäuser Verlag, Basel, 2008, pp. 47–64, DOI 10.1007/978-3-7643-8742-6_3. MR2742620

[AA2] Nicolás Andruskiewitsch and Iván Angiono, *On finite dimensional Nichols algebras of diagonal type*, Bull. Math. Sci. **7** (2017), no. 3, 353–573, DOI 10.1007/s13373-017-0113-x. MR3736568

[AAH] N. Andruskiewitsch, I. Angiono, I. Heckenberger. *On finite GK-dimensional Nichols algebras over abelian groups*. `arXiv:1606.02521`. Mem. Amer. Math. Soc., to appear.

[AS1] Nicolás Andruskiewitsch and Hans-Jürgen Schneider, *Pointed Hopf algebras*, New directions in Hopf algebras, Math. Sci. Res. Inst. Publ., vol. 43, Cambridge Univ. Press, Cambridge, 2002, pp. 1–68, DOI 10.2977/prims/1199403805. MR1913436

[An1] Iván Ezequiel Angiono, *A presentation by generators and relations of Nichols algebras of diagonal type and convex orders on root systems*, J. Eur. Math. Soc. (JEMS) **17** (2015), no. 10, 2643–2671, DOI 10.4171/JEMS/567. MR3420518

[C+] Lisa Carbone, Sjuvon Chung, Leigh Cobbs, Robert McRae, Debajyoti Nandi, Yusra Naqvi, and Diego Penta, *Classification of hyperbolic Dynkin diagrams, root lengths and Weyl group orbits*, J. Phys. A **43** (2010), no. 15, 155209, 30, DOI 10.1088/1751-8113/43/15/155209. MR2608277

[CH] M. Cuntz and I. Heckenberger, *Weyl groupoids with at most three objects*, J. Pure Appl. Algebra **213** (2009), no. 6, 1112–1128, DOI 10.1016/j.jpaa.2008.11.009. MR2498801

[GH] M. Graña and I. Heckenberger, *On a factorization of graded Hopf algebras using Lyndon words*, J. Algebra **314** (2007), no. 1, 324–343, DOI 10.1016/j.jalgebra.2007.02.046. MR2331765

[H1] I. Heckenberger, *The Weyl groupoid of a Nichols algebra of diagonal type*, Invent. Math. **164** (2006), no. 1, 175–188, DOI 10.1007/s00222-005-0474-8. MR2207786

[H2] I. Heckenberger, *Classification of arithmetic root systems*, Adv. Math. **220** (2009), no. 1, 59–124, DOI 10.1016/j.aim.2008.08.005. MR2462836

[HS] I. Heckenberger and H.-J. Schneider, *Root systems and Weyl groupoids for Nichols algebras*, Proc. Lond. Math. Soc. (3) **101** (2010), no. 3, 623–654, DOI 10.1112/plms/pdq001. MR2734956

[Kh] V. K. Kharchenko, *A quantum analogue of the Poincaré-Birkhoff-Witt theorem* (Russian, with Russian summary), Algebra Log. **38** (1999), no. 4, 476–507, 509, DOI 10.1007/BF02671731; English transl., Algebra and Logic **38** (1999), no. 4, 259–276. MR1763385

[K] Victor G. Kac, *Infinite-dimensional Lie algebras*, 3rd ed., Cambridge University Press, Cambridge, 1990. MR1104219

[KL] Günter R. Krause and Thomas H. Lenagan, *Growth of algebras and Gelfand-Kirillov dimension*, Revised edition, Graduate Studies in Mathematics, vol. 22, American Mathematical Society, Providence, RI, 2000. MR1721834

[Ma] Olivier Mathieu, *Classification of simple graded Lie algebras of finite growth*, Proceedings of the International Congress of Mathematicians, Vol. I, II (Kyoto, 1990), Math. Soc. Japan, Tokyo, 1991, pp. 799–806. MR1159266

[Mo] Susan Montgomery, *Hopf algebras and their actions on rings*, CBMS Regional Conference Series in Mathematics, vol. 82, Published for the Conference Board of the Mathematical Sciences, Washington, DC; by the American Mathematical Society, Providence, RI, 1993. MR1243637

[R] Marc Rosso, *Quantum groups and quantum shuffles*, Invent. Math. **133** (1998), no. 2, 399–416, DOI 10.1007/s002220050249. MR1632802

[Y] Hiroyuki Yamane, *Representations of a $\mathbb{Z}/3\mathbb{Z}$-quantum group*, Publ. Res. Inst. Math. Sci. **43** (2007), no. 1, 75–93. MR2317113

FaMAF-CIEM (CONICET), Universidad Nacional de Córdoba, Medina Allende s/n, Ciudad Universitaria, 5000 Córdoba, República Argentina.
*Email address*: `andrus@famaf.unc.edu.ar`

FaMAF-CIEM (CONICET), Universidad Nacional de Córdoba, Medina Allende s/n, Ciudad Universitaria, 5000 Córdoba, República Argentina.
*Email address*: `angiono@famaf.unc.edu.ar`

Philipps-Universität Marburg, Fachbereich Mathematik und Informatik, Hans-Meerwein-Strasse, D-35032 Marburg, Germany.
*Email address*: `heckenberger@mathematik.uni-marburg.de`

Contemporary Mathematics
Volume **728**, 2019
https://doi.org/10.1090/conm/728/14654

# On nonassociative graded-simple algebras over the field of real numbers

Yuri Bahturin and Mikhail Kochetov

ABSTRACT. We extend the loop algebra construction for algebras graded by abelian groups to study graded-simple algebras over the field of real numbers (or any real closed field). As an application, we classify up to isomorphism the graded-simple alternative (nonassociative) algebras and graded-simple finite-dimensional Jordan algebras of degree 2. We also classify the graded-division alternative (nonassociative) algebras up to equivalence.

## 1. Introduction

Let $\mathcal{A}$ be an algebra over a field $\mathbb{F}$ and let $G$ be a group. A *$G$-grading* on $\mathcal{A}$ is a vector space decomposition $\Gamma: \mathcal{A} = \bigoplus_{g\in G} \mathcal{A}_g$ such that $\mathcal{A}_g\mathcal{A}_h \subset \mathcal{A}_{gh}$ for all $g, h \in G$. The subspaces $\mathcal{A}_g$ are called the *homogeneous components*. A *$G$-graded algebra* is an algebra with a fixed $G$-grading. The nonzero elements $x \in \mathcal{A}_g$ are said to be *homogeneous of degree $g$*, which can be written as $\deg x = g$, and the *support* of $\Gamma$ (or of $\mathcal{A}$) is the set $\operatorname{Supp}\Gamma := \{g \in G \mid \mathcal{A}_g \neq 0\}$. A subspace $\mathcal{B}$ in $\mathcal{A}$ (in particular, a subalgebra or an ideal) is called *graded* if $\mathcal{B} = \bigoplus_{g\in G} \mathcal{B}_g$ where $\mathcal{B}_g := \mathcal{B} \cap \mathcal{A}_g$.

Group gradings have been extensively studied for associative, Lie, Jordan, alternative, and other types of algebras. In particular, a classification of gradings is known for many simple finite-dimensional algebras of these types. The situation is especially well understood if the grading group $G$ is abelian and the ground field $\mathbb{F}$ is algebraically closed (see e.g. a recent monograph [**13**] and the references therein).

In the context of $G$-graded algebras, it is natural to consider graded analogs of the standard concepts. For example, a *homomorphism of $G$-graded algebras* $\psi: \mathcal{A} \to \mathcal{B}$ is a homomorphism of algebras that preserves degrees: $\psi(\mathcal{A}_g) \subset \mathcal{B}_g$ for all $g \in G$. In particular, $\mathcal{A}$ and $\mathcal{B}$ are *graded-isomorphic* if there exists an isomorphism of graded algebras $\mathcal{A} \to \mathcal{B}$. A $G$-graded algebra $\mathcal{A}$ is called *graded-simple*, or *simple as a graded algebra*, if $\mathcal{A}^2 \neq 0$ and $\mathcal{A}$ has no graded ideals different from 0 and $\mathcal{A}$. Among these, we are especially interested in studying the *graded-division* algebras, which are defined by the property that every nonzero homogeneous element is

2010 *Mathematics Subject Classification.* Primary 16W50; Secondary 17D05, 17C20, 17C60.

*Key words and phrases.* Graded algebra, loop algebra, alternative algebra, Jordan algebra, division algebra, classification.

The authors acknowledge support by the Natural Sciences and Engineering Research Council (NSERC) of Canada (DG # 227060-2014 and 341792-2013).

invertible (in an appropriate sense). A typical example of a graded-division algebra is the group algebra $\mathbb{F}G$, which is naturally $G$-graded. More generally, the twisted group algebra $\mathbb{F}^\gamma G$ (where $\gamma : G \times G \to \mathbb{F}^\times$ is a 2-cocycle) is a graded-division associative algebra. A graded algebra $\mathcal{A}$ is said to be *graded-central* if the only degree-preserving elements of its centroid are the scalar operators. We recall that the *centroid* $C(\mathcal{A})$ of an algebra $\mathcal{A}$ consists of all elements of $\mathrm{End}_{\mathbb{F}}(\mathcal{A})$ that commute with the operators of left and right multiplication by the elements of $\mathcal{A}$. If $\mathcal{A}$ is unital, we can identify $C(\mathcal{A})$ with the center of $\mathcal{A}$ (i.e., the set of elements that commute and associate with all elements of $\mathcal{A}$).

For abelian $G$, a bridge between simple and graded-simple algebras is given by the *loop algebra* construction in [**1**] (which appeared in [**3**] in a special case). This construction generalizes the so-called multiloop algebras in Lie theory and, if $\mathbb{F}$ is algebraically closed, yields a classification of graded-central-simple (i.e., graded-central and graded-simple) algebras in a given class provided that a classification of gradings is known for central simple algebras in this class (see Section 3 for details). In this paper, we will assume all grading groups abelian, but $\mathbb{F}$ will be the field of real numbers $\mathbb{R}$. We will extend the loop construction to this setting. We do not use the topology of $\mathbb{R}$, so our results are valid over any real closed field $\mathbb{F}$ (with $\mathbb{F}[\sqrt{-1}]$ playing the role of $\mathbb{C}$).

Graded-division associative algebras that are finite-dimensional and graded-central over $\mathbb{R}$ have been classified in [**7**] up to equivalence (see Section 2). That work was preceded by [**6**], where the *division gradings* on finite-dimensional simple real associative algebras (i.e., the gradings that turn them into graded-division algebras) were classified up to equivalence, and [**22**], where these gradings were classified up to isomorphism as well as up to equivalence.

In this paper, we will focus on alternative and Jordan algebras. For the background on these, we refer the reader to [**23**] and [**18**].

Recall that an algebra is said to be *alternative* if the associator $(x, y, z) := (xy)z - x(yz)$ is an alternating function of the variables $x, y, z$ or, equivalently, the left and right alternative identities hold: $x(xy) = x^2y$ and $(yx)x = yx^2$. By Artin's Theorem, this implies that any subalgebra generated by two elements is associative. The definition of inverses in a unital alternative algebra is the same as in the associative case, and the set of invertible elements is closed under multiplication (it is a so-called *Moufang loop*). Kleinfeld proved that any simple alternative algebra is either associative or an octonion algebra over a field.

Group gradings on octonion algebras over any field were described in [**11**] (see also [**13**]). (Here, as in the case of simple Lie algebras, there is no loss of generality in assuming the grading group abelian because the elements of the support always commute.) A classification of gradings up to isomorphism is given in [**15**]. We will use our extension of the loop algebra construction to derive a classification up to isomorphism of graded-simple alternative algebras that are graded-central over $\mathbb{R}$ and not associative (Theorem 4.4). Among these, we will also classify the graded-division algebras up to equivalence (Theorem 4.6 and Corollary 4.8).

Let $\mathbb{F}$ be a field of characteristic different from 2. A *Jordan algebra* over $\mathbb{F}$ is a commutative algebra in which the identity $(xy)x^2 = x(yx^2)$ holds or, equivalently, the operators $L_x$ and $L_{x^2}$ commute for any element $x$, where $L_x(y) := xy$. Any associative algebra $\mathcal{A}$ gives rise to a Jordan algebra, denoted $\mathcal{A}^{(+)}$, which coincides with $\mathcal{A}$ as a vector space, but whose multiplication is given by $x \circ y := \frac{1}{2}(xy + yx)$.

The Jordan algebras that can be embedded into $\mathcal{A}^{(+)}$ for some associative algebra $\mathcal{A}$ are called *special*. Zelmanov proved that any simple Jordan algebra is either special or an Albert algebra (which has dimension 27 over its center).

Abelian group gradings on finite-dimensional simple special Jordan algebras over an algebraically closed field of characteristic 0 were described in [**4, 5**] and classified up to isomorphism and up to equivalence in [**13**] for any characteristic different from 2. The finite-dimensional central simple special Jordan algebras of degree $\geq 3$ over $\mathbb{R}$ can be dealt with following the approach of [**2**], where a classification of gradings for classical central simple Lie algebras over $\mathbb{R}$ was obtained by transferring the problem to associative algebras with involution. Here we will consider degree 2, i.e., Jordan algebras of bilinear forms. We will apply our extension of the loop construction to classify the corresponding graded-simple algebras up to isomorphism (Theorem 5.2), and characterize those that are graded-division algebras (Corollary 5.5).

It should be noted that abelian group gradings on the Albert algebra over $\mathbb{C}$ were classified in [**10**] and over any algebraically closed field of characteristic different from 2 in [**12**] (see also [**13**]). In particular, it has a division grading by the group $\mathbb{Z}_3^3$ (which appeared in [**17**]) if the characteristic is not 3. However, none of the three real forms of the complex Albert algebra admits a division grading, as follows from the classification of fine gradings in [**9**] (since the identity component always contains nontrivial idempotents).

The paper is structured as follows. In Section 2, we review some terminology and facts about gradings. Section 3 is devoted to the loop construction over $\mathbb{R}$: Theorem 3.3 establishes the existence of a "loop model" for a graded-simple algebra and Corollary 3.10 answers the question to what extent it is unique. (The answer turns out to be the same as for the case of "split centroid" in [**1**].) In fact, we also give a description of isomorphisms between graded-simple algebras, which can be conveniently expressed as equivalence of certain groupoids (Theorem 3.9). Sections 4 and 5 deal with our applications of the loop construction to alternative and Jordan algebras, respectively.

Throughout the paper, an unadorned symbol $\otimes$ denotes the tensor product over the ground field, which is usually $\mathbb{R}$. We will also use the following notation for abelian groups: for any $m \in \mathbb{N}$, we have the group homomorphism $[m] : G \to G$ sending $g \mapsto g^m$, and the associated subgroups of $G$: $G_{[m]} = \ker[m]$ and $G^{[m]} = \operatorname{im}[m]$. The order of an element $g \in G$ will be denoted by $o(g)$.

## 2. Background on gradings

**2.1. Gradings induced by a group homomorphism.** Given a $G$-grading $\Gamma : \mathcal{A} = \bigoplus_{g\in G} \mathcal{A}_g$ and a group homomorphism $\alpha : G \to G'$, we can define a $G'$-grading on $\mathcal{A}$ as follows: for any $g' \in G'$, set $\mathcal{A}'_{g'} := \bigoplus_{g\in\alpha^{-1}(g')} \mathcal{A}_g$. Then $\mathcal{A} = \bigoplus_{g'\in G'} \mathcal{A}'_{g'}$ is a $G'$-grading, which will be denoted by ${}^{\alpha}\Gamma$. Thus, if $\alpha$ is fixed, every $G$-graded algebra $\mathcal{A}$ can be considered as a $G'$-graded algebra, which will be denoted by ${}^{\alpha}\mathcal{A}$ or just $\mathcal{A}$ if there is no risk of confusion. In particular, if $G$ is a subgroup of $G'$ then every $G$-graded algebra is also $G'$-graded, with the same support and homogeneous components. More generally, if $\alpha$ is injective on the support then ${}^{\alpha}\Gamma$ has the same components as $\Gamma$, but relabeled through $\alpha$. Otherwise, ${}^{\alpha}\Gamma$ is a so-called *proper coarsening* of $\Gamma$.

If $\psi : \mathcal{A} \to \mathcal{B}$ is a homomorphism of $G$-graded algebras and we consider $\mathcal{A}$ and $\mathcal{B}$ as $G'$-graded through $\alpha$ as above, then $\psi$ is also a homomorphism of $G'$-graded algebras. Thus, $\alpha$ determines a functor $F_\alpha$ from the category of $G$-graded algebras to the category of $G'$-graded algebras: $F_\alpha(\mathcal{A}) = {}^\alpha\mathcal{A}$ and $F_\alpha(\psi) = \psi$.

**2.2. Weak isomorphism and equivalence.** We call a $G$-graded algebra $\mathcal{A}$ and a $G'$-graded algebra $\mathcal{B}$ *graded-weakly-isomorphic* if ${}^\alpha\mathcal{A}$ is graded-isomorphic to $\mathcal{B}$ for some group isomorphism $\alpha : G \to G'$, i.e., there exists an isomorphism of algebras $\psi : \mathcal{A} \to \mathcal{B}$ such that $\psi(\mathcal{A}_g) = \mathcal{B}_{\alpha(g)}$ for all $g \in G$.

There is a more general kind of relabeling: we say that a $G$-graded algebra $\mathcal{A}$ and a $G'$-graded algebra $\mathcal{B}$ are *graded-equivalent* if there exists an isomorphism of algebras $\psi : \mathcal{A} \to \mathcal{B}$ such that, for each $g \in G$, there is $g' \in G'$ such that $\psi(\mathcal{A}_g) = \mathcal{B}_{g'}$.

The support of a division grading on an alternative (in particular, associative) algebra is a subgroup. If, for two such graded-division algebras, we replace the grading groups by the supports, then equivalence is the same as weak isomorphism.

**2.3. Graded modules.** Let $\mathcal{A}$ be a $G$-graded associative algebra over a field $\mathbb{F}$. A $G$-graded vector space $\mathcal{V}$ over $\mathbb{F}$, i.e., a vector space with a fixed decomposition $\mathcal{V} = \bigoplus_{g\in G} \mathcal{V}_g$, is a *graded left $\mathcal{A}$-module* if it is equipped with a left $\mathcal{A}$-action such that $\mathcal{A}_g\mathcal{V}_h \subset \mathcal{V}_{gh}$ for all $g, h \in G$. The concepts of *homogeneous element*, *support*, *isomorphism*, etc., are defined in the same way as for graded algebras.

If $\mathcal{A}$ is a graded-division algebra then every graded $\mathcal{A}$-module is free. More precisely, it admits a basis consisting of homogeneous elements, the $\mathbb{F}$-span $V$ of such a basis is a $G$-graded vector space, and $\mathcal{V} \cong \mathcal{A} \otimes V$ as graded $\mathcal{A}$-modules, where the $G$-grading on the tensor product of $G$-graded vector spaces $U$ and $V$ is defined by setting $\deg(u \otimes v) := \deg(u)\deg(v)$.

## 3. Loop construction

**3.1. Centroid of a graded algebra.** Given a $G$-graded algebra $\mathcal{B}$ over a field $\mathbb{F}$, a linear map $f : \mathcal{B} \to \mathcal{B}$ is said to be homogeneous of degree $g \in G$ if $f(\mathcal{B}_h) \subset \mathcal{B}_{gh}$ for all $h \in G$. In particular, the set of all elements of the centroid $C(\mathcal{B})$ that satisfy this condition will be denoted by $C(\mathcal{B})_g$.

If $\psi : \mathcal{B} \to \mathcal{B}'$ is an isomorphism of $G$-graded algebras then we have an isomorphism of the centroids $C(\psi) : C(\mathcal{B}) \to C(\mathcal{B}')$ defined by $C(\psi)(c) = \psi \circ c \circ \psi^{-1}$ for all $c \in C(\mathcal{B})$. (If the algebras are unital and we identify the centroids with the centers then $C(\psi)$ is just a restriction of $\psi$.) It is clear that $C(\psi)$ maps $C(\mathcal{B})_g$ onto $C(\mathcal{B}')_g$ for any $g \in G$.

Suppose $\mathcal{B}$ is graded-simple. Then, by [**8**, Proposition 2.16], we have $C(\mathcal{B}) = \bigoplus_{g\in G} C(\mathcal{B})_g$, $C(\mathcal{B})$ is a commutative associative graded-division algebra (a *graded-field*), and $\mathcal{B}$ is a graded $C(\mathcal{B})$-module (hence free). Moreover, the graded-simple algebra $\mathcal{B}$ is simple if and only if $C(\mathcal{B})$ is a field [**1**, Lemma 4.2.2]. In any case, the identity component $C(\mathcal{B})_e$ is a field containing $\mathbb{F}$, and $\mathcal{B}$ is a graded-central algebra over $C(\mathcal{B})_e$, with exactly the same homogeneous components, but now viewed as $C(\mathcal{B})_e$-subspaces. For this reason, it is natural to restrict ourselves to the graded-central case.

So, suppose $\mathcal{B}$ is a graded-central-simple algebra over $\mathbb{F}$. Then the homogeneous components of $C(\mathcal{B})$ are one-dimensional: $C(\mathcal{B})_h = \mathbb{F}u_h$ for every $h \in H$ where $H$ is the support of $C(\mathcal{B})$, which is a subgroup of $G$, and $u_h$ is an invertible element.

Thus, $C(\mathcal{B})$ is a twisted group algebra of $H$, with its natural $H$-grading considered as a $G$-grading. Following [**1**], we will say that the centroid is *split* if $C(\mathcal{B})$ is graded-isomorphic to the group algebra $\mathbb{F}H$. This is always the case if $\mathbb{F}$ is algebraically closed [**1**, Lemma 4.3.8], but not if $\mathbb{F}$ is the field of real numbers. For example, $C(\mathcal{B})$ can be the field $\mathbb{C}$ with a nontrivial grading: $\mathbb{C} = \mathbb{C}_e \oplus \mathbb{C}_h$ where $\mathbb{C}_e = \mathbb{R}$, $\mathbb{C}_h = \mathbb{R}i$, and $h$ is an element of order 2 in $G$. The loop construction in [**1**], which we are now going to review, produces graded-central-simple algebras with split centroid. We will extend this construction to cover all possible centroids in the case $\mathbb{F} = \mathbb{R}$ (or any real closed field).

**3.2. Loop algebras with a split centroid.** Let $\pi : G \to \overline{G}$ be an epimorphism of abelian groups and let $H$ be the kernel of $\pi$. As discussed in Subsection 2.1, any $G$-graded algebra $\mathcal{B}$ can be regarded as a $\overline{G}$-graded algebra, by setting $\mathcal{B}_{\overline{g}} = \bigoplus_{g\in\pi^{-1}(\overline{g})} \mathcal{B}_g$ for any $\overline{g} \in \overline{G}$, and this gives us a functor from the category of $G$-graded algebras to the category of $\overline{G}$-graded algebras. The loop construction is the right adjoint of this functor (see [**14**, Remark 3.3]), defined as follows. For a given $\overline{G}$-graded algebra $\mathcal{A}$, consider the tensor product $\mathcal{A}\otimes\mathbb{F}G$. The loop algebra $L_\pi(\mathcal{A})$ is the following subalgebra of $\mathcal{A}\otimes\mathbb{F}G$:

$$L_\pi(\mathcal{A}) = \bigoplus_{g\in G} \mathcal{A}_{\pi(g)} \otimes g,$$

which is naturally $G$-graded: $L_\pi(\mathcal{A})_g = \mathcal{A}_{\pi(g)} \otimes g$. If $\psi : \mathcal{A} \to \mathcal{A}'$ is a homomorphism of $\overline{G}$-graded algebras then the linear map $L_\pi(\psi) : L_\pi(\mathcal{A}) \to L_\pi(\mathcal{A}')$ that sends $a \otimes g \mapsto \psi(a) \otimes g$, for all $a \in \mathcal{A}_{\pi(g)}$ and $g \in G$, is a homomorphism of $G$-graded algebras.

Note that $L_\pi(\mathcal{A})$ is unital if and only if so is $\mathcal{A}$. Also, if $\mathfrak{V}$ is a variety of algebras, i.e., a class defined by polynomial identities, and $\mathbb{F}$ is infinite, then $L_\pi(\mathcal{A})$ belongs to $\mathfrak{V}$ if and only if so does $\mathcal{A}$. Indeed, if $\mathcal{A} \in \mathfrak{V}$ then $\mathcal{A}\otimes\mathbb{F}G \in \mathfrak{V}$ (since $\mathbb{F}G$ is commutative, see e.g. [**16**, p. 10]) and hence $L_\pi(\mathcal{A}) \in \mathfrak{V}$; the converse follows from the fact that $\mathcal{A}$ is a quotient of $L_\pi(\mathcal{A})$ (by means of $a \otimes g \mapsto a$).

One of the assertions of the "Correspondence Theorem" (see [**1**, Theorem 7.1.1]) is that any $G$-graded algebra $\mathcal{B}$ that is graded-central-simple and has a split centroid is graded-isomorphic to $L_\pi(\mathcal{A})$ for some central simple algebra $\mathcal{A}$ equipped with a $\overline{G}$-grading where $\overline{G} = G/H$, $H$ is the support of $C(\mathcal{B})$, and $\pi : G \to \overline{G}$ is the natural homomorphism. Note that the condition on the centroid of $\mathcal{B}$ is necessary because, for any central simple algebra $\mathcal{A}$ with a $\overline{G}$-grading, $L_\pi(\mathcal{A})$ is a graded-central-simple algebra whose centroid can be identified with $L_\pi(\mathbb{F}) = \mathbb{F}H$, where $h \in H$ acts on $L_\pi(\mathcal{A})$ as follows:

$$h(a \otimes g) = a \otimes hg \text{ for all } a \in \mathcal{A}_{\pi(g)},\ g \in G.$$

We will generalize the loop construction to produce $G$-graded algebras with nonsplit centroids over $\mathbb{R}$.

Another assertion of the "Correspondence Theorem" tells us that the $G$-graded algebra $\mathcal{B}$ determines the $\overline{G}$-graded algebra $\mathcal{A}$ up to isomorphism and twist in the following sense. Fix a transversal for the subgroup $H$ in $G$, so we have a section $\xi : \overline{G} \to G$ (which is not necessarily a group homomorphism). This defines a 2-cocycle on $\overline{G}$ with values in $H$, namely,

$$\sigma(\overline{g}_1, \overline{g}_2) = \xi(\overline{g}_1)\xi(\overline{g}_2)\xi(\overline{g}_1\overline{g}_2)^{-1}.$$

Now, given a character $\lambda : H \to \mathbb{F}^\times$ and a $\overline{G}$-graded algebra $\mathcal{A}$, we define $\mathcal{A}^\lambda$ to be $\mathcal{A}$ as a $\overline{G}$-graded space, but with a twisted multiplication:

$$a_1 * a_2 = \lambda(\sigma(\overline{g}_1, \overline{g}_2))a_1 a_2 \text{ for all } a_1 \in \mathcal{A}_{\overline{g}_1},\ a_2 \in \mathcal{A}_{\overline{g}_2}. \tag{1}$$

In the above formula, $\lambda \circ \sigma$ is a 2-cocycle on $\overline{G}$ with values in $\mathbb{F}^\times$, so this is actually a standard cocycle twist of a graded algebra. Note that the isomorhism class of $\mathcal{A}^\lambda$ does not depend on the choice of the transversal. Also, if $\lambda$ extends to a character $G \to \mathbb{F}^\times$ (for example, if $\mathbb{F}$ is algebraically closed or if $H$ is a direct summand of $G$) then $\mathcal{A}^\lambda$ is graded-isomorphic to $\mathcal{A}$.

REMARK 3.1. *As a graded $L_\pi(\mathbb{F})$-module, $L_\pi(\mathcal{A}) \cong L_\pi(\mathbb{F}) \otimes \underline{\mathcal{A}}$ where $\underline{\mathcal{A}}$ is $\mathcal{A}$ regarded as a $G$-graded vector space with $\deg a = \xi(\overline{g})$ for all nonzero $a \in \mathcal{A}_{\overline{g}}$, $\overline{g} \in \overline{G}$. An isomorphism $L_\pi(\mathbb{F}) \otimes \underline{\mathcal{A}} \to L_\pi(\mathcal{A})$ is given by $h \otimes a \mapsto a \otimes h\xi(\overline{g})$, where we have identified $L_\pi(\mathbb{F})$ with $\mathbb{F}H$. In particular, if $\xi$ is a group homomorphism (which can happen if and only if $H$ is a direct summand of $G$) then $\underline{\mathcal{A}} = {}^\xi\!\mathcal{A}$ is a $G$-graded algebra, and $L_\pi(\mathcal{A}) \cong \mathbb{F}H \otimes {}^\xi\!\mathcal{A}$ as graded algebras.*

We are going to generalize, over the field $\mathbb{R}$, the above two assertions of the "Correspondence Theorem" to graded-central-simple algebras whose centroid is not necessarily split. The following result will be useful:

LEMMA 3.2. *If $\mathbb{F}$ is algebraically closed then every graded-automorphism of the centroid $L_\pi(\mathbb{F}) = \mathbb{F}H$ is induced by a graded-automorphism of $L_\pi(\mathcal{A})$.*

PROOF. Any automorphism of $\mathbb{F}H$ as a graded algebra has the form: $\psi_0(h) = \mu(h)h$, for any $h \in H$, where $\mu : H \to \mathbb{F}^\times$ is a character. We extend $\mu$ to a character of $G$ and set $\psi(a \otimes g) = \mu(g)a \otimes g$ for all $g \in G$ and $a \in \mathcal{A}_{\pi(g)}$. This is an automorphism of $L_\pi(\mathcal{A})$ and, for any $c \in \mathbb{F}H$, we have $\psi(c(a \otimes g)) = \psi_0(c)(\psi(a \otimes g))$, which means that $C(\psi) = \psi_0$. □

**3.3. Real loop algebras with a nonsplit centroid.** From now on, assume that the ground field is $\mathbb{R}$ (or, more generally, any real closed field). Let $\pi : G \to \overline{G}$ be an epimorphism of abelian groups, $H = \ker \pi$. We will apply the loop construction from the previous subsection over the algebraic closure $\mathbb{C}$ and then define a suitable Galois descent to $\mathbb{R}$. Denote the generator of $\mathrm{Gal}(\mathbb{C}/\mathbb{R})$ by $\iota$, i.e., $\iota(z) = \bar{z}$ for all $z \in \mathbb{C}$.

Let $\chi : G \to \mathbb{C}^\times$ be a character with values in the unit circle $U \subset \mathbb{C}$ (which is automatic for finite groups). Then the $\mathbb{R}$-linear operator on $\mathbb{C}G = \mathbb{C} \otimes \mathbb{R}G$ defined by $z \otimes g \mapsto \bar{z}\chi(g) \otimes g$ is involutive, so we obtain a semilinear action of $\mathrm{Gal}(\mathbb{C}/\mathbb{R})$ on $\mathbb{C}G$ and hence a real form of $\mathbb{C}G$ as a $G$-graded algebra. (In fact, any such real form is obtained in this way.) The homogeneous component of degree $g$ is $\mathbb{R}u_g$, where $u_g = z_g \otimes g$ and $z_g \in \mathbb{C}^\times$ is any element satisfying $z_g/\bar{z}_g = \chi(g)$. We have

$$u_{g_1} u_{g_2} = \gamma(g_1, g_2) u_{g_1 g_2} \text{ for all } g_1, g_2 \in G,$$

so our real form is the twisted group algebra $\mathcal{D} = \mathbb{R}^\gamma G$ where the 2-cocycle $\gamma : G \times G \to \mathbb{R}^\times$ is given by $\gamma(g_1, g_2) = z_{g_1} z_{g_2} / z_{g_1 g_2}$. We will choose $z_g$ to be in the unit circle, i.e., a square root of $\chi(g)$. Then $\gamma$ takes values in $\{\pm 1\}$.

Now let $\mathcal{A}$ be a $\overline{G}$-graded algebra over $\mathbb{R}$. Similarly to the above, $\mathcal{A} \otimes \mathbb{C} \otimes \mathbb{R}G$ acquires a semilinear action of $\mathrm{Gal}(\mathbb{C}/\mathbb{R})$ defined by $\iota \cdot (a \otimes z \otimes g) = a \otimes \bar{z}\chi(g) \otimes g$, which restricts to a degree-preserving action on $L_\pi(\mathcal{A} \otimes \mathbb{C}) = \bigoplus_{g \in G} \mathcal{A}_{\pi(g)} \otimes \mathbb{C} \otimes \mathbb{R}G$.

As a result, we obtain the following $G$-graded $\mathbb{R}$-form of $L_\pi(\mathcal{A}\otimes\mathbb{C})$:

$$L_\pi^\chi(\mathcal{A}) := \{x \in L_\pi(\mathcal{A}\otimes\mathbb{C}) \mid \iota\cdot x = x\} = \bigoplus_{g\in G} \mathcal{A}_{\pi(g)} \otimes u_g \subset \mathcal{A}\otimes\mathbb{R}^\gamma G.$$

By construction, $L_\pi^\chi(\mathcal{A})\otimes\mathbb{C} \cong L_\pi(\mathcal{A}\otimes\mathbb{C})$. It follows that $L_\pi^\chi(\mathcal{A})$ is unital if and only if so is $\mathcal{A}$, and $L_\pi^\chi(\mathcal{A})$ belongs to a variety $\mathfrak{V}$ if and only if so does $\mathcal{A}$. Indeed, we already know that $L_\pi$ preserves these properties, and they are not affected by field extensions. By the same argument, if $\mathcal{A}$ is central simple then $L_\pi^\chi(\mathcal{A})$ is graded-central-simple.

The action of $\mathrm{Gal}(\mathbb{C}/\mathbb{R})$ passes on to the centroid of $L_\pi(\mathcal{A}\otimes\mathbb{C})$: $(\iota\cdot c)(x) = \iota\cdot c(\iota^{-1}\cdot x)$, for any $x\in L_\pi(\mathcal{A}\otimes\mathbb{C})$ and $c$ in the centroid. Assuming that $\mathcal{A}$ is central simple, we have identified the centroid with $\mathbb{C}H = \mathbb{C}\otimes\mathbb{R}H$. With this identification, we have $\iota\cdot(z\otimes h) = \bar{z}\chi(h)\otimes h$ for all $z\in\mathbb{C}$ and $h\in H$. Therefore, the centroid of $L_\pi^\chi(\mathcal{A})$ can be identified with $\mathcal{D}_H := \bigoplus_{h\in H}\mathcal{D}_h$, which is a graded subalgebra of the real form $\mathcal{D}$ of $\mathbb{C}G$ defined above. Note that $\mathcal{D}_H = L_\pi^\chi(\mathbb{R})$.

We are now ready to extend a part of the "Correspondence Theorem" to the case of nonsplit centroid.

THEOREM 3.3. *Let $\mathcal{B}$ be a $G$-graded algebra that is graded-central-simple over $\mathbb{R}$. Let $H$ be the support of the centroid $C(\mathcal{B})$ and $\pi : G \to \overline{G} = G/H$ be the natural homomorphism. Then, for some character $\chi : G \to U \subset \mathbb{C}^\times$, there exists a central simple algebra $\mathcal{A}$ over $\mathbb{R}$ and a $\overline{G}$-grading on $\mathcal{A}$ such that $\mathcal{B} \cong L_\pi^\chi(\mathcal{A})$ as $G$-graded algebras. Moreover, we can take any character satisfying $L_\pi^\chi(\mathbb{R}) \cong C(\mathcal{B})$.*

PROOF. Since $\mathbb{C}$ is algebraically closed, the centroid of $\mathcal{B}\otimes\mathbb{C}$ is split, hence, by the "Correspondence Theorem" from [**1**], there exists a central simple complex algebra $\widetilde{\mathcal{A}}$ with a $\overline{G}$-grading such that $\mathcal{B}\otimes\mathbb{C} \cong L_\pi(\widetilde{\mathcal{A}})$. Hence, $\mathcal{B}$ is isomorphic to an $\mathbb{R}$-form of $L_\pi(\widetilde{\mathcal{A}})$, and we need to determine all semilinear $\mathrm{Gal}(\mathbb{C}/\mathbb{R})$-actions on the latter (up to isomorphism) to recover $\mathcal{B}$ by Galois decent.

First consider the induced $\mathrm{Gal}(\mathbb{C}/\mathbb{R})$-action on the centroid $\mathbb{C}H$ of $L_\pi(\widetilde{\mathcal{A}})$. As already observed, any automorphism of $\mathbb{C}H$ as a graded $\mathbb{C}$-algebra is of the form $zh \mapsto z\chi(h)h$, where $\chi : H \to \mathbb{C}^\times$ is a character. It follows that any degree-preserving $\mathbb{C}$-antilinear automorphism has the form $zh \mapsto \bar{z}\chi(h)h$. In particular, we have $\iota\cdot(zh) = \bar{z}\chi(h)h$ for some $\chi$, which must take values in the unit circle $U$ because $\iota^2 = \mathrm{id}$. Since $U$ is a divisible abelian group, we can extend $\chi$ to the whole of $G$, with all values still in $U$.

Since the action of $\mathrm{Gal}(\mathbb{C}/\mathbb{R})$ on $L_\pi(\widetilde{\mathcal{A}}) = \bigoplus_{g\in G}\widetilde{\mathcal{A}}_{\pi(g)}\otimes g$ is degree-preserving, we can write, for any $g\in G$, $\iota\cdot(a\otimes g) = \varphi_g(a)\chi(g)\otimes g$ for all $a\in\widetilde{\mathcal{A}}_{\pi(g)}$. It is clear that $\varphi_g : \widetilde{\mathcal{A}}_{\pi(g)} \to \widetilde{\mathcal{A}}_{\pi(g)}$ is a $\mathbb{C}$-antilinear map. Also, $\chi(g)\overline{\chi(g)} = 1$ implies $\varphi_g^2 = \mathrm{id}$. We claim that $\varphi_g$ depends only on $\pi(g)$. Indeed, for all $a\in\widetilde{\mathcal{A}}_{\pi(g)}$ and $h\in H$, we can consider $h$ as an element of the centroid and compute:

$$\begin{aligned}\varphi_{hg}(a)\chi(hg)\otimes hg &= \iota\cdot(a\otimes hg) = \iota\cdot\big(h(a\otimes g)\big) = (\iota\cdot h)\big(\iota\cdot(a\otimes g)\big)\\ &= \chi(h)h\big(\varphi_g(a)\chi(g)\otimes g\big) = \chi(h)\chi(g)\varphi_g(a)\otimes hg.\end{aligned}$$

Thus, $\varphi_{hg} = \varphi_g$, which proves the claim.

Define $\varphi : \widetilde{\mathcal{A}} \to \widetilde{\mathcal{A}}$ by setting $\varphi|_{\widetilde{\mathcal{A}}_{\pi(g)}} = \varphi_g$, for any $g \in G$. Clearly, $\varphi$ is $\mathbb{C}$-antilinear and $\varphi^2 = \mathrm{id}$. Moreover, $\varphi$ is an isomorphism:

$$\begin{aligned}\big(\varphi_g(a)\chi(g)\otimes g\big)\big(\varphi_{g'}(a')\chi(g')\otimes g'\big) &= \big(\iota\cdot(a\otimes g)\big)\big(\iota\cdot(a'\otimes g')\big) = \iota\cdot(aa'\otimes gg') \\ &= \varphi_{gg'}(aa')\chi(gg')\otimes gg',\end{aligned}$$

for all $g, g' \in G, a, a' \in \widetilde{\mathcal{A}}$, so $\varphi_g(a)\varphi_{g'}(a') = \varphi_{gg'}(aa')$, as claimed.

Therefore, $\varphi$ gives a Galois descent for $\widetilde{\mathcal{A}}$. Let $\mathcal{A} = \bigoplus_{\overline{g}\in\overline{G}} \mathcal{A}_{\overline{g}}$ where

$$\mathcal{A}_{\overline{g}} = \{a \in \widetilde{\mathcal{A}}_{\overline{g}} \mid \varphi(a) = a\}.$$

Then $\mathcal{A}\otimes\mathbb{C} \cong \widetilde{\mathcal{A}}$ by means of the mapping $a\otimes z \mapsto az$. Finally, the algebra of fixed points of $\iota$ in $L_\pi(\widetilde{\mathcal{A}})$ is the $\mathbb{R}$-span of the set of all $az_g\otimes g$ where $a \in \mathcal{A}_{\pi(g)}$ and $z_g \in \mathbb{C}^\times$ satisfy $z_g/\bar{z}_g = \chi(g)$. This span is isomorphic to $L^\chi_\pi(\mathcal{A})$ under the isomorphism $L_\pi(\mathcal{A}\otimes\mathbb{C}) \to L_\pi(\widetilde{\mathcal{A}})$ sending $a\otimes z\otimes g$ to $az\otimes g$.

For the last assertion, observe that, in view of Lemma 3.2, the group of automorphisms of the graded $\mathbb{C}$-algebra $L_\pi(\widetilde{\mathcal{A}})$ acts transitively on each isomorphism class of graded $\mathbb{R}$-forms of the centroid $\mathbb{C}H$. This allows us to adjust the $\mathrm{Gal}(\mathbb{C}/\mathbb{R})$-action on $L_\pi(\widetilde{\mathcal{A}})$ so that the induced action on the centroid is given by $\iota\cdot(zh) = \bar{z}\chi(h)h$ where $\chi$ is any character such that the corresponding graded $\mathbb{R}$-form of $\mathbb{C}H$ (spanned by the elements $z_h h$) is isomorphic to $C(\mathcal{B})$. □

The nonzero homogeneous elements of degree $g$ in $L^\chi_\pi(\mathcal{A})$ have the form $a\otimes u_g$ where $0 \neq a \in \mathcal{A}_{\pi(g)}$, so the support of $L^\chi_\pi(\mathcal{A})$ is the inverse image under $\pi$ of the support of $\mathcal{A}$. Remark 3.1 has the following analog:

REMARK 3.4. *As a graded $L^\chi_\pi(\mathbb{F})$-module, $L^\chi_\pi(\mathcal{A}) \cong L^\chi_\pi(\mathbb{F})\otimes\underline{\mathcal{A}}$ where $\underline{\mathcal{A}}$ is $\mathcal{A}$ regarded as a $G$-graded vector space with* $\deg a = \xi(\overline{g})$ *for all nonzero $a \in \mathcal{A}_{\overline{g}}$, $\overline{g} \in \overline{G}$. An isomorphism $L^\chi_\pi(\mathbb{F})\otimes\underline{\mathcal{A}} \to L^\chi_\pi(\mathcal{A})$ is given by $u_h\otimes a \mapsto a\otimes u_h u_{\xi(\overline{g})}$, where we have identified $L^\chi_\pi(\mathbb{F})$ with $\mathcal{D}_H$. In particular, if $\xi$ is a group homomorphism then we can choose $\chi$ to be trivial on the complement $\xi(\overline{G})$ of $H$ in $G$, which entails $L^\chi_\pi(\mathcal{A}) \cong L^\chi_\pi(\mathbb{F})\otimes{}^\xi\!\mathcal{A}$ as graded algebras.*

Since the elements $u_g$ are invertible, we also observe the following:

REMARK 3.5. *If $\mathcal{A}$ is alternative (in particular, associative) or Jordan, then $\mathcal{A}$ is a graded-division algebra with respect to $\overline{G}$ if and only if $L^\chi_\pi(\mathcal{A})$ is a graded-division algebra with respect to $G$.*

EXAMPLE 3.6. *If $\mathcal{B} = \bigoplus_{g\in G}\mathcal{B}_g$ is a graded-field with $\mathcal{B}_e = \mathbb{R}1$ then $\mathcal{B} \cong L^\chi_\pi(\mathbb{R})$ where $H = G$ and $\pi : G \to \{1\}$. If $G$ is finite, we can write it as a direct product of cyclic groups: $G = \langle g_1\rangle\times\cdots\times\langle g_s\rangle$, and choose roots of unity $z_j$, $j = 1,\ldots,s$, satisfying $z_j^2 = \chi(g_j)$. Then*

$$L^\chi_\pi(\mathbb{R}) = \mathbb{R}[u_1]\otimes\cdots\otimes\mathbb{R}[u_s] \text{ where } \deg u_j = g_j \text{ and } u_j^{o(g_j)} = z_j^{o(z_j)} \in \{\pm 1\}.$$

*This is the description of such algebras given in* [**7**].

EXAMPLE 3.7. *Let $\mathcal{C}$ be a real octonion algebra, i.e., either the octonion division algebra $\mathbb{O}$ or the split octonion algebra $\mathbb{O}_s$. Suppose $\pi : G \to \overline{G}$ is an epimorphism of abelian groups and $\mathcal{C}$ is given a $\overline{G}$-grading. Then, for any character $\chi : G \to U$, the $G$-graded algebra $L^\chi_\pi(\mathcal{C})$ is a real alternative nonassociative algebra that is graded-central-simple. Conversely, every such algebra is graded-isomorphic to an algebra of* *the form* $L^\chi_\pi(\mathcal{C})$ *where* $\mathcal{C}$ *is a simple alternative nonassociative algebra with centroid* $\mathbb{R}$*, which is a real octonion algebra by Kleinfeld's Theorem* [**19**, **20**] *(see also* [**23**]*).*

**3.4. Isomorphism problem for real loop algebras.** We have shown how to obtain any $G$-graded algebra $\mathcal{B}$ that is graded-central-simple over $\mathbb{R}$ from a suitable central simple algebra $\mathcal{A}$ equipped with a grading by a quotient group of $G$: $\mathcal{B} \cong L^\chi_\pi(\mathcal{A})$. Our goal is to classify the graded-central-simple algebras up to isomorphism, so we are going to investigate to what extent $\mathcal{A}$ is determined by $\mathcal{B}$, but first we want to fix the parameters $\pi$ and $\chi$. Recall that $\pi$ is the natural homomorphism $G \to \overline{G}$, where $\overline{G} = G/H$ and $H$ is the support of the centroid $C(\mathcal{B})$, so $\pi$ is determined by $\mathcal{B}$. As to $\chi$, it is a homomorphism $G \to U \subset \mathbb{C}^\times$ such that $L^\chi_\pi(\mathbb{R}) \cong C(\mathcal{B})$. Let us see to what extent $\chi$ is determined.

LEMMA 3.8. $L^\chi_\pi(\mathbb{R}) \cong L^{\chi'}_\pi(\mathbb{R})$ *if and only if* $\chi|_{H_{[2]}} = \chi'|_{H_{[2]}}$.

PROOF. Recall that $L^\chi_\pi(\mathbb{R})$ and $L^{\chi'}_\pi(\mathbb{R})$ are the $\mathbb{R}$-forms of $\mathbb{C}H$ given by the $\mathrm{Gal}(\mathbb{C}/\mathbb{R})$-actions coming from $\chi$ and $\chi'$. These $\mathbb{R}$-forms are isomorphic if and only if there exists an automorphism of $\mathbb{C}H$ (as a graded algebra) that sends one onto the other, which happens if and only if there exists a character $\mu : H \to \mathbb{C}^\times$ such that $\overline{\mu(h)}\chi'(h) = \mu(h)\chi(h)$ for all $h \in H$. Normalizing $\mu$, we can assume that $\mu : H \to U$, so the condition becomes $\chi'|_H = \mu^2\chi|_H$, or $\chi'|_H(\chi|_H)^{-1} \in \mathrm{Hom}_\mathbb{Z}(H, U)^{[2]}$. This latter condition is equivalent to $\chi'|_{H_{[2]}} = \chi|_{H_{[2]}}$, because we have the exact sequence

$$\mathrm{Hom}_\mathbb{Z}(H, U) \xrightarrow{[2]} \mathrm{Hom}_\mathbb{Z}(H, U) \xrightarrow{\mathrm{res}} \mathrm{Hom}_\mathbb{Z}(H_{[2]}, U) \to 1$$

as the result of applying the functor $\mathrm{Hom}_\mathbb{Z}(\,.\,, U)$ to the exact sequence

$$1 \to H_{[2]} \to H \xrightarrow{[2]} H.$$

The functor is exact because $U$ is divisible. □

In order to classify graded-central-simple algebras up to isomorphism, we may first sort them according to the isomorphism type of the centroid and then solve the isomorphism problem for two algebras with isomorphic centroids. The first step is achieved by Lemma 3.8, since $L^\chi_\pi(\mathbb{R})$ is the centroid of $L^\chi_\pi(\mathcal{A})$. For the second, we will fix $\pi : G \to \overline{G}$ and $\chi : G \to U$ and determine when $L^\chi_\pi(\mathcal{A})$ and $L^\chi_\pi(\mathcal{A}')$ are isomorphic as graded algebras. To this end, it is convenient to introduce the following notation.

Let $\mathfrak{M}$ be the category whose objects are the central simple algebras over $\mathbb{R}$ equipped with a $\overline{G}$-grading and whose morphisms are the isomorphsms of $\overline{G}$-graded algebras. Let $\mathfrak{N}$ be the category whose objects are the graded-simple $G$-graded algebras with centroid isomorphic to $\mathcal{L}^\chi_\pi(\mathbb{R})$ and whose morphisms are the isomorphisms of $G$-graded algebras. Thus, $\mathfrak{M}$ and $\mathfrak{N}$ are *groupoids* (categories in which all morphisms are invertible).

Then $L^\chi_\pi$ is a functor $\mathfrak{M} \to \mathfrak{N}$. Indeed, we have already seen that, for any object $\mathcal{A}$ of $\mathfrak{M}$, the loop algebra $L^\chi_\pi(\mathcal{A})$ is an object of $\mathfrak{N}$. Now let $\psi : \mathcal{A} \to \mathcal{A}'$ be an isomorphism of $\overline{G}$-graded algebras. Define $L^\chi_\pi(\psi)$ as the restriction of $\psi \otimes \mathrm{id}_{\mathbb{R}^\gamma G}$ to $L^\chi_\pi(\mathcal{A}) \subset \mathcal{A} \otimes \mathbb{R}^\gamma G$, i.e., $L^\chi_\pi(\psi)$ sends $a \otimes u_g \mapsto \psi(a) \otimes u_g$ for all $a \in \mathcal{A}_{\pi(g)}$ and $g \in G$. It is clear that $L^\chi_\pi(\psi)$ is an isomorphism of $G$-graded algebras and that $\psi_1 \neq \psi_2$ implies $L^\chi_\pi(\psi_1) \neq L^\chi_\pi(\psi_2)$, so $L^\chi_\pi$ is a faithful functor $\mathfrak{M} \to \mathfrak{N}$. By Theorem 3.3, it is essentially surjective. However, in general, it fails to be full. To obtain an equivalence of categories, we extend $\mathfrak{M}$ and $L^\chi_\pi$ as follows (cf. [**14**]).

Fix a transversal for $H$ in $G$ and let $\xi : \overline{G} \to G$ be the corresponding section of $\pi$, which determines a 2-cocycle $\sigma : \overline{G} \times \overline{G} \to H$. We define a groupoid $\widetilde{\mathfrak{M}}$ that has the same objects as $\mathfrak{M}$, but more morphisms. Recall that any character $\lambda : H \to \mathbb{R}^\times$ can be used to twist the multiplication of algebras in $\mathfrak{M}$ with the cocycle $\lambda \circ \sigma$: for each $\mathcal{A}$, we denoted by $\mathcal{A}^\lambda$ the $\overline{G}$-graded space $\mathcal{A}$ with the new multiplication $*$ given by Equation (1).

Now let the morphisms in $\widetilde{\mathfrak{M}}$ from $\mathcal{A}$ to $\mathcal{A}'$ be the pairs $(\psi, \lambda)$ where $\lambda$ is a real-valued character on $H$ and $\psi : \mathcal{A}^\lambda \to \mathcal{A}'$ is an isomorphism of $\overline{G}$-graded algebras. Note that the morphisms in $\mathfrak{M}$ can be identified with the pairs that have trivial $\lambda$.

Next we define an extension of $L_\pi^\chi$ to these new morphisms. Let $\widetilde{L_\pi^\chi}(\psi, \lambda) : L_\pi^\chi(\mathcal{A}) \to L_\pi^\chi(\mathcal{A}')$ be the isomorphism of $G$-graded spaces given by

$$a \otimes u_g \mapsto \psi(a)\lambda(g\xi(\pi(g))^{-1}) \otimes u_g \text{ for all } a \in \mathcal{A}_{\pi(g)},\ g \in G. \tag{2}$$

We claim that this is an isomorphism of algebras. Indeed, for any $g_1, g_2 \in G$, $a_1 \in \mathcal{A}_{\pi(g_1)}$ and $a_2 \in \mathcal{A}_{\pi(g_2)}$, we write $g_1 = \xi(\pi(g_1))h_1$, $g_2 = \xi(\pi(g_2))h_2$, for some $h_1, h_2 \in H$, and let $g = g_1 g_2$. Then $u_{g_1} u_{g_2}$ is a scalar multiple of $u_g$ and so $\widetilde{L_\pi^\chi}(\psi, \lambda)$ sends $a_1 a_2 \otimes u_{g_1} u_{g_2}$ to $\psi(a_1 a_2)\lambda(h) \otimes u_{g_1} u_{g_2}$ where

$$h = g\xi(\pi(g))^{-1} = g_1 g_2 \xi(\pi(g_1)\pi(g_2))^{-1} = h_1 h_2 \sigma(\pi(g_1), \pi(g_2)).$$

Since $\psi(a_1 * a_2) = \psi(a_1)\psi(a_2)$, we have

$$\psi(a_1 a_2) = \lambda(\sigma(\pi(g_1), \pi(g_2))^{-1}\psi(a_1)\psi(a_2),$$

so the image of $a_1 a_2 \otimes u_{g_1} u_{g_2}$ is $\psi(a_1)\psi(a_2)\lambda(h_1)\lambda(h_2)$, which is the product of the images of $a_1 \otimes u_{g_1}$ and $a_2 \otimes u_{g_1}$.

It is easy to see that $\widetilde{L_\pi^\chi}(\mathrm{id}_\mathcal{A}, 1) = \mathrm{id}_{\widetilde{L_\pi^\chi}(\mathcal{A})}$ and $\widetilde{L_\pi^\chi}(\psi'\psi, \lambda'\lambda) = \widetilde{L_\pi^\chi}(\psi', \lambda')\widetilde{L_\pi^\chi}(\psi, \lambda)$, so we have defined a functor $\widetilde{L_\pi^\chi} : \widetilde{\mathfrak{M}} \to \mathfrak{N}$.

THEOREM 3.9. *$\widetilde{L_\pi^\chi} : \widetilde{\mathfrak{M}} \to \mathfrak{N}$ is an equivalence of groupoids.*

PROOF. Since $\widetilde{L_\pi^\chi}$ coincides with $L_\pi^\chi$ on the objects, it is essentially surjective by Theorem 3.3. To see that $\widetilde{L_\pi^\chi}$ is faithful, observe that the isomorphism of centroids induced by $\varphi = \widetilde{L_\pi^\chi}(\psi, \lambda) : L_\pi^\chi(\mathcal{A}) \to L_\pi^\chi(\mathcal{A}')$ is given by $u_h \mapsto \lambda(h)u_h$ (recall that the centroids are $L_\pi^\chi(\mathbb{R})$, which is the real form of $\mathbb{C}H$ spanned by the elements $u_h$). Indeed,

$$\begin{aligned}\varphi(u_h(a \otimes u_g)) &= \varphi(a \otimes u_h u_g) = \psi(a)\lambda(hg\xi(\pi(hg))^{-1}) \otimes u_h u_g \\ &= \lambda(h)\psi(a)\lambda(g\xi(\pi(g))^{-1}) \otimes u_h u_g \\ &= \lambda(h)u_h(\varphi(a \otimes u_g))\end{aligned} \tag{3}$$

for all $g \in G$, $h \in H$, $a \in \mathcal{A}_{\pi(g)}$. So, $\lambda$ is uniquely determined by $\varphi$, hence $\psi$ is also uniquely determined.

It remains to show that $\widetilde{L_\pi^\chi}$ is full, so suppose that we have an isomorphism $\varphi : L_\pi^\chi(\mathcal{A}) \to L_\pi^\chi(\mathcal{A}')$ of $G$-graded algebras. For any $g \in G$, write $\varphi(a \otimes u_g) = \varphi_g(a) \otimes u_g$, where $\varphi_g : \mathcal{A}_{\pi(g)} \to \mathcal{A}'_{\pi(g)}$, and take $\lambda : H \to \mathbb{R}^\times$ such that $u_h \mapsto \lambda(h)u_h$ is the isomorphism of centroids induced by $\varphi$. Define $\psi : \mathcal{A} \to \mathcal{A}'$ by $\psi(a) = \varphi_{\xi(\overline{g})}(a)$, for any $\overline{g} \in \overline{G}$ and $a \in \mathcal{A}_{\overline{g}}$. Then $\psi$ is an isomorphism of $\overline{G}$-graded algebras $\mathcal{A}^\lambda \to \mathcal{A}'$, and $\widetilde{L_\pi^\chi}(\psi, \lambda) = \varphi$. The calculations to establish these claims are similar to the above: for the first, put $g_1 = \xi(\overline{g}_1)$ and $g_2 = \xi(\overline{g}_2)$ in the proof that (2) is an isomorphism of algebras, and for the second, use (3) with $g = \xi(\overline{g})$. $\square$

COROLLARY 3.10. *Let $\mathcal{A}$ and $\mathcal{A}'$ be central simple algebras over $\mathbb{R}$ equipped with $\overline{G}$-gradings. Then $L^{\chi}_{\pi}(\mathcal{A})$ and $L^{\chi}_{\pi}(\mathcal{A}')$ are isomorphic as $G$-graded algebras if and only if the $\overline{G}$-graded algebra $\mathcal{A}'$ is isomorphic to a twist of $\mathcal{A}$.* $\square$

## 4. Alternative algebras

As an application of Theorem 3.3 and Corollary 3.10, we will obtain the classification of real $G$-graded algebras that are graded-central-simple and alternative but not associative: as pointed out in Example 3.7, these algebras arise from gradings by the quotient groups of $G$ on the real octonion algebras $\mathbb{O}$ and $\mathbb{O}_s$, and all group gradings on octonion algebras are known. For the classification up to isomorphism, the only remaining question is to calculate the twists, but first we review the relevant facts about octonions.

**4.1. Cayley–Dickson doubling process.** Recall that a *Hurwitz algebra* is a unital composition algebra, i.e., a unital algebra equipped with a multiplicative nonsingular quadratic form, which is called the *norm* and will be denoted by $n$. Except in the case char $\mathbb{F} = 2$, "nonsingular" means that the polar form of $n$, defined by $n(x,y) := n(x+y) - n(x) - n(y)$, is nondegenerate. The *standard involution* of a Hurwitz algebra is given by $\bar{x} := n(x,1)1 - x$.

It is well known that the dimension of a Hurwitz algebra can be only 1, 2, 4 or 8. Hurwitz algebras of dimension 4 are referred to as *quaternion algebras* and those of dimension 8 as *octonion* or *Cayley algebras*.

Given an *associative* Hurwitz algebra $\mathcal{A}$ such that the polar form of $n$ is nondegenerate and any scalar $\alpha \in \mathbb{F}^\times$, let $\mathfrak{CD}(\mathcal{A},\alpha)$ be the direct sum of two copies of $\mathcal{A}$, where we may formally write the element $(x,y)$ as $x + yw$, so $\mathfrak{CD}(\mathcal{A},\alpha) = \mathcal{A} \oplus \mathcal{A}w$. This is a Hurwitz algebra with multiplication and norm given by:

$$(a + bw)(c + dw) = (ac + \alpha\bar{d}b) + (da + b\bar{c})w, \tag{4}$$

$$n\big(a + bw\big) = n(a) - \alpha n(b). \tag{5}$$

For example, the real division algebras of complex numbers, quaternions, and octonions are obtained as $\mathbb{C} = \mathfrak{CD}(\mathbb{R},-1)$, $\mathbb{H} = \mathfrak{CD}(\mathbb{C},-1)$, and $\mathbb{O} = \mathfrak{CD}(\mathbb{H},-1)$. Since $\mathbb{O}$ is not associative, we cannot obtain a Hurwitz algebra by doubling it. We will abbreviate $\mathfrak{CD}(\mathcal{A},\alpha,\beta) := \mathfrak{CD}\big(\mathfrak{CD}(\mathcal{A},\alpha),\beta\big)$, and similarly for $\mathfrak{CD}(\mathbb{F},\alpha,\beta,\gamma)$. For any $\delta \in \mathbb{F}^\times$, the mapping $(x,y) \mapsto (x, \delta^{-1}y)$ is an isomorphism $\mathfrak{CD}(\mathcal{A},\alpha) \to \mathfrak{CD}(\mathcal{A},\alpha\delta^2)$. Hence, over $\mathbb{R}$, the isomorphism class of these algebras depends only on the sign of the parameters $\alpha$, $\beta$ and $\gamma$. It turns out that if any of the parameters is positive, we get the "split complex numbers" $\mathbb{C}_s = \mathbb{R} \times \mathbb{R}$, split quaternions $\mathbb{H}_s \cong M_2(\mathbb{R})$, and split octonions $\mathbb{O}_s$.

Conversely, given any Hurwitz algebra $\mathcal{C}$ with norm $n$ and a subalgebra $\mathcal{A}$ such that the restriction to $\mathcal{A}$ of the polar form of $n$ is nondegenerate, and given any nonisotropic element $w \in \mathcal{A}^\perp$, it follows that $n\big(\mathcal{A},\mathcal{A}w\big) = 0$ and that $\mathcal{A} \oplus \mathcal{A}w$ is a subalgebra of $\mathcal{C}$ isomorphic to $\mathfrak{CD}(\mathcal{A},\alpha)$ with $\alpha = -n(w) = w^2$.

**4.2. Gradings on octonion algebras.** Let $\mathcal{Q}$ be a quaternion subalgebra of a Cayley algebra $\mathcal{C}$ over a field $\mathbb{F}$, and let $w \in \mathcal{Q}^\perp$ with $n(w) = -\gamma \neq 0$. Then

$\mathcal{C} = \mathcal{Q} \oplus \mathcal{Q}w$ is isomorphic to $\mathfrak{CD}(\mathcal{Q}, \gamma)$, which gives a $\mathbb{Z}_2$-grading on $\mathcal{C}$ with $\mathcal{C}_{\bar{0}} = \mathcal{Q}$ and $\mathcal{C}_{\bar{1}} = \mathcal{Q}^\perp = \mathcal{Q}w$.

In its turn, the quaternion subalgebra $\mathcal{Q}$ can be obtained from a 2-dimensional subalgebra $\mathcal{K}$ (either $\mathbb{F} \times \mathbb{F}$ or a separable quadratic field extension of $\mathbb{F}$) as $\mathcal{Q} = \mathcal{K} \oplus \mathcal{K}v \cong \mathfrak{CD}(\mathcal{K}, \beta)$, with $v \in \mathcal{Q} \cap \mathcal{K}^\perp$ and $n(v) = -\beta \neq 0$. Then $\mathcal{C}$ is isomorphic to $\mathfrak{CD}(\mathcal{K}, \beta, \gamma)$, which gives a $\mathbb{Z}_2^2$-grading on $\mathcal{C}$ with $\mathcal{C}_{(\bar{0},\bar{0})} = \mathcal{K}$, $\mathcal{C}_{(\bar{1},\bar{0})} = \mathcal{K}v$, $\mathcal{C}_{(\bar{0},\bar{1})} = \mathcal{K}w$, and $\mathcal{C}_{(\bar{1},\bar{1})} = (\mathcal{K}v)w$.

If $\operatorname{char} \mathbb{F} \neq 2$, then $\mathcal{K}$ can be obtained by doubling $\mathbb{F}$: $\mathcal{K} = \mathbb{F} \oplus \mathbb{F}u \cong \mathfrak{CD}(\mathbb{F}, \alpha)$, with $u \in \mathcal{K} \cap \mathbb{F}^\perp$ and $n(u) = -\alpha \neq 0$, so $\mathcal{C} \cong \mathfrak{CD}(\mathbb{F}, \alpha, \beta, \gamma)$, which gives a $\mathbb{Z}_2^3$-grading on $\mathcal{C}$, with $\deg u = (\bar{1}, \bar{0}, \bar{0})$, $\deg v = (\bar{0}, \bar{1}, \bar{0})$, and $\deg w = (\bar{0}, \bar{0}, \bar{1})$.

The above gradings by $\mathbb{Z}_2^r$, $r = 1, 2, 3$, are called the *gradings induced by the Cayley–Dickson doubling process* [**13**, p. 131] ($r \neq 3$ if $\operatorname{char} \mathbb{F} = 2$).

Finally, over any field, there is a unique (up to isomorphism) split Cayley algebra $\mathcal{C}_s$, and it admits a $\mathbb{Z}^2$-grading, called the *Cartan grading*, whose homogeneous components are the eigenspaces for the action of a maximal torus of $\mathbf{Aut}_{\mathbb{F}}(\mathcal{C}_s)$.

The following is a stronger version of the main result of [**11**], where it was used to classify gradings on Cayley algebras over algebraically closed fields (see also [**13**, Theorem 4.12]).

THEOREM 4.1 ([**15**]). *Any nontrivial grading on a Cayley algebra is, up to equivalence, either a grading induced by the Cayley–Dickson doubling process starting with a Hurwitz division subalgebra, or a coarsening of the Cartan grading on the split Cayley algebra.* □

Starting from this point, a classification of gradings up to isomorphism is obtained in [**15**] for any Cayley algebra. Here we will state the result only for $\mathbb{F} = \mathbb{R}$ and $\mathbb{F} = \mathbb{C}$. To be consistent with our previous notation, we will denote the grading group by $\overline{G}$.

The coarsenings of the Cartan grading on $\mathbb{O}_s$ are induced by arbitrary homomorphisms $\mathbb{Z}^2 \to \overline{G}$ (see Subsection 2.1). Let $\overline{g}_1$, $\overline{g}_2$ and $\overline{g}_3$ be the images of the elements $(1, 0)$, $(0, 1)$, and $(-1, -1)$, respectively, so $\overline{g}_1\overline{g}_2\overline{g}_3 = \overline{e}$ (the identity element of $\overline{G}$). Then the grading is determined by the triple $(\overline{g}_1, \overline{g}_2, \overline{g}_3)$, and two triples yield isomorphic gradings if and only if they are in the same orbit under the action of the group $S_3 \times \mathbb{Z}_2$ (the Weyl group of type $G_2$), where $S_3$ permutes the entries of the triple and the generator of $\mathbb{Z}_2$ inverts them simultaneously. We will denote the resulting graded algebras by $\mathcal{C}(\overline{g}_1, \overline{g}_2, \overline{g}_3)$. None of them is a graded-division algebra (for example, because the identity component contains a nontrivial idempotent).

The remaining nontrivial gradings on $\mathbb{O}$ and $\mathbb{O}_s$ are obtained by arbitrary monomorphisms $\mathbb{Z}_2^r \to \overline{G}$, $1 \leq r \leq 3$, from the gradings induced by the Cayley–Dickson doubling process starting from $\mathbb{H}$ if $r = 1$, from $\mathbb{C}$ if $r = 2$, and from $\mathbb{R}$ if $r = 3$. All of them are division gradings. To include the trivial grading on $\mathbb{O}$, we will allow $r = 0$. The support of the grading is the image of $\mathbb{Z}_2^r$, which we denote by $\overline{T}$. The parameters used in the doubling process determine a character $\mu : \overline{T} \to \{\pm 1\}$ as follows: for any nonzero homogeneous element $x$ of degree $\bar{t}$, we have $x^2 \in -\mu(\bar{t})\mathbb{R}_{>0}$. The trivial characters give gradings on $\mathbb{O}$ and the nontrivial ones on $\mathbb{O}_s$. We will denote the resulting graded-division algebras by $\mathcal{C}(\overline{T}, \mu)$. The algebras corresponding to different pairs $(\overline{T}, \mu)$ are not graded-isomorphic.

A similar classification of division gradings is valid for $\mathbb{H}$ and $\mathbb{H}_s$ on the one hand and $\mathbb{C}$ and $\mathbb{C}_s$ on the other hand. We will denote the resulting graded-division algebras by $\mathcal{Q}(\overline{T},\mu)$ and $\mathcal{K}(\overline{T},\mu)$, respectively.

Over $\mathbb{C}$, there is only one Cayley algebra (up to isomorphism), and the classification of gradings is the same as above, except in the Cayley–Dickson case there is no parameter $\mu$ and $r$ must be equal to 3. We will denote these graded algebras by $\mathcal{C}_{\mathbb{C}}(\overline{g}_1,\overline{g}_2,\overline{g}_3)$ and $\mathcal{C}_{\mathbb{C}}(\overline{T})$. Similarly, we also have $\mathcal{Q}_{\mathbb{C}}(\overline{T})$ with $r=2$ for $M_2(\mathbb{C})$ (the complex quaternion algebra) and $\mathcal{K}_{\mathbb{C}}(\overline{T})$ with $r=1$ for $\mathbb{C}\times\mathbb{C}$.

**4.3. Graded-simple real alternative algebras.** Let $\mathcal{B}$ be a $G$-graded algebra over $\mathbb{R}$ that is graded-simple and alternative but not associative. Assume also that the identity component of the centroid $\mathbb{L}:=C(\mathcal{B})$ is finite-dimensional, so it is either $\mathbb{R}$ or $\mathbb{C}$. As before, let $H$ be the support of $\mathbb{L}$, $\overline{G}=G/H$, and $\pi$ be the natural homomorphism $G\to\overline{G}$.

If $\mathbb{L}_e\cong\mathbb{C}$ then $\mathbb{L}\cong\mathbb{C}H$ and, by the "Correspondence Theorem" from [**1**], we have $\mathcal{B}\cong L_\pi(\mathcal{C})$ where $\mathcal{C}$ is the unique complex Cayley algebra, equipped with a $\overline{G}$-grading, and this grading is determined uniquely up to isomorphism. Thus, in this case the algebras $\mathcal{B}$ can be of two kinds: those with $\mathcal{C}\cong\mathcal{C}_{\mathbb{C}}(\overline{g}_1,\overline{g}_2,\overline{g}_3)$ are classified up to isomorphism by the $(S_3\times\mathbb{Z}_2)$-orbits of the triples $(\overline{g}_1,\overline{g}_2,\overline{g}_3)$ and those with $\mathcal{C}\cong\mathcal{C}_{\mathbb{C}}(\overline{T})$ are classified by the subgroups $\overline{T}\subset\overline{G}$ isomorphic to $\mathbb{Z}_2^3$. The graded-division algebras are the latter.

Now let us assume $\mathbb{L}_e=\mathbb{R}$ and fix a character $\chi:G\to U\subset\mathbb{C}^\times$ such that $\mathbb{L}\cong L_\pi^\chi(\mathbb{R})$. By Theorem 3.3, we have $\mathcal{B}\cong L_\pi^\chi(\mathcal{C})$ where $\mathcal{C}$ is either $\mathbb{O}$ or $\mathbb{O}_s$, equipped with a $\overline{G}$-grading, so $\mathcal{C}\cong\mathcal{C}(\overline{g}_1,\overline{g}_2,\overline{g}_3)$ or $\mathcal{C}\cong\mathcal{C}(\overline{T},\mu)$. To apply Corollary 3.10, we have to classify these $\overline{G}$-graded algebras up to twist. Let $T=\pi^{-1}(\overline{T})$, so $\overline{T}=T/H$. Since $\overline{T}$ is an elementary 2-group (of rank $r$), we have a homomorphism $T\to H$ sending $t\mapsto t^2$, which maps $H$ to $H^{[2]}$ and hence induces a homomorphism

$$\tau:\overline{T}\to H/H^{[2]},\ tH\mapsto t^2H^{[2]}. \tag{6}$$

Note that $\tau$ is the element of $\mathrm{Hom}_{\mathbb{Z}}(\overline{T},H/H^{[2]})$ corresponding to the abelian group extension $H\to T\to\overline{T}$ under the isomorphism $\mathrm{Ext}_{\mathbb{Z}}(\overline{T},H)\cong\mathrm{Hom}_{\mathbb{Z}}(\overline{T},H/H^{[2]})$ (see e.g. [**21**, III.3, Exercise 6]).

LEMMA 4.2. *For any character $\lambda:H\to\mathbb{R}^\times$, we have $\mathcal{C}(\overline{g}_1,\overline{g}_2,\overline{g}_3)^\lambda\cong\mathcal{C}(\overline{g}_1,\overline{g}_2,\overline{g}_3)$ and $\mathcal{C}(\overline{T},\mu)^\lambda\cong\mathcal{C}(\overline{T},\mu')$ where $\mu'(\overline{t})=\mu(\overline{t})\lambda_0(\tau(\overline{t}))$ for all $\overline{t}\in\overline{T}$, and $\lambda_0(hH^{[2]}):=\operatorname{sign}\lambda(h)$ for all $h\in H$.*

PROOF. To compute the twists, we fix a section $\xi:\overline{G}\to G$. It is convenient to take $\xi(\overline{e})=e$, so that the resulting 2-cocycle $\sigma:\overline{G}\times\overline{G}\to H$ is normalized.

The elements $\overline{g}_i$ are the degrees of homogeneous basis elements of one of the Peirce components with respect to a nontrivial homogeneous idempotent (while the other Peirce component gives the elements $\overline{g}_i^{-1}$). Since the degree of the idempotent is $\overline{e}$, it remains an idempotent with respect to the twisted multiplication of Equation (1), and its Peirce components remain the same.

Recall that $\mu(\overline{t})$ is defined by $x^2\in-\mu(\overline{t})\mathbb{R}_{>0}$, for any nonzero homogeneous element $x$ of degree $\overline{t}$. The twisted multiplication gives:

$$x*x=\lambda(\sigma(\overline{t},\overline{t}))x^2=\lambda\big(\xi(\overline{t})^2\xi(\overline{t}^2)^{-1}\big)x^2=\lambda\big(\xi(\overline{t})^2\big)x^2,$$

and the result follows, because $\operatorname{sign}\lambda\big(\xi(\overline{t})^2\big)=\lambda_0\big(\xi(\overline{t})^2H^{[2]}\big)=\lambda_0(\tau(\overline{t}))$. □

LEMMA 4.3. *Let $\overline{T}_0 = \ker\tau$. Then $\mathcal{C}(\overline{T},\mu')$ is graded-isomorphic to a twist of $\mathcal{C}(\overline{T},\mu)$ if and only if $\mu'|_{\overline{T}_0} = \mu|_{\overline{T}_0}$.*

PROOF. We have to let $\lambda_0$ in Lemma 4.2 range over $\mathrm{Hom}_{\mathbb{Z}}(H/H^{[2]},\{\pm1\})$ and consider the resulting $\tau^*(\lambda_0) := \lambda_0\circ\tau$, which are homomorphisms $\overline{T}\to\{\pm1\}$. Consider the exact sequence

$$1\to\overline{T}_0\to\overline{T}\xrightarrow{\tau} H/H^{[2]}.$$

Since these are elementary 2-groups, we may regard them as vector spaces over the field of 2 elements and hence obtain the following exact sequence:

$$\mathrm{Hom}_{\mathbb{Z}}(H/H^{[2]},\{\pm1\})\xrightarrow{\tau^*}\mathrm{Hom}_{\mathbb{Z}}(\overline{T},\{\pm1\})\xrightarrow{\mathrm{res}}\mathrm{Hom}_{\mathbb{Z}}(\overline{T}_0,\{\pm1\})\to 1,$$

which proves the result. □

We summarize the results of this subsection:

THEOREM 4.4. *Let $\mathcal{B}$ be a $G$-graded algebra over $\mathbb{R}$ that is graded-simple and alternative but not associative. Let $\mathbb{L}=C(\mathcal{B})$ and assume that $\mathbb{L}_e$ is finite-dimensional, so it is either $\mathbb{R}$ or $\mathbb{C}$. Let $H$ be the support of $\mathbb{L}$, $\overline{G}=G/H$, and $\pi$ be the natural homomorphism $G\to\overline{G}$. If $\mathbb{L}_e=\mathbb{R}$, pick a character $\chi: G\to U\subset\mathbb{C}^\times$ such that $\mathbb{L}\cong L_\pi^\chi(\mathbb{R})$.*

1. *If $\mathbb{L}_e=\mathbb{R}$ then $\mathcal{B}$ is graded-isomorphic to one of the following:*
   (a) *$L_\pi^\chi(\mathcal{C}(\bar{g}_1,\bar{g}_2,\bar{g}_3))$ for a triple $(\bar{g}_1,\bar{g}_2,\bar{g}_3)\in\overline{G}^3$ with $\bar{g}_1\bar{g}_2\bar{g}_3=\bar{e}$;*
   (b) *$L_\pi^\chi(\mathcal{C}(\overline{T},\mu))$ for a subgroup $\overline{T}\subset\overline{G}$ isomorphic to $\mathbb{Z}_2^r$, $0\le r\le 3$, and a character $\mu:\overline{T}\to\{\pm1\}$.*
2. *If $\mathbb{L}_e\cong\mathbb{C}$ then $\mathcal{B}$ is graded-isomorphic to one of the following:*
   (a) *$L_\pi(\mathcal{C}_{\mathbb{C}}(\bar{g}_1,\bar{g}_2,\bar{g}_3))$ for a triple $(\bar{g}_1,\bar{g}_2,\bar{g}_3)\in\overline{G}^3$ with $\bar{g}_1\bar{g}_2\bar{g}_3=\bar{e}$;*
   (b) *$L_\pi(\mathcal{C}_{\mathbb{C}}(\overline{T}))$ for a subgroup $\overline{T}\subset\overline{G}$ isomorphic to $\mathbb{Z}_2^3$.*

*Two algebras in different items are not graded-isomorphic to one another. Two algebras in item 1(a) or 2(a) are graded-isomorphic if and only if their triples belong to the same $(S_3\times\mathbb{Z}_2)$-orbit. Two algebras in item 2(b) are graded-isomorphic if and only if their subgroups are equal. Finally, two algebras in item 1(b) are graded-isomorphic if and only if their subgroups are equal and their characters have the same restriction to $\overline{T}_0 := \ker\tau$ where $\tau:\overline{T}\to H/H^{[2]}$ is given by Equation (6).* □

**4.4. Graded-division real alternative algebras.** The algebras in items 1(b) and 2(b) of Theorem 4.4 are characterized by the property that they are graded-division algebras (see Remark 3.5). They can be obtained by the Cayley–Dickson doubling process over the graded-field $\mathbb{L}$ as follows. First of all, the formulas (4) and (5) make sense for unital algebras with a quadratic form over any commutative associative ring, in particular over any graded-field $\mathbb{L}$, but we also have to keep track of the gradings. To this end, suppose $\mathcal{A}$ is a graded $\mathbb{L}$-module with an $\mathbb{L}$-bilinear multiplication and a quadratic form $n:\mathcal{A}\to\mathbb{L}$ such that $\mathcal{A}$ is a unital graded algebra and $n(\mathcal{A}_{g_1},\mathcal{A}_{g_2})\subset\mathbb{L}_{g_1g_2}$ for all $g_1,g_2\in G$, so the mapping $x\mapsto\bar{x}:=n(x,1)1-x$ is degree-preserving. Given $k\in G$ and $\alpha\in\mathbb{L}_{k^2}$, Equation (4) makes $\mathcal{B}:=\mathfrak{CD}(\mathcal{A},\alpha)$ a graded algebra if we define

$$\mathfrak{CD}(\mathcal{A},\alpha)_g=\mathcal{A}_g\oplus\mathcal{A}_{gk^{-1}}w \text{ for all } g\in G. \tag{7}$$

Note that $w=(0,1)$ is assigned degree $k$. Moreover, the polar form of the quadratic form $n$ defined by Equation (5), satisfies the condition $n(\mathcal{B}_{g_1}, \mathcal{B}_{g_2}) \subset \mathbb{L}_{g_1 g_2}$ for all $g_1, g_2 \in G$. We will denote the resulting unital graded algebra with a quadratic form by $\mathfrak{CD}(\mathcal{A}, (\alpha, k))$.

For example, the $\overline{G}$-graded algebras $\mathcal{C}(\overline{T}, \mu)$ can be obtained as follows. Take $\mathbb{L} = \mathbb{R}$ with trivial grading. In the case $r = 3$, if $\{\bar{t}_1, \bar{t}_2, \bar{t}_3\}$ is a basis of the elementary 2-group $\overline{T}$, then

$$\mathcal{C}(\overline{T}, \mu) \cong \mathfrak{CD}\big(\mathbb{R}, (-\mu(\bar{t}_1), \bar{t}_1), (-\mu(\bar{t}_2), \bar{t}_2), (-\mu(\bar{t}_3), \bar{t}_3)\big), \tag{8}$$

where we use an abbreviation for the iterated Cayley–Dickson doubling process similar to what we had in Subsection 4.1. The isomorphism (8) remains valid in the case $r < 3$ if we make the convention that $\bar{t}_j = \bar{e}$ for $1 \le j \le 3 - r$, while the $\bar{t}_j$ with $3 - r < j \le 3$ form a basis of $\overline{T}$.

Now let us see how the loop algebra construction interacts with the (graded) Cayley–Dickson doubling process.

Lemma 4.5. *Let $\mathcal{A}$ be a unital $\overline{G}$-graded algebra with a quadratic form over $\mathbb{R}$, $\mathbb{L} = L_\pi^\chi(\mathbb{R})$, $\alpha \in \mathbb{R}^\times$ and $\bar{k} \in \overline{G}$ with $\bar{k}^2 = \bar{e}$. Then*

$$L_\pi^\chi\left(\mathfrak{CD}(\mathcal{A}, (\alpha, \bar{k}))\right) \cong \mathfrak{CD}\big(\, L_\pi^\chi(\mathcal{A}), (\alpha u_k^2, k)\big),$$

*for any $k \in G$ satisfying $\pi(k) = \bar{k}$.*

Proof. In view of (7), applied first with $\overline{G}$ and then with $G$, we have an isomorphism of $G$-graded vector spaces $L_\pi^\chi\left(\mathfrak{CD}(\mathcal{A}, (\alpha, \bar{k}))\right) \to \mathfrak{CD}\big(\, L_\pi^\chi(\mathcal{A}), (\alpha u_k^2, k)\big)$ given by

$$(a + bw) \otimes u_g \mapsto a \otimes u_g + (b \otimes u_g u_k^{-1})w \text{ for all } a \in \mathcal{A}_{\pi(g)},\ b \in \mathcal{A}_{\pi(g)\bar{k}^{-1}},\ g \in G.$$

This is also an isomorphism of $\mathbb{L}$-algebras with quadratic forms because it is the restriction of the mapping $\mathfrak{CD}(\mathcal{A}, \alpha) \otimes \mathcal{D} \to \mathfrak{CD}(\mathcal{A} \otimes \mathcal{D}, \alpha u_k^2)$, where $\mathcal{D} = \mathbb{R}^\gamma G$, obtained by composing the $\mathcal{D}$-algebra isomorphism $\mathfrak{CD}(\mathcal{A}, \alpha) \otimes \mathcal{D} \to \mathfrak{CD}(\mathcal{A} \otimes \mathcal{D}, \alpha)$ sending $(a + bw) \otimes u_g \mapsto a \otimes u_g + (b \otimes u_g)w$ and the $\mathcal{D}$-algebra isomorphism $\mathfrak{CD}(\mathcal{A} \otimes \mathcal{D}, \alpha) \to \mathfrak{CD}(\mathcal{A} \otimes \mathcal{D}, \alpha u_k^2)$ sending $(x, y) \mapsto (x, y u_k^{-1})$, for all $x, y \in \mathcal{A} \otimes \mathcal{D}$. □

Applying this lemma repeatedly, we obtain from (8) the following models for the algebras in item 1(b) of Theorem 4.4:

$$L_\pi^\chi(\mathcal{C}(\overline{T}, \mu)) \cong \mathfrak{CD}\big(\mathbb{L}, (-\mu(\bar{t}_1) u_{t_1}^2, t_1), (-\mu(\bar{t}_2) u_{t_2}^2, t_2), (-\mu(\bar{t}_3) u_{t_3}^2, t_3)\big),$$

for any $t_j \in G$ satisfying $\pi(t_j) = \bar{t}_j$. Note that $u_{t_j}^2 = \gamma(\bar{t}_j, \bar{t}_j) u_{t_j^2} \in \mathbb{L}$.

This can be simplified with a special choice of the elements $t_j$. Indeed, let $r_0$ be the rank of $\overline{T}_0$ and pick the basis of $\overline{T}$ so that the $\bar{t}_j$ with $3 - r < j \le 3 - r + r_0$ form a basis of $\overline{T}_0$. Then the $t_j$ with $3 - r < j \le 3 - r + r_0$ can be chosen to satisfy $t_j^2 = e$, so that the subgroup $T_0$ of $G$ generated by these elements is mapped by $\pi$ isomorphically onto $\overline{T}_0$. Moreover, for these $t_j$, we have $u_{t_j}^2 = \chi(t_j) \in \{\pm 1\}$. Unless $\mathbb{L} \cong \mathbb{R}H$, there exists $h \in H_{[2]}$ such that $\chi(h) = -1$ (see Lemma 3.8), which can be used to adjust the $t_j$ to make $u_{t_j}^2 = \mu(\bar{t}_j)$. If $\mathbb{L} \cong \mathbb{R}H$ then we can take $\chi$ to be trivial, so $u_{t_j}^2 = 1$. Now, for $1 \le j \le 3 - r$, we will take $t_j = e$. As to the remaining case, $3 - r + r_0 < j \le 3$, it does not matter how we choose these $t_j$ because, in view of the isomorphism condition in item 1(b), we may change $\mu(\bar{t}_j)$ arbitrarily without affecting the isomorphism class of the graded algebra. Thus we obtain (cf. Remark 3.4):

$$(9)\qquad L^\chi_\pi(\mathcal{C}(\overline{T},\mu))\cong\begin{cases}\mathcal{C}(T_0,\mu_0)\otimes\mathbb{L} & \text{if } r_0=r;\\ \mathfrak{CD}\big(\mathcal{Q}(T_0,\mu_0)\otimes\mathbb{L},(\pm u_{t_3^2},t_3)\big) & \text{if } r_0=r-1;\\ \mathfrak{CD}\big(\mathcal{K}(T_0,\mu_0)\otimes\mathbb{L},(\pm u_{t_2^2},t_2),(\pm u_{t_3^2},t_3)\big) & \text{if } r_0=r-2;\\ \mathfrak{CD}\big(\mathbb{L},(\pm u_{t_1^2},t_1),(\pm u_{t_2^2},t_2),(\pm u_{t_3^2},t_3)\big) & \text{if } r_0=r-3;\end{cases}$$

where the signs can be chosen arbitrarily, and $\mu_0=\mu\circ\pi|_{T_0}$ if $\mathbb{L}\cong\mathbb{R}H$ and trivial otherwise.

Similar models can be obtained for $L_\pi(\mathcal{C}_\mathbb{C}(\overline{T}))$ in item 2(b) of Theorem 4.4, but we have $\mathbb{L}\cong\mathbb{C}H$, there is no $\mu_0$, $r$ must be 3, and we have $t_j^2$ instead of $\pm u_{t_j^2}$.

We will now obtain a classification of graded-division real alternative algebras up to equivalence. First of all, we replace the grading group $G$ by the support $T$, so equivalence becomes the same as weak isomorphism (see Subsection 2.2), and we have to investigate the effect of the automorphisms of the group $T$ on the graded algebras $\mathcal{B}$ in items 1(b) and 2(b). For any $\alpha\in\operatorname{Aut}(T)$, we have a canonical isomorphism of graded algebras ${}^\alpha C(\mathcal{B})\cong C({}^\alpha\mathcal{B})$. Hence, we fix a representative in each $\operatorname{Aut}(T)$-orbit of the graded-isomorphism classes of graded-fields (over the group $T$) so that we may consider the graded-isomorphism class of $\mathbb{L}=C(\mathcal{B})$ fixed and restrict ourselves to those $\alpha\in\operatorname{Aut}(T)$ that satisfy ${}^\alpha\mathbb{L}\cong\mathbb{L}$. In item 2(b), this latter condition amounts to $\alpha(H)=H$, while in item 1(b) it says more: $\alpha(H)=H$ and $\chi\circ\alpha|_{H_{[2]}}=\chi|_{H_{[2]}}$ (see Lemma 3.8). In any case, $\alpha$ induces an automorphism $\bar\alpha$ of $\overline{T}=T/H$, and we have $\bar\alpha(\overline{T}_0)=\overline{T}_0$.

Consider $\mathcal{B}$ in item 1(b), so $\mathcal{B}\cong L^\chi_\pi(\mathcal{C}(\overline{T},\mu))$. Then ${}^\alpha\mathcal{B}$ is given by the right-hand side of (9) in which $\mathbb{L}$ is replaced by $\mathbb{L}':={}^\alpha\mathbb{L}$, $T_0$ by $T_0':=\alpha(T_0)$, $\mu_0$ by $\mu_0':=\mu_0\circ\alpha^{-1}|_{T_0'}$, and $(\pm u_{t_j^2},t_j)$ by $(\pm u_{t_j^2},\alpha(t_j))$. Note that $\mathbb{L}'_{\alpha(h)}=\mathbb{L}_h=\mathbb{R}u_h$ for all $h\in H$, and there exists an isomorphism of graded algebras $\psi:\mathbb{L}'\to\mathbb{L}$, which must therefore send $u_h$ to either $u_{\alpha(h)}$ or to $-u_{\alpha(h)}$ (depending on $h$). Extending the isomorphism $\mathrm{id}\otimes\psi$ to the Cayley–Dickson doubles, we obtain

$${}^\alpha\mathcal{B}\cong\begin{cases}\mathcal{C}(T_0',\mu_0')\otimes\mathbb{L} & \text{if } r_0=r;\\ \mathfrak{CD}\big(\mathcal{Q}(T_0',\mu_0')\otimes\mathbb{L},(\pm u_{(t_3')^2},t_3')\big) & \text{if } r_0=r-1;\\ \mathfrak{CD}\big(\mathcal{K}(T_0',\mu_0')\otimes\mathbb{L},(\pm u_{(t_2')^2},t_2'),(\pm u_{(t_3')^2},t_3')\big) & \text{if } r_0=r-2;\\ \mathfrak{CD}\big(\mathbb{L},(\pm u_{(t_1')^2},t_1'),(\pm u_{(t_2')^2},t_2'),(\pm u_{(t_3')^2},t_3')\big) & \text{if } r_0=r-3;\end{cases}$$

where $t_j'=\alpha(t_j)$. These elements have the same properties as the $t_j$ in (9), so the net effect of $\alpha$ on the graded-isomorphism class of $\mathcal{B}$ is the replacement of $(T_0,\mu_0)$ by $(T_0',\mu_0')$. Now, $T_0$ can be mapped onto any other complement of $H$ in $\pi^{-1}(\overline{T}_0)$ by an automorphism of the latter that restricts to the identity on $H$ and therefore extends to an automorphism $\alpha$ of $T$ as above. Moreover, any automorphism of $T_0$ extends in the same manner to an automorphism $\alpha$. It follows that there is only one equivalence class of $\mathcal{B}$ when we have fixed $T$ and the graded-isomorphism class of $\mathbb{L}$ other than $\mathbb{R}H$, and two equivalence classes with a fixed $T$ and $\mathbb{L}\cong\mathbb{R}H$: one corresponding to $\mu_0=1$ and the other to $\mu_0\neq1$, where 1 denotes the trivial character.

For $\mathcal{B}$ in item 2(b), i.e., $\mathcal{B}\cong L_\pi(\mathcal{C}_\mathbb{C}(\overline{T}))$, we can use the same arguments or apply the canonical isomorphism ${}^\alpha(L_\pi(\mathcal{A}))\cong L_\pi({}^{\bar\alpha}\mathcal{A})$ defined by $a\otimes g\mapsto a\otimes\alpha(g)$, for any $\overline{G}$-graded algebra $\mathcal{A}$. (This isomorphism could also be applied for item 1(b) in the case of split centroid.)

To summarize our classification, it is convenient to introduce the following notation. Let $T$ be an abelian group with a subgroup $H$ such that $\overline{T} := T/H$ is an elementary 2-group of rank $r \in \{0, 1, 2, 3\}$. Let $\mathcal{O}$ be an orbit of nontrivial characters $H_{[2]} \to \{\pm 1\}$ under the action of the stabilizer of $H$ in $\mathrm{Aut}(T)$. For each such triple $(T, H, \mathcal{O})$, we fix a representative $\chi_0 \in \mathcal{O}$ and extend it to a character $\chi : T \to U \subset \mathbb{C}^\times$. Let

$$\mathcal{DA}(T, H, \mathcal{O}) := L_\pi^\chi(\mathcal{C}(\overline{T}, 1)),$$

where $\pi : T \to \overline{T}$ is the natural homomorphism. For each pair $(T, H)$ as above, fix a character $\mu : \overline{T} \to \{\pm 1\}$ whose restriction to $\overline{T}_0 := \ker \tau$ is nontrivial, where $\tau : \overline{T} \to H/H^{[2]}$ is given by Equation (6), and let

$$\mathcal{DA}(T, H) := L_\pi(\mathcal{C}(\overline{T}, 1)) \text{ and } \mathcal{DA}(T, H)' := L_\pi(\mathcal{C}(\overline{T}, \mu)).$$

(If $\overline{T}_0$ is trivial then only $\mathcal{DA}(T, H)$ is defined.) Finally, for each pair $(T, H)$ with $r = 3$, let

$$\mathcal{DA}_\mathbb{C}(T, H) := L_\pi(\mathcal{C}_\mathbb{C}(\overline{T})).$$

Then we have the following:

THEOREM 4.6. *Let $\mathcal{B}$ be a real graded-division algebra with support $T$ that is alternative but not associative. Let $\mathbb{L} = C(\mathcal{B})$ and assume that $\mathbb{L}_e$ is finite-dimensional, so it is either $\mathbb{R}$ or $\mathbb{C}$, and let $H$ be the support of $\mathbb{L}$.*

1. *If $\mathbb{L}_e = \mathbb{R}$ and $\mathbb{L} \not\cong \mathbb{R}H$, then $\mathcal{B}$ is equivalent to $\mathcal{DA}(T, H, \mathcal{O})$ for a unique orbit $\mathcal{O}$ of nontrivial characters $H_{[2]} \to \{\pm 1\}$ under the action of the stabilizer of $H$ in* $\mathrm{Aut}(T)$.
2. *If $\mathbb{L}_e = \mathbb{R}$ and $\mathbb{L} \cong \mathbb{R}H$, then $\mathcal{B}$ is equivalent to $\mathcal{DA}(T, H)$ or $\mathcal{DA}(T, H)'$, but not both.*
3. *If $\mathbb{L}_e \cong \mathbb{C}$, then $\mathcal{B}$ is equivalent to $\mathcal{DA}_\mathbb{C}(T, H)$.* □

We can make the classification more explicit in the finite-dimensional case by classifying the pairs $(T, H)$ and the triples $(T, H, \chi_0)$ up to isomorphism, where $T$ is a finite abelian group and, as before, $H$ is a subgroup such that $\overline{T} := T/H$ is an elementary 2-group of rank $r \in \{0, 1, 2, 3\}$ and $\chi_0$ is a nontrivial character $H_{[2]} \to \{\pm 1\}$. First of all, we have a canonical decomposition $T = T' \times T''$ where $T'$ is a 2-group and $T''$ has odd order. Then $H = H' \times T''$ and $H_{[2]} = H'_{[2]}$, so the problem reduces to 2-groups.

A subset $\{g_1, \ldots, g_s\}$ of a finite abelian group $G$ will be called a *basis* if the orders of $g_j$ are nontrivial prime powers and $G = \langle g_1 \rangle \times \cdots \times \langle g_s \rangle$. If $G$ is an elementary $p$-group, this is indeed a basis of $G$ as a vector space over the field of $p$ elements. We leave the proof of the following result to the reader.

LEMMA 4.7. *Let $\pi : G \to \overline{G}$ be an epimorphism of finite abelian $p$-groups where $\overline{G}$ is elementary. Then there exists a basis $\{g_1, \ldots, g_s\}$ of $G$ such that the elements $\pi(g_j)$ that are different from the identity form a basis of $\overline{G}$.* □

In particular, our pairs $(T, H)$ are described as follows: there is a basis of $T$ in which $r$ elements are marked, the orders of the marked elements are powers of 2, and $H$ is generated by the unmarked elements and the squares of the marked elements. Up to isomorphism, a pair $(T, H)$ is represented by a tuple of nontrivial prime powers (the orders of the basis elements) in which $r$ powers of 2 are marked: for example, the pair $(\mathbb{Z}_2 \times \mathbb{Z}_2 \times \mathbb{Z}_4, 0 \times \mathbb{Z}_2 \times 2\mathbb{Z}_4)$ is represented by $(\underline{2}, 2, \underline{4})$.

Two pairs are isomorphic if and only if the tuples are the same up to permutation. (The "only if" part can be seen by looking at the quotients $T^{[2^k]}/H^{[2^k]}$ for $k = 1, 2, \ldots$)

Now we want to include $\chi_0$ into consideration. The classification of pairs $(H, \chi_0)$ up to isomorphism is equivalent (in view of Lemma 3.8) to the classifications of graded-fields in Example 3.6 up to equivalence, which was done in [**7**]. Such a pair is described by the tuple consisting of the orders of the basis elements where each power of 2 is given a sign according to the value of $\chi_0$ on the unique element of order 2 in the cyclic group generated by the basis element. Since we consider nontrivial $\chi_0$, at least one entry of the tuple must have negative sign. It turns out that, by changing the basis, one can always achieve that there is exactly one negative sign. This is done using the following moves: if two distinct basis elements $h_i$ and $h_j$ with $o(h_i) = 2^m \le 2^n = o(h_j)$ give negative sign, we replace $h_i$ with $h_i h_j^{2^{n-m}}$, which gives positive sign. Hence, a pair $(H, \chi_0)$ is represented, up to isomorphism, by a tuple of nontrivial prime powers in which exactly one power of 2 is given negative sign: for example, the tuple $(2, -8)$ represents the pair $(\mathbb{Z}_2 \times \mathbb{Z}_8, \chi_0)$ where $\chi_0((\bar{1}, \bar{0})) = 1$ and $\chi_0((\bar{0}, \bar{4})) = -1$. Two pairs are isomorphic if and only if the tuples are the same up to permutation. (The "only if" part can be seen by looking at the restrictions of $\chi_0$ to the subgroups $H^{[2^k]} \cap H_{[2]}$ for $k = 1, 2, \ldots$) Hence, every graded-field in Example 3.6 is equivalent to either $\mathbb{R}H$ (corresponding to the trivial $\chi_0$) or $\mathbb{R}K \otimes (-n)$ where $H \cong K \times \mathbb{Z}_{2^n}$ and the $\mathbb{Z}_{2^{|m|}}$-graded algebra $(m)$ is defined for any $m \in \mathbb{Z}$ as follows:

$$(n) := \mathbb{R}[x]/(x^{2^n} - 1) \text{ and } (-n) := \mathbb{R}[x]/(x^{2^n} + 1) \text{ where } \deg x = \bar{1} \in \mathbb{Z}_{2^n}.$$

Note that $(0) = \mathbb{R}$ (trivially graded) and, if $H$ is a 2-group, $\mathbb{R}H$ is equivalent to the tensor product of graded algebras of the form $(n)$, $n \in \mathbb{N}$.

Finally, for the triples $(T, H, \chi_0)$, we combine the previous arguments: we have a basis of $H$ in which $r_0 \le r$ elements are marked (being the nontrivial squares of the marked basis elements of $T$), and each basis element gives a sign. We can reduce the number of negative signs to one by applying the same moves as above to the basis elements of $T$, except that we are not allowed to replace $t_i$ with $t_i t_j^{2^{n-m}}$ if $t_i$ is in $H$, $t_j$ is not in $H$, and $m = n$. This is not an obstacle, because in this situation we can replace $t_j$ with $t_i t_j$. Therefore, the triples $(T, H, \chi_0)$ are classified by tuples of nontrivial prime powers in which $r$ powers of 2 are marked and one of the powers of 2 (marked or unmarked) is given negative sign. This labeling is arbitrary except that a marked power of 2 can be given negative sign only if it is at least 4.

Now we can describe more explicitly the finite-dimensional graded algebras in Theorem 4.6. For integers $n_1, n_2, n_3$, consider $T = \mathbb{Z}_{2^{|n_1|+1}} \times \mathbb{Z}_{2^{|n_2|+1}} \times \mathbb{Z}_{2^{|n_3|+1}}$ and its standard basis $t_1 = (\bar{1}, \bar{0}, \bar{0})$, $t_2 = (\bar{0}, \bar{1}, \bar{0})$, $t_3 = (\bar{0}, \bar{0}, \bar{1})$. Define the following $T$-graded algebra:

$$\mathbb{R}\{n_1, n_2, n_3\} := \mathfrak{CD}(\mathbb{L}, (x_1, t_1), (x_2, t_2), (x_3, t_3)) \text{ where } \mathbb{L} = (n_1) \otimes (n_2) \otimes (n_3),$$

with $x_j$ being the generator of the $j$-th factor of $\mathbb{L}$, with $\deg x_j = 2t_j$. Thus, the Cayley–Dickson generators satisfy $u^2 = x_1$, $v^2 = x_2, w^2 = x_3$ and are assigned degrees $\deg u = t_1$, $\deg v = t_2$, $\deg w = t_3$. Here we make the convention that $x_j = -1$ if $n_j = 0$. The algebra $\mathbb{R}\{n_1, n_2, n_3\}$ falls into item 1 of Theorem 4.6 if at least one $n_j < 0$ (we may assume without loss of generality that exactly one) and into item 2 otherwise. The pair $(T, H)$ and the orbit of $\chi_0$ are represented by the tuple $(\pm\underline{2}^{|n_1|+1}, \pm\underline{2}^{|n_2|+1}, \pm\underline{2}^{|n_3|+1})$ where the signs are given as follows: plus if $n_j \geq 0$ and minus if $n_j < 0$. We define in a similar manner the graded algebras

$$\mathbb{C}\{n_2, n_3\} := \mathfrak{CD}\big(\mathbb{C} \otimes \mathbb{L}, (x_2, t_2), (x_3, t_3)\big) \text{ where } \mathbb{L} = (n_2) \otimes (n_3);$$
$$\mathbb{H}\{n_3\} := \mathfrak{CD}\big(\mathbb{H} \otimes \mathbb{L}, (x_3, t_3)\big) \text{ where } \mathbb{L} = (n_3).$$

For nonnegative integers $n_1, n_2, n_3$, we also define the graded algebras $\mathbb{R}\{n_1, n_2, n_3\}'$, $\mathbb{C}\{n_2, n_3\}'$ and $\mathbb{H}\{n_3\}'$ as above, but with the convention $x_j = 1$ if $n_j = 0$. They fall into item 2 of Theorem 4.6. Note that in our notation the letters $\mathbb{R}$, $\mathbb{C}$ and $\mathbb{H}$ indicate the identity component. We can also have $\mathbb{O}$ as the identity component. In summary:

COROLLARY 4.8. *Let $\mathcal{B}$ be a finite-dimensional real graded-division algebra that is alternative but not associative.*

1. *If $\mathcal{B}$ is graded-central with a nonsplit centroid, then $\mathcal{B}$ is equivalent to one of the following:*
   (a) $\mathbb{R}K \otimes \mathbb{R}\{n_1, n_2, n_3\}$, $\mathbb{R}K \otimes \mathbb{C}\{n_2, n_3\}$, *or* $\mathbb{R}K \otimes \mathbb{H}\{n_3\}$, *where $K$ is a finite abelian group and exactly one of the integers $n_j$ is negative;*
   (b) $\mathbb{R}K \otimes (-n) \otimes \mathbb{R}\{n_1, n_2, n_3\}$, $\mathbb{R}K \otimes (-n) \otimes \mathbb{C}\{n_2, n_3\}$, $\mathbb{R}K \otimes (-n) \otimes \mathbb{H}\{n_3\}$, *or* $\mathbb{R}K \otimes (-n) \otimes \mathbb{O}$, *where $K$ is a finite abelian group, $n > 0$, and $n_j \geq 0$.*
2. *If $\mathcal{B}$ is graded-central with a split centroid, then $\mathcal{B}$ is equivalent to one of the following:*
   (a) $\mathbb{R}K \otimes \mathbb{R}\{n_1, n_2, n_3\}$, $\mathbb{R}K \otimes \mathbb{C}\{n_2, n_3\}$, $\mathbb{R}K \otimes \mathbb{H}\{n_3\}$, *or* $\mathbb{R}K \otimes \mathbb{O}$, *where $K$ is a finite abelian group and $n_j \geq 0$;*
   (b) $\mathbb{R}K \otimes \mathbb{R}\{n_1, n_2, n_3\}'$, $\mathbb{R}K \otimes \mathbb{C}\{n_2, n_3\}'$, *or* $\mathbb{R}K \otimes \mathbb{H}\{n_3\}'$, *where $K$ is a finite abelian group and $n_j \geq 0$ with at least one being* $0$.
3. *If $\mathcal{B}$ is not graded-central, then $\mathcal{B}$ is equivalent to* $\mathbb{C}K \otimes \mathbb{R}\{n_1, n_2, n_3\}$, *where $K$ is a finite abelian group and $n_j \geq 0$.*

*Two algebras in different items are not equivalent to one another. Two algebras in the same item are equivalent if and only if they have isomorphic abelian groups $K$ and, if applicable, the same value of $n$ and the same $n_j$ up to permutation.* □

## 5. Jordan algebras of degree 2

As another application of Theorem 3.3 and Corollary 3.10, we will classify the finite-dimensional real $G$-graded unital Jordan algebras that are graded-simple and have degree 2 over the center (which is identified with the centroid).

The concept of *degree* is introduced for any finite-dimensional strictly power-associative unital algebra over a field using the so-called *generic minimal polynomial* (see e.g. [**18**, p. 223]), and this can be extended to algebras over direct products of fields: indeed, such an algebra is the direct product of algebras over fields, and we take the maximum degree of the factors.

**5.1. Jordan algebras of symmetric bilinear forms.** Let $V$ be a vector space over a field $\mathbb{F}$ ($\operatorname{char} \mathbb{F} \neq 2$) and let $b : V \times V \to \mathbb{F}$ be a symmetric bilinear form. Then $\mathcal{J}(V, b) := \mathbb{F}1 \oplus V$, with multiplication defined by $uv = b(u, v)1$ for all $u, v \in V$, is a unital Jordan algebra. It is known [**18**, p. 207] that any central simple Jordan algebra of degree 2 over $\mathbb{F}$ is isomorphic to $\mathcal{J}(V, b)$ where $\dim V \geq 2$ and $b$ is nondegenerate. Moreover, the pair $(V, b)$ is determined up to isometry.

As observed in [**4**], all $G$-gradings on these Jordan algebras are obtained from $G$-gradings on $V$ satisfying $b(V_{g_1}, V_{g_2}) = 0$ for all $g_1, g_2 \in G$ with $g_1 g_2 \neq e$.

More generally, if $\mathbb{L}$ is a graded-field, $\mathcal{V}$ is a graded $\mathbb{L}$-module (hence free), and $b : \mathcal{V} \times \mathcal{V} \to \mathbb{L}$ is a symmetric $\mathbb{L}$-bilinear form satisfying $b(\mathcal{V}_{g_1}, \mathcal{V}_{g_2}) \subset \mathbb{L}_{g_1 g_2}$ for all $g_1, g_2 \in G$, then $\mathcal{J}(\mathcal{V}, b) := \mathbb{L}1 \oplus \mathcal{V}$, with multiplication defined as above and $\mathcal{J}(\mathcal{V}, b)_g := \mathbb{L}_g 1 \oplus \mathcal{V}_g$, is a graded Jordan algebra over $\mathbb{L}$.

**5.2. Graded-simple real Jordan algebras of degree 2.** Let $\mathcal{B}$ be a finite-dimensional graded-simple real Jordan algebra, let $\mathbb{L}$ be the centroid of $\mathcal{B}$, and let $H$ be the support of $\mathbb{L}$. Then $H$ is finite, $\mathbb{L}_e$ is either $\mathbb{R}$ or $\mathbb{C}$, and $\mathcal{B} \cong L_\pi(\mathcal{A})$ or $\mathcal{B} \cong L_\pi^\chi(\mathcal{A})$ where $\mathcal{A}$ is a finite-dimensional central simple Jordan algebra over $\mathbb{R}$ or $\mathbb{C}$, respectively, $\pi$ is the natural homomorphism $G \to \overline{G} := G/H$, and $\chi$ is a character $G \to U \subset \mathbb{C}^\times$, as before. Since $\mathcal{A}$ is unital (see e.g. [**18**, p. 201]), so is $\mathcal{B}$, and we can identify $\mathbb{L}$ with the center of $\mathcal{B}$. Disregarding the grading, $\mathbb{L}$ is isomorphic to the direct product of copies of $\mathbb{R}$ and $\mathbb{C}$ (see Example 3.6).

Now assume that $\mathcal{B}$ has degree 2 over $\mathbb{L}$. If $\mathcal{B} \cong L_\pi(\mathcal{A})$ then $\mathbb{L}$ is the group algebra of $H$ and $\mathcal{A}$ is the quotient of $\mathcal{B}$ induced by the augmentation map of $\mathbb{L}$, hence $\mathcal{A}$ also has degree 2. (It cannot be 1, because in this case $\mathcal{A}$ would be the ground field and hence $\mathcal{B} = \mathbb{L}$.) The same conclusion is obtained if $\mathcal{B} \cong L_\pi^\chi(\mathcal{A})$ by extending scalars from $\mathbb{R}$ to $\mathbb{C}$. Therefore, $\mathcal{A} \cong \mathcal{J}(V, b)$ as above, with a $\overline{G}$-grading coming from a $\overline{G}$-grading on $V$ satisfying $b(V_{\bar{g}_1}, V_{\bar{g}_2}) = 0$ for all $\bar{g}_1, \bar{g}_2 \in G$ with $\bar{g}_1 \bar{g}_2 \neq \bar{e}$.

The pair $(V, b)$ is determined up to graded isometry (see [**13**, Theorem 5.42]). Over $\mathbb{C}$, such pairs are parametrized by certain functions $\kappa : \overline{G} \to \mathbb{Z}_{\geq 0}$. Indeed, the condition on $b$ implies that we have an orthogonal decomposition as follows:

$$V = V_{\bar{g}_1} \oplus \cdots \oplus V_{\bar{g}_m} \oplus (V_{\bar{g}'_{m+1}} \oplus V_{\bar{g}''_{m+1}}) \oplus \cdots \oplus (V_{\bar{g}'_l} \oplus V_{\bar{g}''_l}),$$

where all elements of $\overline{G}$ are distinct and, for $1 \leq j \leq m$, we have $\bar{g}_j^2 = \bar{e}$ and the restriction of $b$ to the subspace $V_{\bar{g}_j}$ is nondegenerate, while for $m < j \leq l$, we have $\bar{g}'_j \bar{g}''_j = \bar{e}$ and the subspaces $V_{\bar{g}'_j}$ and $V_{\bar{g}''_j}$ are totally isotropic and in duality through $b$. Therefore, we can choose bases in these subspaces so that $b$ is represented by the following matrix:

$$I_{k_1} \oplus \cdots \oplus I_{k_m} \oplus \begin{bmatrix} 0 & I_{k_{m+1}} \\ I_{k_{m+1}} & 0 \end{bmatrix} \oplus \cdots \oplus \begin{bmatrix} 0 & I_{k_l} \\ I_{k_l} & 0 \end{bmatrix}$$

where $k_j = \dim V_{\bar{g}_j}$ for $1 \leq j \leq m$ and $k_j = \dim V_{\bar{g}'_j} = \dim V_{\bar{g}''_j}$ for $m < j \leq l$ (see [**13**, p. 199]). We define

$$\kappa(\bar{g}) := \dim V_{\bar{g}} \quad \text{for all } \bar{g} \in \overline{G},$$

so the support of $\kappa$ is $\{\bar{g}_1, \ldots, \bar{g}_m, \bar{g}'_{m+1}, \bar{g}''_{m+1}, \ldots, \bar{g}'_l, \bar{g}''_l\}$ and $\kappa$ is *balanced* in the sense that $\kappa(\bar{g}^{-1}) = \kappa(\bar{g})$ for all $\bar{g} \in \overline{G}$. Conversely, every balanced $\kappa$ with finite support corresponds to $(V, b)$ as above. We will denote the corresponding $\overline{G}$-graded Jordan algebra by $\mathcal{J}_{\mathbb{C}}(\kappa)$.

Over $\mathbb{R}$, we have a similar situation, but $b$ will be represented by a matrix of the form

$$I_{p_1, q_1} \oplus \cdots \oplus I_{p_m, q_m} \oplus \begin{bmatrix} 0 & I_{k_{m+1}} \\ I_{k_{m+1}} & 0 \end{bmatrix} \oplus \cdots \oplus \begin{bmatrix} 0 & I_{k_l} \\ I_{k_l} & 0 \end{bmatrix}$$

where $I_{p,q} = \begin{pmatrix} I_p & 0 \\ 0 & -I_q \end{pmatrix}$, so $p_j - q_j$ is the signature of the restriction of $b$ to $V_{\overline{g}_j}$, $1 \leq j \leq m$. Thus, we have one more invariant, namely, the function $\sigma : \overline{G}_{[2]} \to \mathbb{Z}$ defined by

$$\sigma(\overline{g}) := \text{signature}(b|_{V_{\overline{g}}}) \text{ for all } \overline{g} \in \overline{G}_{[2]},$$

which satisfies $|\sigma(\overline{g})| \leq \kappa(\overline{g})$ and $\sigma(\overline{g}) \equiv \kappa(\overline{g}) \pmod 2$ for all $\overline{g} \in \overline{G}_{[2]}$. For a given $\kappa$, we will refer to functions $\sigma$ satisfying these conditions as *signature functions*. Every such pair $(\kappa, \sigma)$ determines $(V, b)$ as above. We will denote the corresponding $\overline{G}$-graded Jordan algebra by $\mathcal{J}(\kappa, \sigma)$.

Similarly to Subsection 4.3, we define a homomorphism

$$\tau : \overline{G}_{[2]} \to H/H^{[2]},\ gH \mapsto g^2 H^{[2]}. \tag{10}$$

Then the cocycle twists of $\mathcal{J}(\kappa, \sigma)$ are calculated similarly to Lemma 4.2:

LEMMA 5.1. *For any character $\lambda : H \to \mathbb{R}^\times$, we have $\mathcal{J}(\kappa, \sigma)^\lambda \cong \mathcal{J}(\kappa, \sigma')$ where $\sigma'(\overline{g}) = \sigma(\overline{g})\lambda_0(\tau(\overline{g}))$ for all $\overline{g} \in \overline{G}_{[2]}$, and $\lambda_0(hH^{[2]}) := \operatorname{sign}\lambda(h)$ for all $h \in H$.* □

Therefore, $\mathcal{J}(\kappa, \sigma')$ is graded-isomorphic to a twist of $\mathcal{J}(\kappa, \sigma)$ if and only if $\sigma' = \sigma\tilde{\mu}$ for some character $\tilde{\mu} : \overline{G}_{[2]} \to \{\pm 1\}$ that restricts to the trivial character on $\ker\tau$ (see the proof of Lemma 4.3). To summarize:

THEOREM 5.2. *Let $\mathcal{B}$ be a $G$-graded finite-dimensional unital Jordan algebra over $\mathbb{R}$ that is graded-simple and has degree 2 over its center $\mathbb{L}$. Let $H$ be the support of $\mathbb{L}$, $\overline{G} = G/H$, and $\pi$ be the natural homomorphism $G \to \overline{G}$. If $\mathbb{L}_e = \mathbb{R}$, pick a character $\chi : G \to U \subset \mathbb{C}^\times$ such that $\mathbb{L} \cong L^\chi_\pi(\mathbb{R})$.*

1. *If $\mathbb{L}_e = \mathbb{R}$ then $\mathcal{B}$ is graded-isomorphic to $L^\chi_\pi(\mathcal{J}(\kappa, \sigma))$ for a unique balanced function $\kappa : \overline{G} \to \mathbb{Z}_{\geq 0}$ with finite support and a signature function $\sigma : \overline{G}_{[2]} \to \mathbb{Z}$, which is determined up to multiplication by the characters $\overline{G}_{[2]} \to \{\pm 1\}$ with trivial restriction to $\ker\tau$, where $\tau : \overline{G}_{[2]} \to H/H^{[2]}$ is given by Equation (10).*
2. *If $\mathbb{L}_e \cong \mathbb{C}$ then $\mathcal{B}$ is graded-isomorphic to $L_\pi(\mathcal{J}_\mathbb{C}(\kappa))$ for a unique balanced function $\kappa : \overline{G} \to \mathbb{Z}_{\geq 0}$ with finite support.* □

The graded algebras in Theorem 5.2 can be obtained as Jordan algebras of symmetric bilinear forms over $\mathbb{L}$. Indeed, consider the $G$-graded $\mathbb{L}$-module $\mathcal{V} = L^\chi_\pi(V)$ and define a symmetric $\mathbb{L}$-bilinear form $B : \mathcal{V} \times \mathcal{V} \to \mathbb{L}$ as follows:

$$B(v_1 \otimes u_{g_1}, v_2 \otimes u_{g_2}) := b(v_1, v_2) \otimes u_{g_1} u_{g_2} \text{ for all } v_1 \in V_{\pi(g_1)},\ v_2 \in V_{\pi(g_2)},\ g_1, g_2 \in G.$$

Since $b(v_1, v_2) = 0$ unless $g_1 g_2 \in H$, the right-hand side is indeed in $\mathbb{L}$, and $B$ satisfies $B(\mathcal{V}_{g_1}, \mathcal{V}_{g_2}) \subset \mathbb{L}_{g_1 g_2}$. Hence, we can define the $G$-graded Jordan algebra $\mathcal{J}(\mathcal{V}, B)$, and the canonical isomorphism $(\mathbb{R}1 \oplus V) \otimes \mathbb{R}^\gamma G \cong \mathbb{R}^\gamma G \oplus (V \otimes \mathbb{R}^\gamma G)$ restricts to an isomorphism $L^\chi_\pi(\mathcal{J}(V, b)) \cong \mathcal{J}(\mathcal{V}, B)$.

Now choose $g_j \in G$ such that $\pi(g_j) = \overline{g}_j$ for $1 \leq j \leq m$ and $g'_j \in G$ such that $\pi(g'_j) = \overline{g}'_j$ for $m < j \leq l$. (If $\overline{g}_j = \overline{e}$, we take $g_j = e$.) Then $\mathcal{V}$ can be identified with $V \otimes \mathbb{L}$ as a graded $\mathbb{L}$-module through the isomorphism sending, for each $1 \leq j \leq m$, $v \otimes u_h \mapsto v \otimes u_{g_j} u_h$ for all $v \in V_{\overline{g}_j}$, $h \in H$, and for each $m < j \leq l$, $v \otimes u_h \mapsto v \otimes u_{g'_j} u_h$ for all $v \in V_{\overline{g}'_j}$, $h \in H$, and $v \otimes u_h \mapsto v \otimes u_{(g'_j)^{-1}} u_h$ for all $v \in V_{\overline{g}''_j}$, $h \in H$. Then any choice of bases in $V_{\overline{g}_j}$, $V_{\overline{g}'_j}$ and $V_{\overline{g}''_j}$ gives us an $\mathbb{L}$-basis of $\mathcal{V}$. We can choose the bases so that $B$ is represented by the matrix

$$I_{p_1,q_1} \otimes \delta_1 u_{h_1} \oplus \cdots \oplus I_{p_m,q_m} \otimes \delta_m u_{h_m} \oplus \begin{bmatrix} 0 & I_{k_{m+1}} \\ I_{k_{m+1}} & 0 \end{bmatrix} \otimes 1 \oplus \cdots \oplus \begin{bmatrix} 0 & I_{k_l} \\ I_{k_l} & 0 \end{bmatrix} \otimes 1$$

where $\delta_j = \gamma(\overline{g}_j, \overline{g}_j) \in \{\pm 1\}$ and $h_j = g_j^2$.

If we consider these graded algebras up to equivalence rather than isomorphism, some further simplifications can be made. First of all, it is convenient to change the grading group as follows. Let $\widetilde{G}$ be the abelian group generated by $H$ and the new symbols $\tilde{g}_1, \ldots, \tilde{g}_m, \tilde{g}'_{m+1}, \tilde{g}''_{m+1}, \ldots, \tilde{g}'_l, \tilde{g}''_l$ subject to the following relations: for $1 \leq j \leq m$, we impose $\tilde{g}_j^2 = h_j$ if $\overline{g}_j \neq \bar{e}$ and $\tilde{g}_j = e$ otherwise, while for $m < j \leq l$, we impose $\tilde{g}'_j \tilde{g}''_j = e$. Then $H$ is a subgroup of $\widetilde{G}$ and we have $\widetilde{G}/H \cong \mathbb{Z}_2^m \times \mathbb{Z}^{l-m}$ if $\overline{g}_j \neq \bar{e}$ for all $j$ and $\widetilde{G}/H \cong \mathbb{Z}_2^{m-1} \times \mathbb{Z}^{l-m}$ otherwise. We define a $\widetilde{G}$-grading on $\mathcal{V}$ by declaring, for each $1 \leq j \leq m$, $\mathcal{V}_{g_j h}$ to be the component of degree $\tilde{g}_j h$ for all $h \in H$, and similarly for $m < j \leq l$. The $\mathbb{L}$-bilinear form $B$ is compatible with this new grading, so we obtain a $\widetilde{G}$-grading on $\mathcal{J}(\mathcal{V}, B)$, which is equivalent to the original $G$-grading.

REMARK 5.3. *We can say more: the $G$-grading on $\mathcal{J}(\mathcal{V}, B)$ is induced from the $\widetilde{G}$-grading by the homomorphism $\alpha : \widetilde{G} \to G$ extending the identity on $H$ and sending $\tilde{g}_j \mapsto g_j$, $\tilde{g}'_j \mapsto g'_j$, and $\tilde{g}''_j \mapsto (g'_j)^{-1}$. It follows that $\widetilde{G}$ is the* universal abelian group *(see e.g.* [**13**, §1.2]*) of the grading on $\mathcal{J}(\mathcal{V}, B)$.*

Now replace $G$ by $\widetilde{G}$ and drop the tilde. If $\mathbb{L}$ is nonsplit, there exists $h \in H$ such that $u_h^2 = -1$, and we can use it to adjust our $\mathbb{L}$-basis of $\mathcal{V}$ to make $\delta_j = 1$ or $\delta_j = -1$ arbitrarily, but note that this also changes $g_j$, so we cannot apply this if $\overline{g}_j = \bar{e}$. Therefore, we can make the matrix of $B$ look as follows:

$$I_{p_1,q_1} \otimes u_{h_1} \oplus \cdots \oplus I_{p_m,q_m} \otimes u_{h_m} \oplus \begin{bmatrix} 0 & I_{k_{m+1}} \\ I_{k_{m+1}} & 0 \end{bmatrix} \otimes 1 \oplus \cdots \oplus \begin{bmatrix} 0 & I_{k_l} \\ I_{k_l} & 0 \end{bmatrix} \otimes 1$$

where we may assume $p_j \geq q_j$ if $\overline{g}_j \neq \bar{e}$. If $\mathbb{L} \cong \mathbb{R}H$ then we may assume $u_h = h$, so $\delta_j = 1$, but note that the isomorphism condition in Theorem 5.2 allows us to interchange $p_j$ and $q_j$ arbitrarily if $\overline{g}_j \notin \ker \tau$. Therefore, we can make the matrix of $B$ look as above, where we may assume $p_j \geq q_j$ if $\overline{g}_j \notin \ker \tau$.

**5.3. Graded-division Jordan algebras of degree 2.** In a unital Jordan algebra $\mathcal{J}$, an inverse of an element $a$ is $b \in \mathcal{J}$ such that $ab = 1$ and $a^2 b = a$. The existence of such $b$ is equivalent to the invertibility of the operator $U_a := 2L_a^2 - L_{a^2}$. If this is the case, the element $b$ is unique and given by $b = U_a^{-1}(a)$ (see e.g. [**18**, p. 52]), so the usual notation $a^{-1}$ is used. $\mathcal{J}$ is called a *division algebra* if every nonzero element is invertible. It is clear that if $\mathcal{J}$ is $G$-graded and $a \in \mathcal{J}_g$ is invertible then $a^{-1} \in \mathcal{J}_{g^{-1}}$. As stated in the introduction, $\mathcal{J}$ is a *graded-division algebra* if every nonzero homogeneous element is invertible. Note, however, that, for an invertible element $a$, the operator $L_a$ is not necessarily invertible, and the set of invertible elements need not be closed under multiplication.

The Jordan algebra of a symmetric bilinear form, $\mathcal{J}(V, b)$, is a division algebra if and only if $b(v, v)$ is not a square in $\mathbb{F}$, for any $0 \neq v \in V$ (see [**18**, Exercise 2, p. 54]). Clearly, in the case $\mathbb{F} = \mathbb{R}$, this is equivalent to $b$ being negative, i.e., $b(v, v) < 0$, for any $0 \neq v \in V$.

PROPOSITION 5.4. *Suppose we have a $G$-grading on the Jordan algebra $\mathcal{J} = \mathcal{J}(V, b)$ over any field $\mathbb{F}$, $\operatorname{char}\mathbb{F} \neq 2$. Then $\mathcal{J}$ is a graded-division algebra if and only if the following conditions hold:*

(i) *all components $V_g$ are nonisotropic (i.e., the restriction of the quadratic form $b(v, v)$ to $V_g$ does not represent $0$);*

(ii) *for every $0 \neq v \in V_e$, $b(v, v)$ is not a square in $\mathbb{F}$.*

*Condition (i) implies that all elements of the support have order $\leq 2$.*

PROOF. Since $\mathcal{J}_e = \mathbb{F}1 \oplus V_e = \mathcal{J}(V_e, b_e)$, where $b_e$ is the restriction of $b$ to $V_e$, the nonzero elements of $\mathcal{J}_e$ are invertible if and only if condition (ii) holds, which also implies condition (i) for $g = e$. Now suppose $0 \neq v \in \mathcal{J}_g$ for some $g \neq e$. Then $v \in V_g$ and hence $v^2 = b(v, v)1$. If $b(v, v) = 0$ then $v$ is not invertible because $v^2 b = 0 \neq v$ for any $b \in \mathcal{J}$. Conversely, if $b(v, v) \neq 0$ then $v^{-1} = \frac{1}{b(v,v)} v$. Therefore, condition (i) is equivalent to the invertibility of all nonzero elements of $\mathcal{J}_g$. If $g^2 \neq e$ then $V_g$ is totally isotropic. $\square$

Recalling Remark 3.5 and applying the above proposition to our $\mathcal{J}(\kappa, \sigma)$ and $\mathcal{J}_{\mathbb{C}}(\kappa)$, with $\overline{G}$ playing the role of $G$, we obtain the following:

COROLLARY 5.5. *Let $\mathcal{B}$ be as in Theorem 5.2. Then $\mathcal{B}$ is a graded-division algebra if an only if one of the following holds:*

1. *$\mathbb{L}_e = \mathbb{R}$, the support of $\kappa$ is contained in $\overline{G}_{[2]}$, $|\sigma(\overline{g})| = \kappa(\overline{g})$ for all $\overline{g} \in \overline{G}_{[2]}$, and $\sigma(\overline{e}) = -\kappa(\overline{e})$;*
2. *$\mathbb{L}_e \cong \mathbb{C}$, the support of $\kappa$ is contained in $\overline{G}_{[2]}$, $\kappa(\overline{g}) \in \{0, 1\}$ for all $\overline{g} \in \overline{G}_{[2]}$, and $\kappa(\overline{e}) = 0$.* $\square$

Note that in item 2, the conditions imply that all nonzero homogeneous components of $\mathcal{B}$ are 1-dimensional, whereas in item 1, they can have arbitrary dimensions. This is in contrast with alternative (in particular, associative) graded-division algebras, where all nonzero components must have the same dimension (which, if finite, must be 1, 2, 4 or 8 over $\mathbb{R}$).

## Acknowledgments

The authors are grateful to Alberto Elduque and Efim Zelmanov for fruitful discussions.

## References

[1] Bruce Allison, Stephen Berman, John Faulkner, and Arturo Pianzola, *Realization of graded-simple algebras as loop algebras*, Forum Math. **20** (2008), no. 3, 395–432, DOI 10.1515/FORUM.2008.020. MR2418198

[2] Yuri Bahturin, Mikhail Kochetov, and Adrián Rodrigo-Escudero, *Gradings on classical central simple real Lie algebras*, J. Algebra **506** (2018), 1–42, DOI 10.1016/j.jalgebra.2018.02.036. MR3800069

[3] Yu. A. Bahturin, S. K. Sehgal, and M. V. Zaicev, *Group gradings on associative algebras*, J. Algebra **241** (2001), no. 2, 677–698, DOI 10.1006/jabr.2000.8643. MR1843319

[4] Yuri Bahturin and Ivan Shestakov, *Gradings of simple Jordan algebras and their relation to the gradings of simple associative algebras*, Comm. Algebra **29** (2001), no. 9, 4095–4102, DOI 10.1081/AGB-100105990. Special issue dedicated to Alexei Ivanovich Kostrikin. MR1857029

[5] Y. A. Bahturin, I. P. Shestakov, and M. V. Zaicev, *Gradings on simple Jordan and Lie algebras*, J. Algebra **283** (2005), no. 2, 849–868, DOI 10.1016/j.jalgebra.2004.10.007. MR2111225

[6] Yuri Bahturin and Mikhail Zaicev, *Simple graded division algebras over the field of real numbers*, Linear Algebra Appl. **490** (2016), 102–123, DOI 10.1016/j.laa.2015.10.021. MR3429035

[7] Yuri Bahturin and Mikhail Zaicev, *Graded division algebras over the field of real numbers*, J. Algebra **514** (2018), 273–309, DOI 10.1016/j.jalgebra.2018.08.015. MR3853066
[8] Georgia Benkart and Erhard Neher, *The centroid of extended affine and root graded Lie algebras*, J. Pure Appl. Algebra **205** (2006), no. 1, 117–145, DOI 10.1016/j.jpaa.2005.06.007. MR2193194
[9] A. J. Calderón Martín, C. Draper, and C. Martín González, *Gradings on the real forms of the Albert algebra, of* $\mathfrak{g}_2$*, and of* $\mathfrak{f}_4$, J. Math. Phys. **51** (2010), no. 5, 053516, 21, DOI 10.1063/1.3326238. MR2666994
[10] Cristina Draper Fontanals and Cándido Martín González, *Gradings on the Albert algebra and on* $\mathfrak{f}_4$, Rev. Mat. Iberoam. **25** (2009), no. 3, 841–908. MR2590049
[11] Alberto Elduque, *Gradings on octonions*, J. Algebra **207** (1998), no. 1, 342–354, DOI 10.1006/jabr.1998.7474. MR1643126
[12] Alberto Elduque and Mikhail Kochetov, *Gradings on the exceptional Lie algebras* $F_4$ *and* $G_2$ *revisited*, Rev. Mat. Iberoam. **28** (2012), no. 3, 773–813, DOI 10.4171/RMI/691. MR2949619
[13] Alberto Elduque and Mikhail Kochetov, *Gradings on simple Lie algebras*, Mathematical Surveys and Monographs, vol. 189, American Mathematical Society, Providence, RI; Atlantic Association for Research in the Mathematical Sciences (AARMS), Halifax, NS, 2013. MR3087174
[14] Alberto Elduque and Mikhail Kochetov, *Graded simple modules and loop modules*, Groups, rings, group rings, and Hopf algebras, Contemp. Math., vol. 688, Amer. Math. Soc., Providence, RI, 2017, pp. 53–85. MR3649168
[15] Alberto Elduque and Mikhail Kochetov, *Gradings on the simple real Lie algebras of types* $G_2$ *and* $D_4$, J. Algebra **512** (2018), 382–426, DOI 10.1016/j.jalgebra.2018.06.035. MR3841529
[16] Antonio Giambruno and Mikhail Zaicev, *Polynomial identities and asymptotic methods*, Mathematical Surveys and Monographs, vol. 122, American Mathematical Society, Providence, RI, 2005. MR2176105
[17] Robert L. Griess Jr., *A Moufang loop, the exceptional Jordan algebra, and a cubic form in* 27 *variables*, J. Algebra **131** (1990), no. 1, 281–293, DOI 10.1016/0021-8693(90)90176-O. MR1055009
[18] Nathan Jacobson, *Structure and representations of Jordan algebras*, American Mathematical Society Colloquium Publications, Vol. XXXIX, American Mathematical Society, Providence, R.I., 1968. MR0251099
[19] Erwin Kleinfeld, *Simple alternative rings*, Ann. of Math. (2) **58** (1953), 544–547, DOI 10.2307/1969753. MR0058581
[20] Erwin Kleinfeld, *Alternative nil rings*, Ann. of Math. (2) **66** (1957), 395–399, DOI 10.2307/1969897. MR0087638
[21] Saunders Mac Lane, *Homology*, Classics in Mathematics, Springer-Verlag, Berlin, 1995. Reprint of the 1975 edition. MR1344215
[22] Adrián Rodrigo-Escudero, *Classification of division gradings on finite-dimensional simple real algebras*, Linear Algebra Appl. **493** (2016), 164–182, DOI 10.1016/j.laa.2015.11.025. MR3452733
[23] K. A. Zhevlakov, A. M. Slin′ko, I. P. Shestakov, and A. I. Shirshov, *Rings that are nearly associative*, Pure and Applied Mathematics, vol. 104, Academic Press, Inc. [Harcourt Brace Jovanovich, Publishers], New York-London, 1982. Translated from the Russian by Harry F. Smith. MR668355

Department of Mathematics and Statistics, Memorial University of Newfoundland, St. John's, NL, A1C5S7, Canada
*Email address*: bahturin@mun.ca

Department of Mathematics and Statistics, Memorial University of Newfoundland, St. John's, NL, A1C5S7, Canada
*Email address*: mikhail@mun.ca

Contemporary Mathematics
Volume **728**, 2019
https://doi.org/10.1090/conm/728/14655

# Nonsemisimple Hopf algebras of dimension $8p$ with the Chevalley property

Margaret Beattie, Gastón A. García, Siu-Hung Ng, and Jolie Roat

Abstract. Let $p$ be an odd prime and $\Bbbk$ an algebraically closed field of characteristic zero. We classify nonsemisimple Hopf algebras over $\Bbbk$ of dimension $8p$ with the Chevalley property, and give partial results on nonsemisimple Hopf algebras of dimension 24.

## 1. Introduction

It was conjectured by Kaplansky [**K**] that there are only finitely many isomorphism classes of finite-dimensional Hopf algebras over $\Bbbk$ for any given dimension. This conjecture was shown to be false by counterexamples of dimension $p^4$ with $p$ an odd prime [**AS1**,**BDG**,**Ge2**], but it is true for dimension 16 [**GV**]. However, the module categories over Hopf algebras of dimension 32 yield infinitely many inequivalent finite tensor categories [**EG**]. In particular, there are infinitely isomorphism classes of Hopf algebras of dimension 32, see [**B**], [**Gr3**].

In the past two decades, Hopf algebras in a dimension which is factorized with *few* primes have been classified (cf. [**BG2**]). As a consequence, the number of isomorphsim classes of Hopf algebras of dimensions less than 32 but not equal to 24 is known to be finite. Semisimple Hopf algebras of dimension 24 with a $C^*$-structure were classified in [**IK**], but the general classification without $C^*$-structure remains open. In the nonsemisimple case, pointed Hopf algebras of dimension 24 are classified and there are finitely many of them up to isomorphism. However, not all the 24-dimensional Hopf algebras are pointed.

The family of Hopf algebras with the Chevalley property consists of Hopf algebras whose coradical forms a Hopf subalgebra. For example, pointed Hopf algebras have the Chevalley property. In fact, there exist $8p$-dimensional Hopf algebras with the Chevalley property which are not pointed for each odd prime $p$, see Proposition 3.14. In this paper, in addition to some new general results, we will provide a complete classification of nonsemisimple Hopf algebras of dimension $8p$ with $p$ an odd prime, which have the Chevalley property. Even though this result brings us one step closer to the complete classification of 24-dimensional nonsemisimple Hopf algebras, it remains unknown whether there exists any Hopf algebra $H$ of dimension 24 such that neither $H$ nor $H^*$ has the Chevalley property. Nevertheless, some

2010 *Mathematics Subject Classification.* 16T05.

The second author was partially supported by ANPCyT-Foncyt, CONICET, and Secyt (UNLP). The third author was partially supported by the NSF grant DMS 1501179.

new observations of 24-dimensional Hopf algebras are presented in the last section of the paper which could be essential to the complete classification.

The paper is organized as follows. In Section 2 we introduce several definitions and conventions that will be needed along the paper. We also provide some descriptions on Hopf algebras with respect to their coalgebra structure and study pointed Hopf algebras over the group given by the semidirect product $\Gamma_{4p} = C_p \rtimes_\chi C_4$. In Section 3 we collect some facts on nonsemisimple Hopf algebras of dimension $8p$ and prove our main result in Theorem 3.16. The last section is devoted to the study of nonsemisimple Hopf algebras of dimension 24.

## 2. Preliminaries

**2.1. Definitions and Conventions.** We work over an algebraically closed field $\Bbbk$ of characteristic zero. Let $H$ be a finite-dimensional Hopf algebra over $\Bbbk$ with antipode $\mathcal{S}$. We use Sweedler's notation $\Delta(h) = h_1 \otimes h_2$ for the comultiplication in $H$, but dropping the summation symbol, see [**S**]. The readers are referred to [**Mo**] for details on the basic definitions of Hopf algebras.

The *coradical* $H_0$ of $H$ is the sum of all simple subcoalgebras of $H$. In particular, if $G(H)$ denotes the group of *group-like elements* of $H$, we have $\Bbbk G(H) \subseteq H_0$. We write $H_0 = \bigoplus_{d\geq 1} H_{0,d}$, where $H_{0,d}$ denotes the direct sum of the simple subcoalgebras of $H$ of dimension $d^2$. By the Nichols-Zoeller Theorem, its dimension is divisible by the order of the group of group-like elements (see [**AN**]).

We say that a Hopf algebra $H$ is *pointed* if $H_0 = \Bbbk G(H)$, and it is called *basic* if all the simple $H$-modules are 1-dimensional. Since $H$ is finite-dimensional, $H$ is basic if and only if $H^*$ is pointed.

We denote by $\{H_i\}_{i\geq 0}$ the *coradical filtration* of $H$. If $H_0$ is a Hopf subalgebra of $H$, then

$$\operatorname{gr} H = \bigoplus_{n\geq 0} \operatorname{gr} H(n)$$

is the associated graded Hopf algebra, where $\operatorname{gr} H(n) = H_n/H_{n-1}$ for $n \geq 0$ and $H_{-1} = 0$. Let $\pi : \operatorname{gr} H \to H_0$ be the homogeneous projection. Then

$$R = (\operatorname{gr} H)^{\operatorname{co}\pi} := \{r \in \operatorname{gr} H \mid (\operatorname{id} \otimes \pi)\Delta(r) = r \otimes 1\}$$

is called the *diagram* of $H$, which is a braided Hopf algebra in the category ${}^{H_0}_{H_0}\mathcal{YD}$ of left Yetter-Drinfeld modules over $H_0$, and it is a graded subalgebra of $\operatorname{gr} H$. The linear subspace $R(1)$ of homogeneous elements of $R$ of degree one, with the braiding from ${}^{H_0}_{H_0}\mathcal{YD}$, is called the *infinitesimal braiding* of $H$ and coincides with the subspace of primitive elements

$$\mathcal{P} = \{r \in R \mid \Delta_R(r) = r \otimes 1 + 1 \otimes r\},$$

where $\Delta_R$ denotes the comultiplication of $R$. It turns out that the Hopf algebra $\operatorname{gr} H$ is the Radford biproduct (or Magid bosonization) $\operatorname{gr} H \simeq R\# H_0$ (cf. [**Ra**, **Ma**]). The subalgebra of $R$ generated by $R(1) = V$ is isomorphic to the Nichols algebra $\mathfrak{B}(V)$. If $H_0 = \Bbbk G$ is a group algebra, we denote the category of Yetter-Drinfeld modules simply by ${}^G_G\mathcal{YD}$. Specifically, a left Yetter-Drinfeld module over $\Bbbk G$ is a finite-dimensional $G$-graded $\Bbbk$-linear space $M = \oplus_{h\in G} M_h$ with a left $\Bbbk G$-module $(M, \cdot)$ structure such that

$$\delta(g \cdot m) = ghg^{-1} \otimes g \cdot m, \qquad \forall\, m \in M_h, g, h \in G.$$

Here $\delta$ is left $\Bbbk G$-comodule structure on $M$ given by the $G$-grading, namely with $\delta(m) = h \otimes m$ for $m \in M_h$.

Yetter-Drinfeld modules over $G$ are completely reducible and the irreducible ones are parametrized by pairs $(\mathcal{O}, \rho)$, where $\mathcal{O}$ is a conjugacy class of $G$ and $(\rho, V)$ is an irreducible representation of the centralizer $C_G(\sigma)$ of a fixed element $\sigma \in \mathcal{O}$: Then the irreducible Yetter-Drinfeld module $M(\mathcal{O}, \rho)$ is the induced $\Bbbk G$-module $\Bbbk G \otimes_{\Bbbk C_G(\sigma)} V$. Let $1 = g_1, \dots, g_n$ be a complete set of left coset representatives of $C_G(\sigma)$ in $G$. Then $\sigma_i = g_i \sigma g_i^{-1}$, $i = 1, \dots, n$, are all the elements of $\mathcal{O}$ and $M(\mathcal{O}, \rho) = \bigoplus_{i \leq 1 \leq n} g_i \otimes V$. The $\Bbbk G$-comodule structure on $M(\mathcal{O}, \rho)$ is given by $\delta(gv_i) = \sigma_i \otimes gv_i$ for all $i = 1, \dots, n$, where $g_i v$ denotes the element $g_i \otimes v \in g_i \otimes V$. In particular, $\dim M(\mathcal{O}, \rho) = |\mathcal{O}| \dim V$. We denote the corresponding Nichols algebra by $\mathfrak{B}(\mathcal{O}, \rho)$. For the definition of Nichols algebras and basic facts, we refer the reader to [**A**].

Throughout this paper, the cyclic group of order $\ell$ is denoted by $C_\ell$. For $n \in \mathbb{N}$, $n \geq 3$, the dihedral group of order $2n$ is denoted by $\mathbb{D}_n$; we will use the presentation

$$\mathbb{D}_n = \langle a, y \mid a^2 = y^n = 1,\ aya = y^{-1} \rangle.$$

The dicyclic group of order $4n$, denoted by $\mathrm{Dic}_n$, has the presentation

$$\mathrm{Dic}_n = \langle x, y \mid x^4 = y^n = 1,\ xyx^{-1} = y^{-1} \rangle.$$

Let $N$ be a positive integer and let $q \in \Bbbk$ be a primitive $N$-th root of unity. The *Taft algebra* $T_q$ is the $\Bbbk$-algebra generated by the elements $g$ and $x$ satisfying the relations $x^N = 0 = 1 - g^N$, $gx = qxg$. It is a Hopf algebra with its coalgebra structure determined by $g$ being group-like and $x$ a $(1, g)$-primitive, *i.e.*, $\Delta(g) = g \otimes g$ and $\Delta(x) = x \otimes 1 + g \otimes x$. It is self-dual and pointed of dimension $N^2$. If $N = 2$ so that $q = -1$, then $T_{-1}$ is called the *Sweedler algebra* and will be denoted by $H_4$ throughout this article.

Recall that a tensor category $\mathcal{C}$ over $\Bbbk$ has the Chevalley property if the tensor product of any two simple objects is semisimple. We shall say that a Hopf algebra $H$ has the *Chevalley property* if the category $\mathrm{Comod}\,(H)$ of $H$-comodules does. In particular, $H$ has the Chevalley property if and only if the coradical is a Hopf subalgebra.

REMARK 2.1. In some literature (for example [**NN**]), $H$ is said to have the Chevalley property if $H$-mod does. In this case, $H$ has the Chevalley property if the Jacobson radical of $H$ is a Hopf ideal.

**2.2. On simple coalgebras, group-likes and skew-primitives.** In this subsection we collect some structural results that will be needed later.

Recall that $H$ is a free left and a free right $H^*$-module of rank 1 via the $H^*$-actions $\rightharpoonup$ and $\leftharpoonup$ defined by

$$f \rightharpoonup h = f(h_2)h_1 \quad \text{and} \quad h \leftharpoonup f = f(h_1)h_2, \qquad \text{for all } f \in H^* \text{ and } h \in H.$$

Thus, a finite-dimensional $\Bbbk$-linear space $V$ is a right (resp. left) $H$-comodule if and only if $V$ is a left (resp. right) $H^*$-module. If $\sigma$ is an algebra automorphism of $H^*$, then $V$ can be turned into another left $H^*$-module ${}_\sigma V$ with the $H^*$-action twisted by $\sigma$, namely

$$f \cdot v := \sigma(f)v, \qquad \text{for } f \in H^* \text{ and } v \in V.$$

LEMMA 2.2. Let $C$ be a 4-dimensional simple coalgebra over $\Bbbk$ and $T : C \to C$ a coalgebra automorphism with finite order. If $\mathrm{Tr}(T) = 0$, then $\mathrm{ord}(T) = 2$.

PROOF. Let $n = \operatorname{ord}(T)$. By [**Š97**, Theorem 1.4 (b)], there is a primitive $n$-th root $\omega \in \Bbbk$ and a basis $\{e_{ij}\}_{i,j=1,2}$ of $C$ such that $T(e_{ij}) = \omega^{i-j} e_{ij}$. Thus, $\operatorname{Tr}(T) = (1+\omega)(1+\omega^{-1})$ which is zero, if and only if, $\omega = -1$. Therefore, $n = 2$. □

LEMMA 2.3. Let $H$ be a finite-dimensional Hopf algebra that admits a decomposition of coalgebras $H = C \oplus C'$ with $C$ simple. Then $C$ is invariant under $\mathcal{S}^{-2} \circ L(a)$, where $a$ is the distinguished group-like element of $H$ and $L(a)$ denotes the left multiplication operator on $H$ by $a$. Moreover, $\operatorname{Tr}_C(\mathcal{S}^{-2} \circ L(a)) = 0$ provided $a$ is nontrivial. If, in addition, $\dim C = 4$, then $\mathcal{S}^4|_C = L(a^2)|_C$.

PROOF. Let $I$ be a simple right coideal of $H$ such that $\Delta(I) \subset I \otimes C$. Then $C \cong \dim(I) I$ as right $H$-comodules and hence left $H^*$-modules. In particular, $I$ is a simple projective $H^*$-module. By [**Ng**, Lemma 1.1], we have that $I \cong \Bbbk_{\widehat{a^{-1}}} \otimes I^{\vee\vee}$, where $I^\vee$ denotes the left dual of $I$ in the category of finite-dimensional $H^*$-modules, and $\hat{x} \in H^{**}$ denotes the image of $x \in H$ under the natural isomorphism of $H \simeq H^{**}$. In particular, $I \cong {}_\sigma I$ where $\sigma = (\mathcal{S}^*)^2 \circ R(\widehat{a^{-1}})$ (cf. [**Ng**, Lemma 1.2]). Therefore, the annihilator ideal $\operatorname{ann}_{H^*} I$ is invariant under the automorphism $\sigma$. Note that $\sigma = \tau^*$ where $\tau = \mathcal{S}^2 \circ L(a^{-1})$ is a coalgebra automorphism. Since $\operatorname{ann}_{H^*} I = C^\perp$, we find $\sigma(C^\perp) = C^\perp$ and hence $C$ is invariant under $\tau$.

Since $I \cong {}_\sigma I$ as $H^*$-modules, $C \cong {}_\sigma C$ as $H^*$-modules. To complete the proof, by [**Ng**, Lemma 1.2], it suffices to show that $\tau^{-1} : C \to {}_\sigma C$ is a $H^*$-module homomorphism. For $f \in H^*$ and $c \in C$,

$$\sigma(f) \rightharpoonup \tau^{-1}c = \sigma(f)(\tau^{-1}c_2)\tau^{-1}c_1 = f(\tau(\tau^{-1}c_2))\tau^{-1}c_1 = f(c_2)\tau^{-1}c_1 = \tau^{-1}(f \rightharpoonup c)\,.$$

Therefore, $\tau^{-1} \in \operatorname{Hom}_{H^*}(C, {}_\sigma C)$. The last assertion follows then by Lemma 2.2. □

LEMMA 2.4. Let $H$ be a finite-dimensional Hopf algebra and $A$ a Hopf subalgebra. Suppose $H$ admits a subcoalgebra decomposition $H = A \oplus C$ for some subcoalgebra $C$ of $H$. Then $H$ and $A$ have a common distinguished group-like element.

PROOF. Let $\lambda$ be a nonzero right integral of $H^*$. Then, for all $h \in H$, $h \leftharpoonup \lambda = \lambda(h)1$ and $\lambda \rightharpoonup h = \lambda(h)a$ where $a$ is the distinguished group-like element of $H$. Note that for $c \in C$, we have $\lambda(c) = 0$. Otherwise, $\lambda(c)1 = \lambda(c_1)c_2 \in C$ and hence $1 \in C$, which contradicts that $C \cap A = 0$. Therefore, $\lambda|_A \neq 0$ and it is a right integral of $A$. Let $h \in A$ such that $\lambda(h) \neq 0$. Then

$$a = \lambda(h)^{-1}\lambda(h_2)h_1 \in A\,.$$

Thus, $a$ is the distinguished group-like element of $A$. □

The following proposition is a refined version of [**BG**, Proposition 2.17] and [**HN**, Proposition 1.3].

PROPOSITION 2.5. Let $\pi : H \to A$ be a Hopf algebra epimorphism and assume $\dim H = 2 \dim A$. Then $H^{co\pi} = \Bbbk\{1, x\}$ with $x$ a $(1, g)$-primitive element, $g \in G(H)$ and $2 \mid \operatorname{ord} g$. Moreover, if $x$ is a trivial skew-primitive, *i.e.*, a nonzero scalar multiple of $1 - g$, then $\operatorname{ord} g = 2$ and $H$ fits into an exact sequence of Hopf algebras $\Bbbk C_2 \hookrightarrow H \twoheadrightarrow A$. If $x$ is nontrivial, then $x^2 = 0$ and $H$ contains a Hopf subalgebra of dimension $2 \operatorname{ord}(g)$; in particular, $4 \mid \dim H$. □

**2.3. On pointed Hopf algebras over $\Gamma_{4p}$.** Our main results will involve Hopf algebras whose coradical is a semisimple Hopf algebra of dimension $4p$ for $p$ an odd prime. To this end, we need some results on the simple representations of the group $\Gamma_{4p}$. Recall that for $p$ a prime congruent to 1 mod 4, $\Gamma_{4p}$ is defined to be the group given by the semidirect product $\Gamma_{4p} = C_p \rtimes_\chi C_4$ where $C_p = \langle y\rangle$, $C_4 = \langle x\rangle$ are the cyclic groups of order $p$ and 4, respectively, and $\chi(x)(y) = y^\ell$, where $\ell = \frac{p-1}{2}$. Moreover, $\Gamma_{4p}$ has the presentation

$$\Gamma_{4p} = \langle x, y : \; x^4 = 1 = y^p, \; xy = y^\ell x\rangle,$$

and it has $4 + \frac{p-1}{4}$ conjugacy classes and hence $\Gamma_{4p}$ has $4 + \frac{p-1}{4}$ simple representations. Among them, 4 are one-dimensional, say $\alpha_0, \ldots, \alpha_3$, which are given the quotient $\Gamma_{4p}/C_p$. The remaining $\frac{p-1}{4}$ simple representations are 4-dimensional, say $(\beta_1, V_1), \ldots, (\beta_{\frac{p-1}{4}}, V_{\frac{p-1}{4}})$.

Denote by $\omega$ a primitive 4th root of unity and by $q$ a primitive $p$th root of unity.

| **Conjugacy class** $\mathcal{O}$ | **Centralizer** | **Irrepn.** $\rho$ | $\dim M(\mathcal{O}, \rho)$ |
|---|---|---|---|
| $e$ | $\Gamma_{4p}$ | $\alpha_j$, $0 \le j \le 3$ | 1 |
| | | $\beta_j$, $1 \le j \le \frac{p-1}{4}$ | 4 |
| $\mathcal{O}_{x^m} = \{x^m, yx^m, \ldots, y^{p-1}x^m\}$, $\mid \mathcal{O}_{x^m} \mid = p$, $0 < m < 4$ | $\langle x\rangle \simeq C_4$ | $\chi_j$, $\chi_j(x) = \omega^j$ $0 \le j \le 3$ | $p \ge 5$ |
| $\mathcal{O}_{y^k} = \{y^k, y^{k\ell}, y^{k\ell^2}, y^{k\ell^3}\}$, $\mid \mathcal{O}_{y^k} \mid = 4$, $1 < k \le \frac{p-1}{4}$ | $\langle y\rangle \simeq C_p$ | $\psi_s$, $\psi_s(y) = q^s$ $0 \le s < p$ | 4 |

TABLE 1. $\Gamma_{4p}$, with $p \equiv 1 \mod 4$.

LEMMA 2.6. *Assume either $\mathcal{O}$ is trivial or $\mathcal{O} = \mathcal{O}_{y^k}$ for $1 < k \le \frac{p-1}{4}$. Then $\dim \mathfrak{B}(\mathcal{O}, \rho)$ is infinite for $\rho$ any simple representation of $\Gamma_{4p}$ or simple representation of $C_p$, respectively.*

PROOF. Assume first that $\mathcal{O} = \{e\}$ and let $(\rho, V)$ be a simple representation of $\Gamma_{4p}$. Then $M(\mathcal{O}, \rho) = V$ is the $\Bbbk G$-module $V$ with the trivial $G$-grading. This implies that $c(v \otimes w) = w \otimes v$ for all $v, w \in V$; in particular, $c(v \otimes v) = v \otimes v$. Consequently, $\dim \mathfrak{B}(\mathcal{O}, \rho)$ is infinite, since it contains the polynomial algebra in $v$.

Assume now that

$$\mathcal{O} = \mathcal{O}_{y^k} = \{y^k, y^{k\ell}, y^{k\ell^2}, y^{k\ell^3}\} = \{y^k, xy^kx^{-1}, x^2y^kx^{-2}, x^3y^kx^{-3}\}$$

for some $k \ne 0$. Since $C_{\Gamma_{4p}}(y^k) = \langle y\rangle \simeq C_p$, all simple representations are one-dimensional, say $\psi_0, \ldots, \psi_{p-1}$ with $\psi_s(y) = q^s$. Then $M(\mathcal{O}_{y^k}, \psi_s) = \bigoplus_{j=0}^3 x^j \otimes \Bbbk 1$. If we denote $w_j = x^j \otimes 1$, we have that

$$x \cdot w_j = w_{j+1}, \qquad y \cdot w_j = q^{s\ell^{4-j}} w_j, \qquad \text{and} \qquad \delta(w_j) = y^{k\ell^j} \otimes w_j \qquad \text{for all } j \in \mathbb{Z}_4.$$

In particular, the formula for the braiding equals

$$c(w_r \otimes w_t) = y^{k\ell^r} \cdot w_t \otimes w_r = q^{ks\ell^{r+4-t}} w_t \otimes w_r \qquad \text{for all } 0 \le r, t < 4;$$

that is, $M(\mathcal{O}_{y^k}, \psi_s)$ is a braided vector space of diagonal type. If $s = 0$, then $c(w_r \otimes w_t) = w_t \otimes w_r$, which implies that $\dim \mathfrak{B}(\mathcal{O}_{y^k}, \psi_0) = \infty$. Assume $s \ne 0$ and denote $\xi = q^{ks}$. Then by [**H**, Theorem 17], we have that $\dim \mathfrak{B}(\mathcal{O}_{y^k}, \psi_s) = \infty$ for all $0 < s < p$. □

REMARK 2.7. For $0 < m < 4$, consider the conjugacy class $\mathcal{O} = \mathcal{O}_{x^m} = \{x^m, yx^my^{-1}, \dots, y^{p-1}x^my^{1-p}\}$, and let $\chi_k$ be the one-dimensional representation of $\Gamma_{4p}$ given by $\chi_k(x) = \omega^k$ for $0 < k < 4$. Then, $M(\mathcal{O}_{x^m}, \chi_k) = \bigoplus_{j=0}^{p-1} y^j \otimes \Bbbk 1$; in particular, $\dim M(\mathcal{O}_{x^m}, \chi_k) = p$. If we write $v_j = y^j \otimes 1$ for $0 \le j < p$, we have that

$$x \cdot v_j = \omega^k v_{j\ell+1}, \quad y \cdot v_j = v_{j+1}, \quad \text{and} \quad \delta(v_j) = y^{j(1-\ell^m)}x^m \otimes v_j \quad \text{for all } j \in \mathbb{Z}_p.$$

In particular, the braiding equals

$$c(v_r \otimes v_t) = y^{r(1-\ell^m)}x^m \cdot v_t \otimes v_r = \omega^{mk} v_{t\ell^m + r(1-\ell^m)} \otimes v_r \qquad \text{for all } 0 \le r, t < p.$$

Consequently, if $mk \equiv 0 \mod 4$, we have that $c(v_0 \otimes v_0) = x^m \cdot v_0 \otimes v_0 = v_0 \otimes v_0$, which implies that $\dim \mathfrak{B}(\mathcal{O}_{x^m}, \chi_k) = \infty$. If $mk \not\equiv 0 \mod 4$, pointed Hopf algebras with group of points $\Gamma_{4p}$ do occur, see [**Gr4**], [**AG**].

## 3. Nonsemisimple Hopf algebras $H$ of dimension $8p$

Throughout this section, we assume that $H$ is a nonsemisimple Hopf algebra of dimension $8p$, with $p$ an odd prime unless specified otherwise. For each $\beta \in G(H^*)$, the 1-dimensional $H$-module associated with $\beta$ is denoted by $\Bbbk_\beta$. In particular, $\Bbbk_\varepsilon$ is the trivial 1-dimensional $H$-module. We denote the Jacobson radical of $H$ by $J(H)$, and the projective cover of an $H$-module $V$ by $P(V)$.

**3.1. Nonsemisimple Hopf algebras $H$ of dimension $8p$ with $|G(H)| = 2p$.** In this subsection we improve upon the results of [**BG2**] for $H$ a nonsemisimple nonpointed Hopf algebra of dimension $8p$ and group likes of order $2p$, $p$ an odd prime. We begin by recalling the well-known structure of the pointed Hopf algebras of dimension $4p$.

REMARK 3.1. For $p$ an odd prime, nonsemisimple pointed Hopf algebras of dimension $4p$ were described in [**AN**, A.1] and for $p = 2$ the pointed Hopf algebras of dimension 8 are well-known and were described for example in [**Ş97**]. In the first case, all such Hopf algebras $H$ have $G(H) \simeq C_{2p}$. The pointed Hopf algebras of dimension 8 can have group-likes of order 2 or 4 and, if order 4, can have group of group-likes isomorphic to the Klein 4-group or to the cyclic group of order 4. In either case ($p$ odd or $p = 2$), there are 4 nonisomorphic Hopf algebras with group-likes of order $2p$ as listed below. Here $g, h$ are group-like elements, $x$ is skew-primitive and $\lambda$ is a primitive $2p$th root of unity.

$$\mathcal{A}(-1,0) := \Bbbk\langle g, x \mid g^{2p} - 1 = x^2 = gx + xg = 0\rangle, \quad \Delta(x) = x \otimes 1 + g \otimes x.$$

$$\mathcal{A}(-1,0)^* := \Bbbk\langle g, x \mid g^{2p} - 1 = x^2 = gx - \lambda xg = 0\rangle, \quad \Delta(x) = x \otimes 1 + g^p \otimes x.$$

$$\mathcal{A}(-1,1) := \Bbbk\langle g, x \mid g^{2p} - 1 = x^2 - g^2 + 1 = gx + xg = 0\rangle, \quad \Delta(x) = x \otimes 1 + g \otimes x.$$

$$H_4 \otimes \Bbbk C_p := \Bbbk\langle g, h, x | h^2 - 1 = g^p - 1 = x^2 = gh - hg = hx + xh = gx - xg = 0\rangle,$$
$$\Delta(x) = x \otimes 1 + h \otimes x.$$

In the case $p = 2$, we write $\mathcal{A}_8(-1,1)$ to emphasize it is 8-dimensional. Note that $\mathcal{A}_8(-1,1) = A_4''$ in [**Ş97**] notation.

The dual of $\mathcal{A}(-1,1)$ is nonpointed and as a coalgebra is isomorphic to $H_4 \oplus \mathcal{M}^*(2,\Bbbk)^{p-1}$. On the other hand, the Hopf algebras $\mathcal{A}(-1,0)$ and $\mathcal{A}(-1,1)$ do not have Hopf subalgebras isomorphic to $H_4$ but $\mathcal{A}(-1,0)^*$ and $H_4 \otimes \Bbbk C_p$ do. In all four cases, $\mathcal{S}^4 = \mathrm{id}$. The distinguished group-like elements of $\mathcal{A}(-1,0)$, $\mathcal{A}(-1,0)^*$ and $\mathcal{A}(-1,1)$ are respectively $g$, $g$ and $g^p$ in the above convention.

The following is known from [**BG2**].

PROPOSITION 3.2. [**BG2**, Propositions 4.5, 4.6, 4.7] Let $H$ be a nonsemisimple nonpointed Hopf algebra of dimension $8p$ and group-likes of order $2p$, $p$ an odd prime. Then:

($a$) $H$ contains a pointed Hopf subalgebra $A$ of dimension $4p$. As a coalgebra $H \simeq A \oplus \mathcal{M}^*(2,\Bbbk)^p$ and $H_0 \simeq \Bbbk C_{2p} \oplus \mathcal{M}^*(2,\Bbbk)^p$ and so has dimension $6p$.
($b$) $H^*$ does not have a Hopf subalgebra isomorphic to $H_4$.
($c$) $|G(H^*)|$ is 2 or 4.
($d$) Suppose either
  ($i$) $A^*$ is nonpointed (so that $A$ is isomorphic to $\mathcal{A}(-1,1)$), or
  ($ii$) $A^*$ is pointed and $H^*$ has a nontrivial skew-primitive,
  then $H^*$ has a Hopf subalgebra isomorphic to $\mathcal{A}_8(-1,1)$.

Now we strengthen the results above.

LEMMA 3.3. Let $H$ be an $8p$-dimensional nonsemisimple Hopf algebra with $|G(H)| = 2p$ such that $H$ and $H^*$ are nonpointed. Then $H$ contains a pointed normal Hopf subalgebra $A$, isomorphic to either $\mathcal{A}(-1,1)$ or $\mathcal{A}(-1,0)$, and fits into the exact sequence $1 \to A \to H \to \Bbbk C_2 \to 1$, which has the dual exact sequence $1 \to \Bbbk C_2 \to H^* \xrightarrow{\pi} A^* \to 1$. Furthermore, $G(H^*) \simeq C_4$ and $H^*$ has a Hopf subalgebra $Y$ isomorphic to $\mathcal{A}_8(-1,1)$ with image in $A^*$ isomorphic to $H_4$.

PROOF. Let $A$ be the Hopf subalgebra of $H$ given by Proposition 3.2 ($a$), and write $\pi : H^* \to A^*$ for the induced surjection. By Proposition 2.5, $R = H^{*\operatorname{co}\pi}$ is either a group algebra of dimension 2 or it is generated by a nontrivial $(1,g)$-primitive $y$ with $y^2 = 0$ and $2 \mid \operatorname{ord}(g)$. Assume the latter case. As $R$ is left-normal, it follows that $h_1 y \mathcal{S}(h_2) \in \Bbbk y$ for all $h \in H^*$. Thus, $hy = h_1 y \mathcal{S}(h_2) h_3 = cyk$ with $c \in \Bbbk$ and $k \in H^+$, which implies that $H^* y \subset y H^*$. Therefore, $yH^* = R^+H^*$ is a Hopf ideal of $H^*$ and $H^*/R^+H^* \cong A^*$. In particular, $\dim R^+H^* = 4p$. As $y^2 = 0$, it follows that $R^+H^* \subset J(H^*)$. But since $\dim H_0 = 6p$, we have that $\dim J(H^*) = 2p$, a contradiction. Therefore, $R$ is a 2-dimensional normal Hopf subalgebra of $H^*$, and hence $A$ is a normal Hopf subalgebra of $H$.

Let $e_0, \ldots, e_{2p-1}$ be the primitive idempotents of $K = \Bbbk G(H)$. Since $H$ is a free $K$-module, $\dim He_i = 4$. Therefore, $\dim P \leq 4$ for all principal projective $H$-modules $P$. Suppose $H^*$ has only trivial skew-primitive elements. Then $P(\Bbbk)$ must have a simple constituent $V$ with $\dim V > 1$. This implies $\dim P(V^*) > 4$, giving a contradiction and so showing that $H^*$ has a nontrivial skew-primitive. Thus by Proposition 3.2 ($d$), $H^*$ contains a Hopf subalgebra $Y$ isomorphic to $\mathcal{A}_8(-1,1)$. By the structure of $\mathcal{A}_8(-1,1)$ and by Proposition 3.2 ($c$), $G(H^*) \simeq \Bbbk C_4$.

Write $1 \to \Bbbk\langle\beta\rangle \to H^* \to A^* \to 1$ with $\beta$ a group-like element of order 2. Suppose that $Y$ is generated by the $(1,\alpha)$-primitive $y$, where $\alpha$ has order 4. Then $\beta = \alpha^2$ and $\pi(\beta) = 1$. Thus the image of $G(H^*)$ under $\pi$ is $\Bbbk C_2 \simeq \Bbbk\langle\pi(\alpha)\rangle$. Since $y \notin H^{*co\pi}$, then $\pi(y) \neq 0$ so that $\pi(y)$ is a nontrivial $(1,\pi(\alpha))$-primitive. and $\pi(A_8(-1,1))$ is isomorphic to $H_4$. Then $A^*$ must contain a Hopf subalgebra isomorphic to $H_4$, so that by Remark 3.1, $A$ is isomorphic to either $\mathcal{A}(-1,1)$ or $\mathcal{A}(-1,0)$. □

We end this subsection with the following theorems.

THEOREM 3.4. Let $H$ be an $8p$-dimensional nonsemisimple Hopf algebra with $|G(H)| = 2p$ such that both $H$ and $H^*$ are nonpointed. Then $\operatorname{ord}(\mathcal{S}^2) = 2p$.

PROOF. By Lemma 3.3 and Remark 3.1, $H$ contains a pointed $4p$-dimensional Hopf subalgebra $A$ generated by a group-like element $g$ of order $2p$ and a $(1, g)$-primitive element $x$, and $H = A \oplus H_{0,2}$ as coalgebras. Note that, $\mathcal{S}^4|_A = \mathrm{id}_A$ and by Lemma 2.4, $g$ is the distinguished group-like element of $H$. Let $D$ be a simple subcoalgebra of dimension 4. By Lemmas 2.2 and 2.3, there exists a matrix basis $\{e_{11}, e_{22}, e_{21}, e_{12}\}$ of $D$ such that $\mathcal{S}^2(e_{ij}) = (-1)^{i-j} g e_{ij}$. Since $G(H)$ acts on the 4-dimensional simple subcoalgebras of $H$, $g^p D = D$ and so $\mathcal{S}^{2p}(D) = D$. If $gD = D$, then $D$ is a left $(\Bbbk\langle g\rangle, H)$-Hopf module and hence $D$ must be free over $\Bbbk\langle g\rangle$. This is not possible since $\dim D = 4$. Therefore, $gD \neq D$. In particular, $\mathcal{S}^2(D) \cap D = 0$ and $p \mid \operatorname{ord}(\mathcal{S}^2) \mid 4p$.

Suppose $\mathcal{S}^{2p}|_D = \mathrm{id}$. Then the subalgebra $K$ of $H$ generated by $\Bbbk\langle g\rangle \oplus D$ is a sub-bialgebra (and hence a Hopf subalgebra) of dimension $> 4p$. Therefore, $K = H$ and hence $\mathcal{S}^{2p} = \mathrm{id}$. This implies $\mathcal{S}^{2p}|_A = \mathrm{id}$. However, since $A$ is a pointed nonsemisimple Hopf algebra of dimension $4p$, $\operatorname{ord}(\mathcal{S}^2|_A) = 2$, a contradiction. Therefore, $\operatorname{ord}(\mathcal{S}^2|_{H_{0,2}}) = 2p = \operatorname{ord}(\mathcal{S}^2)$. □

THEOREM 3.5. Let $H$ be a Hopf algebra of dimension $2^n p$ over $\Bbbk$, where $n \geq 0$ and $p$ is an odd prime. If $H$ has a semisimple Hopf subalgebra of dimension $2^n$, then $H$ is semisimple.

PROOF. Suppose $H$ has a semisimple Hopf subalgebra $K$ of dimension $2^n$, and let $\mathcal{S}$ be the antipode of $H$. We consider three cases to show that $H$ must be semisimple.

First, suppose that $H$ contains a group-like element $a$ of order $p$. Define $\overline{K}$ to be the $\Bbbk$-subalgebra of $H$ generated by $C = K \oplus \Bbbk\langle a\rangle$. Then, $\overline{K}$ is a Hopf subalgebra and $\dim \overline{K} > 2^n$, and so $\overline{K} = H$. Further, since $C$ is the direct sum of two semisimple Hopf algebras as a vector space, $\mathcal{S}^2|_C = \mathrm{id}_C$ and so $S^2 = \mathrm{id}_H$. Hence $H$ must be semisimple.

Next suppose that $H^*$ contains a group-like element $\alpha$ of order $p$. Let $\pi : H^* \to K^*$ be the Hopf algebra surjection dual to the inclusion of $K$ in $H$. Since $\operatorname{ord}(\pi(\alpha)) \mid \dim K$, $\pi(\alpha) = 1$ and so $\alpha \in (H^*)^{\operatorname{co}\pi}$. Thus $\Bbbk\langle\alpha\rangle \subset (H^*)^{\operatorname{co}\pi}$. However, since $\dim (H^*)^{\operatorname{co}\pi} = p$, we actually have $\Bbbk\langle\alpha\rangle = (H^*)^{\operatorname{co}\pi}$ and thus the exact sequence

$$1 \to \Bbbk C_p \to H^* \to K^* \to 1$$

of Hopf algebras. Since $\Bbbk C_p$ and $K^*$ are semisimple, so is $H^*$ and thus so is $H$.

Finally suppose that neither $H$ nor $H^*$ has any group-like element of order $p$. By Radford's formula, we have $(\mathcal{S}^4)^{2n} = \mathrm{id}_H$. In particular, $\operatorname{ord}(\mathcal{S}^2)$ is a 2-power. Let $e$ be the normalized integral of $K$. Since $H$ is a free right $K$-module, $\dim He = p$.

Suppose $H$ is not semisimple. It follows from [**RS**, Lemma 3] that $\operatorname{Tr}(\mathcal{S}^2 \circ R(e)) = 0$ and hence $\operatorname{Tr}(\mathcal{S}^2|_{He}) = 0$. However, we know $\dim He = p$ and the order of $\mathcal{S}^2$ is a 2-power, so by [**HN**, Lemma 1.8], $\operatorname{Tr}(\mathcal{S}^2|_{He}) \neq 0$, a contradiction! Therefore, $H$ is semisimple. □

**3.2. Nonsemisimple Hopf algebras of dimension $8p$ with the Chevalley Property.** Throughout this subsection, we assume that $H$ is a Hopf algebra of dimension $8p$, $p$ an odd prime. If $H$ has the Chevalley property, by the

Nichols-Zoeller Theorem, the dimension $n$ of $H_0$ must be a divisor of $8p$, *i.e.*, $n \in \{2, 4, 8, p, 2p, 4p, 8p\}$. If $H$ is not pointed, since the only semisimple Hopf algebras of dimension $2, 4, p$ are group algebras, it follows that $n \in \{8, 2p, 4p\}$.

Assume $L$ is a semisimple Hopf algebra of dimension 8. By [**Mk1**], $L$ is isomorphic to one of the following Hopf algebras:

- $\Bbbk C_8$, $\Bbbk(C_2 \times C_4)$, $\Bbbk(C_2 \times C_2 \times C_2)$ (abelian case),
- $\Bbbk Q_8$, $\Bbbk\mathbb{D}_4$, where $Q_8$ is the quaternion group (nonabelian case),
- $\Bbbk^{Q_8}$, with $\Bbbk^{Q_8} = \Bbbk C_4 \oplus \mathcal{M}^*(2, \Bbbk)$ as coalgebras (dual of non-abelian case),
- $\Bbbk^{\mathbb{D}_4}$, with $\Bbbk^{\mathbb{D}_4} = \Bbbk C_4 \oplus \mathcal{M}^*(2, \Bbbk)$ as coalgebras (dual of non-abelian case), or
- $B_8 = \Bbbk(C_2 \times C_2) \oplus \mathcal{M}^*(2, \Bbbk)$ as coalgebras, see the definition in Remarks 3.7 and 3.8.

Also recall that every Hopf algebra of dimension $2p$ is a group algebra or isomorphic to $\Bbbk^{\mathbb{D}_p}$, the dual of the group algebra on the dihedral group of order $2p$.

PROPOSITION 3.6. Let $H$ be a nonsemisimple nonpointed Hopf algebra of dimension $8p$. If $H$ has the Chevalley property, then $\dim H_0 = 4p$.

PROOF. Since $H$ is nonsemisimple and nonpointed, we have that $\dim H_0$ cannot be $1, 2, 4, p$ or $8p$. Further, by Theorem 3.5, we also have that $\dim H_0 \neq 8$. We show that $2p$ is also impossible.

Assume $\dim H_0 = 2p$. Since $H$ is not pointed, we must have that $H_0 \cong \Bbbk^{\mathbb{D}_p}$. Then $\operatorname{gr} H \cong R\#\Bbbk^{\mathbb{D}_p}$ with $R$ a connected graded Hopf algebra in ${}^{\Bbbk^{\mathbb{D}_p}}_{\Bbbk^{\mathbb{D}_p}}\mathcal{YD}$. Let $V = R(1)$ be the vector space of homogeneous elements of degree one and consider the subalgebra $\mathfrak{B}(V)$ of $R$ generated by $V$; it is a Nichols algebra of dimension less than or equal to 4. Since ${}^{\Bbbk^{\mathbb{D}_p}}_{\Bbbk^{\mathbb{D}_p}}\mathcal{YD} \simeq {}^{\mathbb{D}_p}_{\mathbb{D}_p}\mathcal{YD}$ as braided tensor categories, $\mathfrak{B}(V)$ corresponds to a Nichols algebra $\mathfrak{B}(W)$ of dimension less or equal than 4 in ${}^{\mathbb{D}_p}_{\mathbb{D}_p}\mathcal{YD}$. For dihedral groups $\mathbb{D}_n$ with $n$ odd, by [**AHS**, Theorem 4.8] the only possible finite-dimensional Nichols algebra over a simple module in ${}^{\mathbb{D}_n}_{\mathbb{D}_n}\mathcal{YD}$ comes from the conjugacy class $\{a, ay, ay^2, ..., ay^{n-1}\}$ in $\mathbb{D}_n$ and the braiding induced from the category. So, if $W \in {}^{\mathbb{D}_p}_{\mathbb{D}_p}\mathcal{YD}$ and $\mathfrak{B}(W)$ is finite-dimensional, then $W$ must contain a simple module isomorphic to this simple module of dimension $p$. As $\mathfrak{B}(W)$ is graded and the first two homogeneous components satisfy that $\mathfrak{B}(W)^0 = \Bbbk$, $\mathfrak{B}(W)^1 = W$, we have that $\dim \mathfrak{B}(W) > p$. But moreover, if the top degree of $\mathfrak{B}(W)$ is $m$, then by the Poincaré duality $\dim \mathfrak{B}(W)^{m-i} = \dim \mathfrak{B}(W)^i$. So, at least, $\dim \mathfrak{B}(W) \geq 2 + p$, which is impossible. □

In the remainder of this subsection we first recall the construction of a family of Hopf algebras of dimension $8p$ due to [**CDMM**], and prove that any nonsemisimple, nonpointed and nonbasic Hopf algebra of dimension $8p$ with the Chevalley property is isomorphic to a member of this family.

REMARK 3.7. For $p \equiv 3 \mod 4$ the semisimple Hopf algebras of dimension $4p$ are:

- $\Bbbk\Gamma$ with $\Gamma$ an abelian group of order $4p$;
- $\Bbbk\mathbb{D}_{2p}$ and $\Bbbk^{\mathbb{D}_{2p}}$, where $\Bbbk^{\mathbb{D}_{2p}} \cong \Bbbk[C_2 \times C_2] \oplus \mathcal{M}^*(2, \Bbbk)^{p-1}$ as coalgebras;
- $\Bbbk \operatorname{Dic}_p$ and $\Bbbk^{\operatorname{Dic}_p}$, where $\Bbbk^{\operatorname{Dic}_p} \cong \Bbbk C_4 \oplus \mathcal{M}^*(2, \Bbbk)^{p-1}$ as coalgebras;
- $A_{4p}$, with $A_{4p} = \Bbbk C_4 \oplus \mathcal{M}^*(2, \Bbbk)^{p-1}$ as coalgebras and $A_{4p} \simeq A_{4p}^*$ (nontrivial self-dual case);

- $B_{4p}$, with $B_{4p} = \Bbbk(C_2 \times C_2) \oplus \mathcal{M}^*(2, \Bbbk)^{p-1}$ as coalgebras and $B_{4p} \simeq B_{4p}^*$.

Here, $A_{4p}$ denotes the $\Bbbk$-algebra generated by the elements $a$, $s_+$ and $s_-$ satisfying the relations

$$a^2 = 1, \qquad as_\pm = s_\pm a, \qquad s_\pm^2 = 1, \qquad (s_+ s_-)^p = 1. \tag{3.1}$$

Let $e_0 = (1+a)/2$ and $e_1 = (1-a)/2$. Then $A_{4p}$ is a Hopf algebra with

$$\Delta(a) = a \otimes a, \qquad \varepsilon(a) = 1, \qquad \mathcal{S}(a) = a$$

$$\Delta(s_\pm) = s_\pm \otimes e_0 s_\pm + s_\mp \otimes e_1 s_\pm, \qquad \varepsilon(s_\pm) = 1, \qquad \mathcal{S}(s_\pm) = e_0 s_\pm + e_1 s_\mp.$$

On the other hand, $B_{4p}$ is the Hopf algebra defined in the same way, but replacing the last relation in (3.1) by $(s_+ s_-)^p = a$. For more details, see [**Mk2**, Definition 3.3]. These Hopf algebras were first constructed by Gelaki, see [**Ge1**].

If $p \equiv 1 \mod 4$ then there is an additional nonabelian group of order $4p$ given by the semidirect product $\Gamma_{4p} = C_p \rtimes_\chi C_4$, see Subsection 2.3. If $C_p = \langle y \rangle$ and $C_4 = \langle x \rangle$ then $\chi(x)(y) = y^{(p-1)/2}$.

REMARK 3.8. If $p = 2$, the definition of $B_{4p}$ above still makes sense and $B_8$ is the semisimple Hopf algebra of dimension 8 due to Kac and Paljutkin.

Let $N > 1$ be an integer and $q \in \Bbbk$ a primitive $N$th root of 1. Recall that for $n \in \mathbb{N}$

$$(n)_q = \frac{q^n - 1}{q - 1}, \qquad (n)_q! = (n)_q (n-1)_q \cdots (1)_q.$$

Let $R_q$ denote the graded algebra $\Bbbk[y]/(y^N)$ endowed with a coalgebra structure such that

$$d_0 = 1, \qquad d_1 = y, \qquad d_2 = \frac{y^2}{(2)_q!}, \qquad \ldots, \qquad d_{N-1} = \frac{y^{N-1}}{(N-1)_q!},$$

form a divided power sequence in the sense that $\Delta(d_n) = \sum_{i=0}^{n} d_i \otimes d_{n-i}$ for $0 \leq n < N$. The homogeneous component $R_q(n)$ is $\Bbbk y^n (= \Bbbk d_n)$ and is 0 if $n \geq N$.

DEFINITION 3.9. [**CDMM**, Definition 2.1] A Yetter-Drinfeld (or YD) datum for $R_q$, is a triple $(L, g, \chi)$ which consists of a Hopf algebra $L$ with bijective antipode, a group-like element $g$ in $L$ and an algebra map $\chi : L \to \Bbbk$ such that $\chi(g) = q$ and

$$(\chi \rightharpoonup h)g = g(h \leftharpoonup \chi) \text{ for all } h \in L.$$

Using the YD-datum $(L, g, \chi)$ for $R_q$, one can define the $L$-action and $L$-coaction on $R_q$ by:

$$h \cdot y^n = \chi^n(h) y^n \quad \text{and} \quad \delta(y^n) = g^n \otimes y^n, \quad \text{for } n = 0, \ldots, N-1,$$

so that $R_q$ is a graded braided Hopf algebra in ${}^L_L\mathcal{YD}$. Conversely, it can be seen that any structure that makes $R_q$ into a graded braided Hopf algebra in ${}^L_L\mathcal{YD}$ arises uniquely from a YD datum for $R_q$, see [**AS2**]. Note that by the above definition $q \neq 1$ and so $g \neq 1$. Thus to have a YD datum, it is necessary to have $g \neq 1$ in the center of $G(L)$.

EXAMPLES 3.10. (*a*) Let $\Bbbk^{\mathrm{Dic}_p}$ be the function algebra on the dicyclic group of order $4p$. By [**CDMM**], this commutative Hopf algebra is generated by the elements $a$, $x$ satisfying:

$$a^2 = 1, \quad \varepsilon(a) = 1, \quad \Delta(a) = a \otimes a, \quad \mathcal{S}(a) = a,$$

$$x^{2p} = 1, \quad \varepsilon(x) = 1, \quad \Delta(x) = x \otimes e_0 x + a x^{-1} \otimes e_1 x, \quad \mathcal{S}(x) = e_0 x^{-1} + e_1 x.$$

Here, $e_0 = (1+a)/2$, $e_1 = (1-a)/2$ and $g = (e_0 + \sqrt{-1}e_1)x^p$ is a group-like element of order 4 that generates $G(\Bbbk^{\mathrm{Dic}_p})$.

A YD-Datum for $R_{-1} = \Bbbk[y]/(y^2)$ is given by the triple $(\Bbbk^{\mathrm{Dic}_p}, g, \chi)$, where $\chi : \Bbbk^{\mathrm{Dic}_p} \to \Bbbk$ is the algebra map defined by $\chi(g) = -1$, $\chi(a) = 1$ and $\chi(x) = -1$.

$(b)$ Consider the semisimple Hopf algebra $A_{4p}$ from Remark 3.7. There exist two YD-Datum for $R_{-1} = \Bbbk[y]/(y^2)$, see [**CDMM**, Theorem 4.3], which are isomorphic to each other. They are given by the pairs $(s_+(p), \chi_2)$ and $(s_+(p)a, \chi_3)$, where $s_+(p) = s_+s_-s_+s_- \cdots$ ($p$-times) $\in G(A_{4p})$ and $\chi_2$, $\chi_3$ are defined by $\chi_2(a) = -1$, $\chi_2(s_\pm) = 1$ and $\chi_3(a) = -1 = \chi_3(s_\pm)$.

LEMMA 3.11. Let $H$ be a nonsemisimple Hopf algebra of dimension $2n$ with antipode $\mathcal{S}$. Assume that $H$ has a semisimple Hopf subalgebra $K$ of dimension $n$. Then,

- $(a)$ $H$ has the Chevalley property with $K = H_0$, $\mathrm{ord}(\mathcal{S}^2) = 2$ and $n$ must be even.
- $(b)$ Consider the graded Hopf algebra $\mathrm{gr}\, H = R\#K$. Then $\dim R = 2$, $R \cong R_{-1}$ as $\Bbbk$-algebras, and $R$ admits a YD-datum structure $(K, g, \chi)$.

PROOF. $(a)$ Denote by $\pi : H^* \to K^*$ the adjoint of the inclusion of $K$ in $H$ and set $R = (H^*)^{co\,\pi}$. Then by Proposition 2.5, $R = \Bbbk\{1, x\}$, with $x$ a $(1, g)$-primitive element, $g \in G(H^*)$ and $2 \mid \mathrm{ord}(g)$. If $x$ is trivial, then $H^*$ fits into an exact sequence of Hopf algebras $0 \to \Bbbk C_2 \to H^* \to K^* \to 0$, which implies that $H^*$ is semisimple, a contradiction. Thus, $x$ must be nontrivial. In such a case, we know that $x^2 = 0$ and $4 \mid \dim H^*$. Thus, $n$ must be even.

Further, as $R$ is left-normal, it follows that $h_1 r \mathcal{S}(h_2) \in R$ for all $h \in H^*$ and $r \in R$. Moreover, one has that $\varepsilon(h_1 x \mathcal{S}(h_2)) = \varepsilon(x)\varepsilon(h) = 0$. Since $R^+ = \Bbbk x$, we conclude $h_1 x \mathcal{S}(h_2) \in \Bbbk x$. Thus, $hx = h_1 x \mathcal{S}(h_2) h_3 = cxh$ where $c$ is a scalar, implying $H^* x \subset x H^*$. But in such a case, $(R^+ H^*)^2 = (xH^*)^2 = xH^*xH^* = x^2 H^* = 0$ so $R^+ H^* \subseteq J(H^*)$. As $K$ is semisimple, it holds that $J(H^*) \subset R^+ H^*$, from which follows that $J(H^*) = R^+ H^*$. Then, $K^* \cong H^*/R^+H^*$ and hence $K \cong (H^*/J(H^*))^* = H_0$.

Since $H$ is nonsemisimple, $0 = \mathrm{Tr}(\mathcal{S}^2)$. Moreover, as $\mathcal{S}^2|_K = \mathrm{id}$, $\mathcal{S}^2$ induces a linear automorphism $\overline{\mathcal{S}}^2$ on $H/K$ and we have

$$0 = \mathrm{Tr}(\mathcal{S}^2) = \mathrm{Tr}(\mathcal{S}^2|_K) + \mathrm{Tr}(\overline{\mathcal{S}}^2) = n + \mathrm{Tr}(\overline{\mathcal{S}}^2).$$

Therefore, $\mathrm{Tr}(\overline{\mathcal{S}}^2) = -n$. Note that $\dim(H/K) = n$ and $\overline{\mathcal{S}}^2$ is of finite order. This forces $\overline{\mathcal{S}}^2 = -\mathrm{id}$ and so the eigenvalues of $\mathcal{S}^2$ are $\pm 1$. Thus, we have $\mathcal{S}^4 = \mathrm{id}$. Since $H$ is nonsemisimple, $\mathrm{ord}(\mathcal{S}^2) = 2$.

$(b)$ Since $H$ is not semisimple, $\mathrm{gr}\, H$ is not semisimple. Moreover, as $R = (\mathrm{gr}\, H)^{\mathrm{co}\,\pi}$ with $\pi : \mathrm{gr}\, H \twoheadrightarrow K$ the canonical projection, we have that $\dim R = 2$. Then by Proposition 2.5, $R$ has basis $\{1, x\}$ where $x$ is a nontrivial $(1, g)$-primitive and $x^2 = 0$. Then we must have that $R \simeq R_{-1}$ and $x \in R(1)$. Since $R(1) \in {}^K_K\mathcal{YD}$, there exists an algebra map $\chi : K \to \Bbbk$ such that

$$h \cdot x = \chi(h)x \text{ for all } h \in K.$$

Then by [**AS2**, Lemma 8.1] and the following remark we have that $(K, g, \chi)$ is a YD-Datum for $R_{-1}$ (see Definition 3.9).

□

DEFINITION 3.12. (cf. [**CDMM**]) Define $\mathcal{H}_{8p}$ as the $\Bbbk$-algebra generated by the elements $a, x, z$ satisfying the relations

$$a^2 = 1, \qquad x^{2p} = 1, \qquad z^2 = a - 1,$$
$$ax = xa, \qquad az = za, \qquad xz = -zx.$$

It is a Hopf algebra with the coalgebra structure and antipode given by

$$\Delta(a) = a \otimes a, \qquad \Delta(x) = x \otimes e_0 x + ax^{-1} \otimes e_1 x, \qquad \Delta(z) = g \otimes z + z \otimes 1$$
$$\mathcal{S}(a) = a, \qquad \mathcal{S}(x) = e_0 x^{-1} + e_1 x, \qquad \mathcal{S}(z) = -g^{-1} z$$
$$\varepsilon(a) = 1, \qquad \varepsilon(x) = 1, \qquad \varepsilon(z) = 0,$$

where $g = (e_0 + \sqrt{-1} e_1) x^p$. Clearly, $\mathcal{H}_{8p}$ contains $\Bbbk^{\mathrm{Dic}_p}$ as a Hopf subalgebra and $g$ is a group-like element of order 4 that generates $G(\mathcal{H}_{8p})$. Moreover, by [**CDMM**, Theorem 4.2] we have that $\mathcal{H}_{8p}$ is a lifting of a two-dimensional Nichols algebra over $\Bbbk^{\mathrm{Dic}_p}$.

REMARK 3.13. Note that the definition above is denoted $\mathcal{H}_{8p}(1)$ in [**CDMM**]. There, the authors define the Hopf algebra $\mathcal{H}_{8p}(\alpha)$ as above except that $z^2 = \alpha(a-1)$. Since we are working over an algebraically closed field of characteristic 0, $\mathcal{H}_{8p}(\alpha)$ is isomorphic to $\mathcal{H}_{8p}(1)$ for $\alpha \neq 0$, and we denote $\mathcal{H}_{8p}(1)$ simply by $\mathcal{H}_{8p}$. Notice that $\mathcal{H}_{8p}(0) \simeq R_{-1} \# \Bbbk^{\mathrm{Dic}_p}$.

PROPOSITION 3.14. *The Hopf algebra $\mathcal{H}_{8p}$ is a nonsemisimple, nonpointed, nonbasic Hopf algebra of dimension $8p$ with coradical $\Bbbk^{\mathrm{Dic}_p}$. Moreover, $(\mathcal{H}_{8p}^*)_0 = \Bbbk C_{2p} \oplus \mathcal{M}^*(2, \Bbbk)^p$.*

PROOF. By [**CDMM**, Theorem 4.2], $\mathcal{H}_{8p}$ is a non-trivial lifting of a quantum line $R_{-1}$ over the semisimple Hopf algebra $\Bbbk^{\mathrm{Dic}_p}$. Hence it is nonsemisimple and nonpointed. Since the diagram $R_{-1}$ is given by the YD-Datum $(\Bbbk^{\mathrm{Dic}_p}, g, \chi)$, with $\chi : \Bbbk^{\mathrm{Dic}_p} \to \Bbbk$ an algebra map defined by $\chi(g) = -1$, $\chi(a) = 1$ and $\chi(x) = -1$, we have that $\dim R_{-1} = 2$ and consequently, $\dim \mathcal{H}_{8p} = 8p$.

To finish the proof, we find all simple representations of $\mathcal{H}_{8p}$. Since not all simple representations are one-dimensional, the algebra is nonbasic.

Let $\xi$ be a primitive $2p$th root of unity. Since $\Bbbk^{\mathrm{Dic}_p}$ is a commutative algebra, all simple representations are one-dimensional. They are given by the family of $\Bbbk$-vector spaces $(V_{i,j} = \Bbbk v_{i,j})_{i \in \mathbb{Z}_2, j \in \mathbb{Z}_{2p}}$ with the action determined by $a \cdot v_{i,j} = (-1)^i v_{i,j}$ and $x \cdot v_{i,j} = \xi^j v_{i,j}$.

The one-dimensional representations of $\mathcal{H}_{8p}$ are given by the vector spaces $W_j = \Bbbk w_j$ with $j \in \mathbb{Z}_{2p}$. The action is determined by $a \cdot w_j = w_j$, $x \cdot w_j = \xi^j w_j$ and $z \cdot w_j = 0$. Clearly, $W_j \simeq W_\ell$ if and only if $\ell = j$ in $\mathbb{Z}_{2p}$.

The two-dimensional representations of $\mathcal{H}_{8p}$ are given by the vector spaces $U_i = \Bbbk u_1^i + \Bbbk u_2^i$ with $i \in \mathbb{Z}_{2p}$. The action is determined by

$$a \cdot u_j^i = -u_j^i \qquad x \cdot u_j^i = (-1)^{j-1} \xi^i u_j^i, \qquad z \cdot u_1^i = u_2^i, \quad \text{and} \quad z \cdot u_2^i = -2u_1^i,$$

for $j = 1, 2$. We show now that $U_i$ is simple for all $i \in \mathbb{Z}_{2p}$ and $U_i \simeq U_j$ if and only if $i = j$ or $i = j + p$ in $\mathbb{Z}_{2p}$. Clearly, $U_i = V_{1,i} \oplus V_{1,i+p}$ as $\Bbbk^{\mathrm{Dic}_p}$-modules with $V_{1,i} = \Bbbk u_1^i$ and $V_{1,i+p} = \Bbbk u_2^i$.

Let $i \in \mathbb{Z}_{2p}$ and assume that $U_i$ is not simple. Since it is two-dimensional, it must contain a one-dimensional simple module, say spanned by $v = \beta u_1^i + \gamma u_2^i$.

Thus, $a \cdot v = -v$ and $x \cdot v = \mu v$ for some $2p$th root of unity $\mu$. Since $x \cdot v = \beta\xi^i u_1^i - \gamma\xi^i u_2^i$, it follows that $\mu = \xi^i$ and $v = \beta u_1^i$ or $\mu = -\xi^i = \xi^{i+p}$ and $v = \gamma u_2^i$. But this is impossible, because $z$ would not stabilize $v$.

Let $\varphi : U_i \to U_j$ be a morphism of $\mathcal{H}_{8p}$-modules. Since both modules are simple, we have that $\varphi = 0$ or $\varphi$ is an isomorphism. Assume that $\varphi \neq 0$. Then $\varphi(u_1^i)$ must be an eigenvector of $x$ in $U_j$ of eigenvalue $\xi^i$, since the eigenvalues of $x$ in $U_j$ are $\pm\xi^j$, the claim follows. Consequently, we have at least $p$ isomorphism classes of two-dimensional simple modules.

Then $\dim(\mathcal{H}_{8p}^*)_0 \geq 6p$ with group-likes of order at least $2p$. By [**BG**, Proposition 3.2], $\mathcal{H}_{8p}^*$ has a nontrivial skew-primitive element and so the dimension of the linear space spanned by the nontrivial skew-primitive elements is at least $2p$. Thus, $(\mathcal{H}_{8p}^*)_0 = \Bbbk C_{2p} \oplus \mathcal{M}^*(2, \Bbbk)^p$ as claimed. □

THEOREM 3.15. *Let $H$ be a nonsemisimple, nonpointed, nonbasic Hopf algebra of dimension $8p$ with the Chevalley property. Then $H$ is isomorphic either to $R_{-1}\#A_{4p}$ or $\mathcal{H}_{8p}$.*

PROOF. By Proposition 3.6, we know that $\dim H_0 = 4p$. If we denote $A = \operatorname{gr} H$, then we have that $A = R\#H_0$ and by Lemma 3.11, $R = R_{-1}$ admits a YD-datum given by $(H_0, g, \chi)$ for some $g \in G(H_0)$ and $\chi : H_0 \to \Bbbk$. If $p \equiv 3 \mod 4$, we have four possibilities (up to isomorphism) for $H_0$: $A_{4p}$, $B_{4p}$, $\Bbbk^{\mathbb{D}_{2p}}$ and $\Bbbk^{\mathrm{Dic}_p}$. If $p \equiv 1 \mod 4$, there is an additional possibility given by $\Bbbk^{\Gamma_{4p}}$.

First we show that the cases $B_{4p}$, $\Bbbk^{\mathbb{D}_{2p}}, \Bbbk^{\Gamma_{4p}}$ are impossible. By [**CDMM**, Theorem 4.4], there is no YD-datum for $R$ when $H_0 = B_{4p}$ and so the first case is impossible. For $H_0 = \Bbbk^{\mathbb{D}_{2p}}$, by [**CDMM**, Theorem 4.1] there is one YD-datum but it admits no lifting. This implies that $H = A = R_{-1}\#\Bbbk^{\mathbb{D}_{2p}}$ from which follows that $H^* \simeq R_{-1}^*\#\Bbbk\mathbb{D}_{2p}$ is pointed, a contradiction. Finally, let $H_0 = \Bbbk^{\Gamma_{4p}}$ and assume that there exists a braided Hopf algebra $R$ of dimension 2 such that $A = R\#\Bbbk^{\Gamma_{4p}}$. Then $A^* \simeq R^*\#\Bbbk\Gamma_{4p}$, where $R^*$ is a braided Hopf algebra in ${}^{\Gamma_{4p}}_{\Gamma_{4p}}\mathcal{YD}$. Since this category is semisimple, $R^*$ is a sum of simple modules. But since the simple modules in ${}^{\Gamma_{4p}}_{\Gamma_{4p}}\mathcal{YD}$ have dimensions $1, 4$ or $p$ by Table 1, $R^*$ must be the sum of two simple one-dimensional modules of the form $M(\mathcal{O}, \rho)$ with $\mathcal{O} = \{1\}$. This is impossible, because $R^*$ would contain the Nichols algebra generated by these modules and the later are infinite-dimensional by Lemma 2.6.

If $H_0 = A_{4p}$, then by [**CDMM**, Theorem 4.3] $R$ admits a unique (up to isomorphism) YD-datum and it has no lifting. Thus $H = A = R_{-1}\#A_{4p}$; in this case, $H$ is a self-dual Hopf algebra. If $H_0 = \Bbbk^{\mathrm{Dic}_p}$, then by [**CDMM**, Theorem 4.2], $A = R_{-1}\#\Bbbk^{\mathrm{Dic}_p}$ has a nontrivial lifting isomorphic to $\mathcal{H}_{8p}$. □

We end this section with the classification of Hopf algebras of dimension $8p$ with the Chevalley property. The determination of the pointed Hopf algebras in this case follows essentially from the results of Nichols, Graña ([**Ni**], [**Gr1**, **Gr2**]), and several authors that contributed to the solution of the abelian case, among them Andruskiewitsch, Schneider, Heckenberger and Angiono, see for example [**AA**].

THEOREM 3.16. *Let $H$ be a nonsemisimple Hopf algebra of dimension $8p$ with the Chevalley property.*

(*a*) *Assume $H$ is pointed. Then $H$ is isomorphic to one of the Hopf algebras constructed in* [**Gr1**].

(*b*) Assume neither $H$ nor $H^*$ is pointed. Then $H$ is isomorphic either to $R_{-1}\#A_{4p}$ or $\mathcal{H}_{8p}$.

PROOF. (*a*) Assume $H$ is pointed. Then $|G(H)| \in \{2,4,8,2p,4p\}$. The cases $|G(H)| = 2$ and 8 are impossible by [**Ni**, Theorem 4.2.1] and [**Gr1**, Corollary 4.1], respectively, because $\dim H$ would be a power of 2. The case $|G(H)| = 4$ is also impossible for the same reason, although the result is due to several authors working on pointed Hopf algebras over abelian groups. For details see for example [**AA**]. Finally, if $|G(H)| = 2p$ or $4p$, then $H$ is isomorphic to one of the Hopf algebras constructed in [**Gr1**], since $\frac{\dim H}{|G(H)|} < 32$.

(*b*) Assume $H$ is neither pointed nor basic. Then $H$ is isomorphic either to $R_{-1}\#A_{4p}$ or $\mathcal{H}_{8p}$ by Theorem 3.15. □

## 4. Nonsemisimple Hopf algebras of dimension 24

We end the paper with some structural results on nonsemisimple Hopf algebras of dimension 24. Throughout this section $H$ will be a nonsemisimple Hopf algebra of dimension 24. We begin with the following lemma.

LEMMA 4.1. If $G(H)$ is nontrivial, then $H$ has a nontrivial skew-primitive element.

PROOF. It suffices to prove the same statement for $H^*$. Note that $H^*$ having the Chevalley property is equivalent to the $\Bbbk$-linear full subcategory generated by the simple $H$-modules being a fusion subcategory of $H$-mod.

Suppose $G(H^*)$ is nontrivial and $H^*$ has only trivial skew-primitive elements. Then $H$ has a nontrivial 1-dimension $H$-module, and $\operatorname{Ext}(\Bbbk, \Bbbk_\beta) = \operatorname{Ext}(\Bbbk_\beta, \Bbbk) = 0$ for all $\beta \in G(H^*)$. Thus, $P(\Bbbk)/J^2P(\Bbbk)$ has a simple $H$-submodule $V$ such that $\dim V \geq 2$.

(i) We first observe that $\dim V = 2$ otherwise $\dim V \geq 3$ and so

$$\dim H \geq |G(H^*)| \dim P(\Bbbk) + \dim V \dim P(V) \geq 2\cdot 5 + 3\cdot 7 > 24.$$

(ii) We claim that if $W$ is another simple $H$-module with $\dim W > 1$, then $W$ is projective with $\dim W = 2$. Since

$$\begin{aligned}\dim H &\geq |G(H^*)| \dim P(\Bbbk) + \dim V \dim P(V) + \dim W \dim P(W)\\ &\geq 2\cdot 4 + 2\cdot 5 + \dim W \dim P(W)\,.\end{aligned}$$

This forces $\dim W \dim P(W) \leq 6$, and hence $\dim W = \dim P(W) = 2$. In particular, $W$ is projective.

(iii) Since $V$ is the unique simple $H$-module with $\dim V > 1$ that is not projective, $V^* \cong V$ and $\Bbbk_\beta \otimes V \cong V \cong V \otimes \Bbbk_\beta$ for all $\beta \in G(H^*)$. Since $P(\Bbbk_\beta) \cong \Bbbk_\beta \otimes P(\Bbbk)$, $V$ is a simple $H$-submodule of $P(\Bbbk_\beta)/J^2P(\Bbbk_\beta)$. This implies that $\sum_{\beta\in G(H^*)} \Bbbk_\beta$ is an $H$-submodule of $P(V)/J^2P(V)$, and hence

$$\dim P(V) \geq |G(H^*)| + 2\dim V = |G(H^*)| + 4 \geq 6\,.$$

(iv) $|G(H^*)| = 2$ for otherwise $\dim P(V) \geq 7$ and so

$$\dim H \geq |G(H^*)| \dim P(\Bbbk) + 2\dim P(V) \geq 3\cdot 4 + 2\cdot 7 > 24.$$

(v) We now claim that $V$ is the unique simple $H$-module with $\dim V > 1$. Suppose not. By (ii), there is a 2-dimensional simple projective $H$-module $W$.

The inequality

$$\dim H \geq 2\dim P(\Bbbk) + 2\dim P(V) + (\dim W)^2 \geq 2\cdot 4 + 2\cdot 6 + 2\cdot 2 = 24$$

implies that $W$ is the unique simple projective $H$-module,

$$\dim P(\Bbbk) = 4 \quad \text{and} \quad \dim P(V) = 6.$$

Therefore, $W \cong W^*$ and $\Bbbk_\alpha \otimes W \cong W \cong W \otimes \Bbbk_\alpha$, where $\alpha \in G(H^*)$ is the nontrivial group-like element. Since $W$ is projective, so is $W \otimes W^*$. Therefore, $W \otimes W^*$ is a direct sum of indecomposable projective $H$-modules. However, $W \otimes W^*$ maps surjectively onto $\Bbbk$ and $\Bbbk_\alpha$ which implies $P(\Bbbk) \oplus P(\Bbbk_\alpha)$ is a summand of $W \otimes W^*$. This is absurd as $\dim(W \otimes W^*) = \dim P(\Bbbk) = \dim P(\Bbbk_\alpha) = 4$.

(vi) Let $m$ be the multiplicity of $V$ as a composition factor of $P(\Bbbk)$. Then, by (iii), $m$ is also the multiplicity of $V$ as a composition factor of $P(\Bbbk_\alpha)$. Since $\Bbbk$ is algebraically closed of characteristic zero, $\dim P(V)$ is the multiplicity of $V$ in the composition factors of $H$ (cf. [**CR**, Theorem 54.19]). By (v), we have

$$H = P(\Bbbk) \oplus P(\Bbbk_\alpha) \oplus 2P(V).$$

As $V$ is composition factor of $P(\Bbbk)$, we have that $m > 0$. Since $\dim \operatorname{Soc}(P(V)) = \dim V = 2$, $\operatorname{Soc}(P(V)) = V$ and hence $[P(V) : V] \geq 2$. Therefore, $\dim P(V) \geq 2m + 4$. Thus, the inequality

$$\dim H = 2\dim P(\Bbbk) + 2\dim P(V) \geq 2(2+2m) + 2(2m+4) = 4(2m+3)$$

forces $m = 1$. In particular, $\dim P(\Bbbk) = 4$. Now we have

$$\dim H = 2\cdot 4 + 2\dim P(V),$$

and so $\dim P(V) = 8$.

(vii) Since $P(\Bbbk_\beta) \otimes V$ is projective for any $\beta \in G(H^*)$ and it maps surjectively onto $V$, $P(V)$ is a summand of $P(\Bbbk_\beta) \otimes V$. This implies $P(\Bbbk_\beta) \otimes V \cong P(V)$ by comparing dimensions. Similarly, $V \otimes P(\Bbbk_\beta) \cong P(V)$.

(viii) We now prove that $H^*$ having the Chevalley property means that the $\Bbbk$-linear abelian full subcategory generated by $\Bbbk, \Bbbk_\alpha$ and $V$ is closed under the tensor product of $H$-mod. Since

$$\dim \operatorname{Hom}_H(P(\Bbbk_\beta), V \otimes V) = \dim \operatorname{Hom}_H(V^* \otimes P(\Bbbk_\beta), V) = \dim \operatorname{Hom}_H(P(V), V) = 1,$$

the multiplicity of the $\Bbbk_\beta$ as a composition factor of $V \otimes V$ is 1. The dual basis map $\mathrm{db} : \Bbbk \to V \otimes V^*$ and the evaluation map $\mathrm{ev} : V^* \otimes V \to \Bbbk$ are nonzero $H$-module maps, so are $\Bbbk_\beta \otimes \mathrm{db}$ and $\Bbbk_\beta \otimes \mathrm{ev}$ for all $\beta \in G(H^*)$. Therefore, both $\Bbbk$ and $\Bbbk_\alpha$ are summands of $V \otimes V$, and hence $V \otimes V \cong V \oplus \Bbbk \oplus \Bbbk_\alpha$. By (iii), the $\Bbbk$-linear abelian full subcategory $\mathcal{D}$ generated by $\Bbbk, \Bbbk_\alpha$ and $V$ is a fusion subcategory of $H$-mod.

(ix) By the reconstruction theorem, $\mathcal{D}$ is monoidally equivalent to $\overline{H}$-mod for some semisimple quotient Hopf algebra $\overline{H}$ of $H$. Since $\dim \overline{H} = 1 + 1 + 2^2 = 6$, it follows from the classification of 6-dimensional Hopf algebras that $\overline{H} \cong \Bbbk\mathbb{S}_3$. Thus, $\Bbbk^{\mathbb{S}_3} \cong (H^*)_0$ but this contradicts that the smallest Hopf algebras with the Chevalley property and the coradical isomorphic to $\Bbbk^{\mathbb{S}_3}$ is of dimension 72 [**AV**]. This completes the proof of this lemma. □

COROLLARY 4.2. *If $H$ and $H^*$ do not have the Chevalley property, then both $H$ and $H^*$ have a nontrivial skew-primitive element.*

PROOF. By [**BG**, Lemma 4.25], $|G(H)|$ and $|G(H^*)|$ are even. The statement then follows immediately from Lemma 4.1. □

PROPOSITION 4.3. Let $H$ be a nonbasic Hopf algebra of dimension 24 without the Chevalley property. Then $H$ is generated by its coradical.

PROOF. Let $K$ be the Hopf subalgebra of $H$ generated by the coradical $H_0$. As $H$ is nonpointed, $\dim H_0 > 4$. Since 6-dimensional Hopf algebras are semisimple, $\dim K = 8, 12$ or 24. Since $H$ does not have the Chevalley property, then $K \neq H_0$ and $K$ is nonsemisimple and nonpointed. Consider the standard filtration associated to $K$ and write $\operatorname{gr} H = R\#K$, see [**AC**].

Assume $\dim K = 8$. In such a case, $\dim R = 3$. Let $V$ be an indecomposable Yetter-Drinfeld submodule of $R(1)$ and consider the subalgebra $L$ of $R$ generated by $V$. Note that $L$ is not necessarily a Nichols algebra since it might contain primitive elements in degree bigger than one. In any case, there is a projection $L \twoheadrightarrow \mathfrak{B}(V)$ and consequently $\dim L \geq \dim \mathfrak{B}(V)$. This implies by [**GG**, Theorem A] that $V$ is simple and $\dim V = 1$ or 2. The latter case is impossible since $\dim \mathfrak{B}(V) > 3$. In the former case, by [**GG**, Proposition 3.2] it follows that $\dim L = 2$, which is impossible, since $\operatorname{gr} H$ would contain a Hopf subalgebra of dimension 16.

Assume now that $\dim K = 12$. In such a case, $\dim R = 2$. By the classification result of Natale [**Na**] (see also [**CN**]), there is a unique Hopf algebra (up to isomorphism) whose coradical is not a Hopf subalgebra. It is the dual of a pointed Hopf algebra; denote it by $\mathcal{C}$. As before, let $V$ be an indecomposable Yetter-Drinfeld submodule of $R(1)$ and consider the subalgebra $L$ of $R$ generated by $V$. Since $\dim R(0) = 1$ and so $\dim R(1) = 1$. Let $x$ be a nonzero element of $R(1)$. Then $x^2 = 0$ and so $R \cong R_{-1}$. In particular, $R^* \cong R$ as algebras and coalgebras. Thus by [**X**, Theorem B], $H$ is isomorphic to bosonization $\bigwedge \Bbbk_\chi \# \mathcal{C}$. This implies that $H^* \simeq \bigwedge \Bbbk_\chi \# \mathcal{C}^*$ and whence $H$ would be basic, a contradiction. □

## References

[A] Nicolás Andruskiewitsch, *An introduction to Nichols algebras*, Quantization, geometry and noncommutative structures in mathematics and physics, Math. Phys. Stud., Springer, Cham, 2017, pp. 135–195. MR3751453

[AA] Nicolás Andruskiewitsch and Iván Angiono, *On finite dimensional Nichols algebras of diagonal type*, Bull. Math. Sci. **7** (2017), no. 3, 353–573, DOI 10.1007/s13373-017-0113-x. MR3736568

[AC] Nicolás Andruskiewitsch and Juan Cuadra, *On the structure of (co-Frobenius) Hopf algebras*, J. Noncommut. Geom. **7** (2013), no. 1, 83–104, DOI 10.4171/JNCG/109. MR3032811

[AG] Nicolás Andruskiewitsch and Matías Graña, *From racks to pointed Hopf algebras*, Adv. Math. **178** (2003), no. 2, 177–243, DOI 10.1016/S0001-8708(02)00071-3. MR1994219

[AHS] Nicolás Andruskiewitsch, István Heckenberger, and Hans-Jürgen Schneider, *The Nichols algebra of a semisimple Yetter-Drinfeld module*, Amer. J. Math. **132** (2010), no. 6, 1493–1547. MR2766176

[AN] Nicolás Andruskiewitsch and Sonia Natale, *Counting arguments for Hopf algebras of low dimension*, Tsukuba J. Math. **25** (2001), no. 1, 187–201, DOI 10.21099/tkbjm/1496164220. MR1846876

[AS1] N. Andruskiewitsch and H.-J. Schneider, *Lifting of quantum linear spaces and pointed Hopf algebras of order $p^3$*, J. Algebra **209** (1998), no. 2, 658–691, DOI 10.1006/jabr.1998.7643. MR1659895

[AS2] Nicolás Andruskiewitsch and Hans-Jürgen Schneider, *Hopf algebras of order $p^2$ and braided Hopf algebras of order $p$*, J. Algebra **199** (1998), no. 2, 430–454, DOI 10.1006/jabr.1997.7175. MR1489920

[AV] Nicolás Andruskiewitsch and Cristian Vay, *Finite dimensional Hopf algebras over the dual group algebra of the symmetric group in three letters*, Comm. Algebra **39** (2011), no. 12, 4507–4517, DOI 10.1080/00927872.2011.616429. MR2863448

[B] M. Beattie, *An isomorphism theorem for Ore extension Hopf algebras*, Comm. Algebra **28** (2000), no. 2, 569–584, DOI 10.1080/00927870008826844. MR1736747

[BDG] M. Beattie, S. Dăscălescu, and L. Grünenfelder, *On the number of types of finite-dimensional Hopf algebras*, Invent. Math. **136** (1999), no. 1, 1–7, DOI 10.1007/s002220050302. MR1681117

[BG] Margaret Beattie and Gastón Andrés García, *Techniques for classifying Hopf algebras and applications to dimension* $p^3$, Comm. Algebra **41** (2013), no. 8, 3108–3129, DOI 10.1080/00927872.2012.692004. MR3169445

[BG2] Margaret Beattie and Gastón Andrés García, *Classifying Hopf algebras of a given dimension*, Hopf algebras and tensor categories, Contemp. Math., vol. 585, Amer. Math. Soc., Providence, RI, 2013, pp. 125–152, DOI 10.1090/conm/585/11615. MR3077235

[CDMM] C. Călinescu, S. Dăscălescu, A. Masuoka, and C. Menini, *Quantum lines over non-cocommutative cosemisimple Hopf algebras*, J. Algebra **273** (2004), no. 2, 753–779, DOI 10.1016/j.jalgebra.2003.08.006. MR2037722

[CN] Yi-Lin Cheng and Siu-Hung Ng, *On Hopf algebras of dimension* $4p$, J. Algebra **328** (2011), 399–419, DOI 10.1016/j.jalgebra.2010.08.029. MR2745573

[CR] Charles W. Curtis and Irving Reiner, *Representation theory of finite groups and associative algebras*, Pure and Applied Mathematics, Vol. XI, Interscience Publishers, a division of John Wiley & Sons, New York-London, 1962. MR0144979

[EG] Pavel Etingof and Shlomo Gelaki, *On families of triangular Hopf algebras*, Int. Math. Res. Not. **14** (2002), 757–768, DOI 10.1155/S1073792802110014. MR1891233

[GG] G. A. García and J. M. J. Giraldi, On Hopf algebra over quantum subgroups. Preprint: `arXiv:1605. 03995`.

[GV] Gastón Andrés García and Cristian Vay, *Hopf algebras of dimension 16*, Algebr. Represent. Theory **13** (2010), no. 4, 383–405, DOI 10.1007/s10468-009-9128-7. MR2660853

[Ge1] Shlomo Gelaki, *Quantum groups of dimension* $pq^2$, Israel J. Math. **102** (1997), 227–267, DOI 10.1007/BF02773801. MR1489107

[Ge2] Shlomo Gelaki, *Pointed Hopf algebras and Kaplansky's 10th conjecture*, J. Algebra **209** (1998), no. 2, 635–657, DOI 10.1006/jabr.1998.7513. MR1659891

[Gr1] Matías Graña, *On Nichols algebras of low dimension*, New trends in Hopf algebra theory (La Falda, 1999), Contemp. Math., vol. 267, Amer. Math. Soc., Providence, RI, 2000, pp. 111–134, DOI 10.1090/conm/267/04267. MR1800709

[Gr2] Matías Graña, *A freeness theorem for Nichols algebras*, J. Algebra **231** (2000), no. 1, 235–257, DOI 10.1006/jabr.2000.8363. MR1779599

[Gr3] Matías Graña, *Pointed Hopf algebras of dimension 32*, Comm. Algebra **28** (2000), no. 6, 2935–2976, DOI 10.1080/00927870008827002. MR1757439

[Gr4] Matías Graña, Finite dimensional Nichols algebras of non-diagonal group type, zoo of examples available at `http://mate.dm.uba.ar/~matiasg/zoo.html`.

[H] I. Heckenberger, *Classification of arithmetic root systems*, Adv. Math. **220** (2009), no. 1, 59–124, DOI 10.1016/j.aim.2008.08.005. MR2462836

[HN] Michael Hilgemann and Siu-Hung Ng, *Hopf algebras of dimension* $2p^2$, J. Lond. Math. Soc. (2) **80** (2009), no. 2, 295–310, DOI 10.1112/jlms/jdp026. MR2545253

[IK] Masaki Izumi and Hideki Kosaki, *Kac algebras arising from composition of subfactors: general theory and classification*, Mem. Amer. Math. Soc. **158** (2002), no. 750, 198, DOI 10.1090/memo/0750. MR1903399

[K] Irving Kaplansky, *Bialgebras*, Lecture Notes in Mathematics, Department of Mathematics, University of Chicago, Chicago, Ill., 1975. MR0435126

[Ma] Shahn Majid, *Cross products by braided groups and bosonization*, J. Algebra **163** (1994), no. 1, 165–190, DOI 10.1006/jabr.1994.1011. MR1257312

[Mk1] Akira Masuoka, *Semisimple Hopf algebras of dimension* $6, 8$, Israel J. Math. **92** (1995), no. 1-3, 361–373, DOI 10.1007/BF02762089. MR1357764

[Mk2] Akira Masuoka, *Cocycle deformations and Galois objects for some cosemisimple Hopf algebras of finite dimension*, New trends in Hopf algebra theory (La Falda, 1999), Contemp. Math., vol. 267, Amer. Math. Soc., Providence, RI, 2000, pp. 195–214, DOI 10.1090/conm/267/04271. MR1800713

[Mo] Susan Montgomery, *Hopf algebras and their actions on rings*, CBMS Regional Conference Series in Mathematics, vol. 82, Published for the Conference Board of the Mathematical Sciences, Washington, DC; by the American Mathematical Society, Providence, RI, 1993. MR1243637

[Na] Sonia Natale, *Hopf algebras of dimension 12*, Algebr. Represent. Theory **5** (2002), no. 5, 445–455, DOI 10.1023/A:1020504123567. MR1935855

[Ng] Siu-Hung Ng, *Hopf algebras of dimension pq. II*, J. Algebra **319** (2008), no. 7, 2772–2788, DOI 10.1016/j.jalgebra.2007.08.003. MR2397407

[Ni] Warren D. Nichols, *Bialgebras of type one*, Comm. Algebra **6** (1978), no. 15, 1521–1552, DOI 10.1080/00927877808822306. MR0506406

[NN] Cris Negron and Siu-Hung Ng, *Gauge invariants from the powers of antipodes*, Pacific J. Math. **291** (2017), no. 2, 439–460, DOI 10.2140/pjm.2017.291.439. MR3703451

[Ra] David E. Radford, *The structure of Hopf algebras with a projection*, J. Algebra **92** (1985), no. 2, 322–347, DOI 10.1016/0021-8693(85)90124-3. MR778452

[RS] David E. Radford and Hans-Jürgen Schneider, *On the even powers of the antipode of a finite-dimensional Hopf algebra*, J. Algebra **251** (2002), no. 1, 185–212, DOI 10.1006/jabr.2001.9127. MR1900281

[Ş97] Dragoş Ştefan, *The set of types of $n$-dimensional semisimple and cosemisimple Hopf algebras is finite*, J. Algebra **193** (1997), no. 2, 571–580, DOI 10.1006/jabr.1996.6991. MR1458803

[S] Moss E. Sweedler, *Hopf algebras*, Mathematics Lecture Note Series, W. A. Benjamin, Inc., New York, 1969. MR0252485

[X] R. Xiong, On Hopf algebras over the unique 12-dimensional Hopf algebra without the dual Chevalley property. Preprint: arXiv:1712.00826.

Mount Allison University, Sackville, NB E4L 1E6, Canada
*Email address*: mbeattie@mta.ca

Departamento de Matemática, Facultad de Ciencias Exactas, Universidad Nacional de La Plata. CONICET. Casilla de Correo 172, (1900) La Plata, Argentina
*Email address*: ggarcia@mate.unlp.edu.ar

Department of Mathematics, Louisiana State University, Baton Rouge, LA 70803, USA
*Email address*: rng@math.lsu.edu

J.R.: Mathematics Department, SUNY Cortland, Cortland, NY 13045, USA
*Email address*: jolie.roat@cortland.edu

Contemporary Mathematics
Volume **728**, 2019
https://doi.org/10.1090/conm/728/14656

# Pointed braided tensor categories

Costel-Gabriel Bontea and Dmitri Nikshych

Abstract. We classify finite pointed braided tensor categories admitting a fiber functor in terms of bilinear forms on symmetric Yetter-Drinfeld modules over abelian groups. We describe the groupoid formed by braided equivalences of such categories in terms of certain metric data, generalizing the well-known result of Joyal and Street [**JS93**] for fusion categories. We study symmetric centers and ribbon structures of pointed braided tensor categories and examine their Drinfeld centers.

## 1. Introduction

In this paper, we work over an algebraically closed field $k$ of characteristic 0. All tensor categories are assumed to be $k$-linear and finite. All Hopf algebras and modules over them are defined over $k$ and are assumed to be finite dimensional.

A tensor category is called *pointed* if all its simple objects are invertible. An example of such a category is the category of finite dimensional corepresentations $\mathsf{Corep}(H)$ of a pointed Hopf algebra $H$. Any pointed tensor category admitting a fiber functor is equivalent to some $\mathsf{Corep}(H)$. The classification of pointed Hopf algebras having abelian group of group-like elements is nearing its completion, see [**A14**].

This paper deals with classification of *braided* pointed tensor categories. Such a classification is well known in the semisimple case, i.e., for fusion categories. It was proved by Joyal and Street [**JS93**] that the 1-categorical truncation of the 2-category of braided fusion categories is equivalent to the category of pre-metric groups, i.e., pairs $(\Gamma, q)$, where $\Gamma$ is a finite abelian group and $q : \Gamma \to k^\times$ is a quadratic form. In particular, braidings on pointed fusion categories are in bijection with abelian 3-cocycles. In the presence of a fiber functor such cocycles are precisely bicharacters on abelian groups. Explicitly, a braided fusion category having a fiber functor is equivalent to $\mathcal{C}(\Gamma, r_0) := \mathsf{Corep}(k[\Gamma], r_0)$, where the $r$-form $r_0$ is given by a bicharacter on $\Gamma$.

In this work we extend the above results to non-semisimple braided tensor categories. We classify co-quasitriangular pointed Hopf algebras up to tensor equivalence of their corepresentation categories, thereby obtaining a classification of braided tensor categories having a fiber functor.

The work of the second author was partially supported by the National Security Agency grant H98230-16-1-0008 and the National Science Foundation grant DMS-1801198.

THEOREM 1.1. *Let $\mathcal{C}$ be a pointed braided tensor category having a fiber functor. Then $\mathcal{C}$ is completely determined by a finite abelian group $\Gamma$, a bicharacter $r_0 : \Gamma \times \Gamma \to k^\times$, an object $V \in \mathcal{Z}_{sym}(\mathcal{C}(\Gamma, r_0))_-$, and a morphism $r_1 : V \otimes V \to k$. More precisely,*

$$\mathcal{C} \cong \mathcal{C}(\Gamma, r_0, V, r_1) := \mathsf{Corep}(\mathfrak{B}(V)\#k[\Gamma], r), \tag{1}$$

*where $r|_{\Gamma\times\Gamma} = r_0$ and $r|_{V\otimes V} = r_1$.*

Here the symmetric center $\mathcal{Z}_{sym}(\mathcal{C}(\Gamma, r_0))$ has a canonical (possibly trivial) grading by $\mathbb{Z}/2\mathbb{Z}$:

$$\mathcal{Z}_{sym}(\mathcal{C}(\Gamma, r_0)) = \mathcal{Z}_{sym}(\mathcal{C}(\Gamma, r_0))_+ \oplus \mathcal{Z}_{sym}(\mathcal{C}(\Gamma, r_0))_-,$$

where $\mathcal{Z}_{sym}(\mathcal{C}(\Gamma, r_0))_+$ denotes the maximal Tannakian subcategory.

Theorem 1.1 is proved in Section 4, where details of the construction of $r$ can be found. Quasitriangular structures on $\mathfrak{B}(V)\#k[\Gamma]$ were explicitly described by Nenciu in [**Ne04**] in terms of generators of $\Gamma$ and a basis of $V$. The classification of co-quasitriangular structures can be obtained by duality. Theorem 1.1 says that *every* pointed co-quasitriangular Hopf algebra is equivalent to the above by a 2-cocycle deformation. Also, our description of $r$-forms is given in invariant terms and avoids the use of bases and generators.

We describe the symmetric center of $\mathcal{C}(\Gamma, r_0, V, r_1)$ and show that a pointed braided tensor category is not factorizable unless it is semisimple. We also show that this category is always ribbon and classify its ribbon structures.

We obtain a parameterization of pointed braided tensor categories similar to the parameterization of braided fusion categories by quadratic forms [**JS93**]. Namely, we introduce the groupoid of *metric data* $(\mathcal{P}, V, r)$, where $\mathcal{P}$ is a braided fusion category admitting a fiber functor, $V$ is an object in $\mathcal{Z}_{sym}(\mathcal{P})_-$, and $r : V \otimes V \to k$ is an alternating morphism.

THEOREM 1.2. *The groupoid of isomorphism classes of equivalences of pointed braided tensor categories having a fiber functor is equivalent to the groupoid of metric data.*

Theorem 1.2 is proved in Section 7. The alternating form $r$ can be canonically recovered from the restriction of the squared braiding on two-dimensional objects of a category, see Remark 7.4.

The Drinfeld center of $\mathcal{C}(\Gamma, r_0, V, r_1)$ is not pointed when $V \neq 0$. We show that when $V \cong V^*$ the trivial component of the universal grading of $\mathcal{Z}(\mathcal{C}(\Gamma, r_0, V, r_1))$ is pointed and corresponds to a certain braided vector space with a symplectic bilinear form (the Drinfeld double of $V$), see Section 3.2 and Theorem 8.5 for details.

The paper is organized as follows.

Section 2 contains background material about braided tensor categories, co-quasitriangular Hopf algebras and their twisting deformations by 2-cocycles.

In Section 3 we discuss quantum linear spaces of symmetric type and their bosonizations $\mathfrak{B}(V)\#k[\Gamma]$. We use Mombelli's classification of Galois objects for quantum linear spaces [**Mo11**] to obtain a classification of 2-cocycles on such bosonizations in terms of $V$ (equivalently, we obtain a classification of fiber functors on corresponding corepresentation categories), see Proposition 3.9. We also compute the second invariant cohomology group of $\mathfrak{B}(V)\#k[\Gamma]$ in Proposition 3.13.

In Section 4 we prove Theorem 1.1.

In Section 5 we study the symmetric center of $\mathcal{C}(\Gamma, r_0, V, r_1)$. It turns out that this center is always non-trivial if $V \neq 0$.

In Section 6 we show that a pointed braided tensor category always has a ribbon structure and classify all such structures, improving the result of [**Ne04**].

In Section 7 we prove Theorem 1.2. The correspondence between pointed braided tensor categories and metric data is useful, in particular, for computing the groups of autoequivaleces, see Corollary 7.5.

Finally, Section 8 contains a description of the Drinfeld center of the pointed braided tensor category $\mathcal{C}(\Gamma, r_0, V, r_1)$ when $V$ is self-dual. This category is no longer pointed if $V \neq 0$, but it has a faithful grading with a pointed trivial component. We describe the structure of this component and show that it is the corepresentation category of the bosonization of the Drinfeld double of $V$.

## 2. Preliminaries

**2.1. Finite tensor categories and Hopf algebras.** We assume familiarity with basic results of the theory of finite tensor categories [**EGNO15**], and the theory of Hopf algebras [**M93, R11**].

For a Hopf algebra $H$ we denote by $\Delta$, $\varepsilon$, $S$ the comultiplication, counit, and antipode of $H$, respectively. We make use of Sweedler's summation notation: $\Delta(x) = x_{(1)} \otimes x_{(2)}$, $x \in H$. We denote by $*$ the convolution, i.e., the multiplication in the dual Hopf algebra. By $H^{\text{cop}}$ we denote the co-opposite Hopf algebra of $H$, i.e., the algebra $H$ with co-multiplication $\Delta^{\text{op}}(x) = x_{(2)} \otimes x_{(1)}$, $x \in H$. We denote $G(H)$ the group of group-like elements of $H$ (i.e., elements $g \in H$ such that $\Delta(g) = g \otimes g$). Also, we denote $H^+ = \text{Ker}(\varepsilon)$.

Let $\mathsf{Rep}(H)$ and $\mathsf{Corep}(H)$ be tensor categories of left $H$-modules and right $H$-comodules, respectively, over a Hopf algebra $H$. Note that there is a canonical tensor equivalence between $\mathsf{Corep}(H)$ and $\mathsf{Rep}(H^*)$. In general, a tensor category $\mathcal{C}$ is equivalent to the co-representation category of some Hopf algebra $H$ if and only if there exists a fiber functor (i.e., an exact faithful tensor functor) $F : \mathcal{C} \to \mathsf{Vec}$, where $\mathsf{Vec}$ is the tensor category of $k$-vector spaces.

A tensor category is *pointed* if all of its simple objects are invertible with respect to the tensor product. A Hopf algebra $H$ is *pointed* if $\mathsf{Corep}(H)$ is pointed. The classification of finite dimensional pointed Hopf algebras is still an open problem, though important progress has been made so far (see [**A14**] and the references therein). The best understood class is that of pointed Hopf algebras with abelian coradical [**AS10**]. It was shown by Angiono [**An13**] that such a Hopf algebra $H$ is generated by its group-like elements and skew-primitive elements.

**2.2. Braided tensor categories and co-quasitriangular Hopf algebras.** A *braiding* on a finite tensor category $\mathcal{C}$ is a natural isomorphism

$$c_{X,Y} : X \otimes Y \to Y \otimes X, \qquad X, Y \in \mathcal{C},$$

satisfying the hexagon axioms. A *braided* tensor category is a pair consisting of a tensor category and a braiding on it.

A *co-quasitriangular* Hopf algebra is a pair $(H, r)$, where $H$ is a Hopf algebra and $r : H \otimes H \to k$ is a convolution invertible linear map, called an *$r$-form*, satisfying the following conditions:

$$x_{(1)}y_{(1)}r(y_{(2)}, x_{(2)}) = r(y_{(1)}, x_{(1)})y_{(2)}x_{(2)}, \tag{2}$$
$$r(x, yz) = r(x_{(1)}, z)r(x_{(2)}, y), \tag{3}$$
$$r(xy, z) = r(x, z_{(1)})r(y, z_{(2)}), \tag{4}$$

for all $x,\ y,\ z \in H$.

REMARK 2.1. Let $r : H \otimes H \to k$ be a linear map and let $\varphi_r : H \to H^{*\mathrm{cop}}$ be defined by $\varphi_r(x) = r(x, -)$, for all $x \in H$. Then $r$ satisfies (3), respectively (4), if and only if $\varphi_r$ is a coalgebra map, respectively an algebra map.

There is a bijective correspondence between the set of $r$-forms on a Hopf algebra $H$ and the set of braidings on $\mathsf{Corep}(H)$. The braiding corresponding to $r : H \otimes H \to k$ is given by

$$c_{U,V} : U \otimes V \to V \otimes U, \quad u \otimes v \mapsto \sum r(u_{(1)}, v_{(1)})v_{(0)} \otimes u_{(0)}, \tag{5}$$

where $U,\ V$ are $H$-comodules and $u \in U,\ v \in V$.

We denote by $\mathsf{Corep}(H,\ r)$ the braided tensor category $\mathsf{Corep}(H)$ with braiding given by $r$.

**2.3. Ribbon categories and ribbon elements.** A *ribbon* tensor category is a braided tensor category $\mathcal{C}$ together with a *ribbon structure* on it, i.e., an element $\theta \in \mathsf{Aut}(\mathrm{id}_{\mathcal{C}})$ such that

$$\theta_{X \otimes Y} = (\theta_X \otimes \theta_Y) \circ c_{Y,X} \circ c_{X,Y} \tag{6}$$
$$(\theta_X)^* = \theta_{X^*} \tag{7}$$

for all $X,\ Y \in \mathcal{C}$.

If $(H, r)$ is a co-quasitriangular Hopf algebra then ribbon structures on $\mathsf{Corep}(H, r)$ are in bijection with *ribbon elements* of $(H, r)$, i.e., convolution invertible central elements $\alpha \in H^*$ such that $\alpha \circ S = \alpha$ and

$$\alpha(xy) = \alpha(x_{(1)})\alpha(y_{(1)})(r_{21} * r)(x_{(2)}, y_{(2)})$$

for all $x,\ y \in H$. The ribbon structure associated to the ribbon element $\alpha$ is

$$\theta_V : V \to V, \quad v \mapsto \sum \alpha(v_{(1)})v_{(0)}.$$

The ribbon elements of $(H, r)$ can be determined in the following way (see [**R94**, Proposition 2] where the result appears in dual form). Let $\eta : H \to k$, $\eta(h) = r(h_{(2)}, S(h_{(1)}))$, $h \in H$, be the Drinfeld element of $(H, r)$. Then $\eta^{-1}(h) = r(S^2(h_{(2)}), h_{(1)})$, for all $h \in H$, the element $(\eta \circ S) * \eta^{-1}$ is a group-like element of $H^*$, and the map

$$\gamma \mapsto \gamma * \eta$$

establishes a one-to-one correspondence between the set of group-like elements $\gamma \in H^*$ satisfying $\gamma^2 = (\eta \circ S) * \eta^{-1}$ and $S^2_{H^*}(p) = \gamma^{-1} * p * \gamma$, for all $p \in H^*$, and the set of ribbon elements of $(H, r)$.

**2.4. Pointed braided fusion categories.** Let $\mathcal{C}$ be a pointed braided fusion category. The isomorphism classes of simple objects of $\mathcal{C}$ form a finite abelian group $\Gamma$. The braiding determines a function $c : \Gamma \times \Gamma \to k^\times$ and the function $q : \Gamma \to k^\times$, $q(g) = c(g, g)$, $g \in \Gamma$, is a *quadratic form* on $\Gamma$, i.e., the symmetric function

$$b(g,h) = \frac{q(gh)}{q(g)q(h)}, \qquad g, h \in \Gamma,$$

is a bicharacter on $\Gamma$ and $q(g^{-1}) = q(g)$ for all $g \in \Gamma$. The result of [**JS93**] (see also [**DGNO10**, Appendix D]) says, in particular, that the assignment

$$\mathcal{C} \mapsto (\Gamma,\, q)$$

determines an equivalence between the 1-categorical truncation of the 2-category of pointed braided fusion categories and the category of pre-metric groups. The objects of the latter category are finite abelian groups equipped with a quadratic form, and morphisms are group homomorphisms preserving the quadratic forms.

A pointed braided fusion category associated to $(\Gamma,\, q)$ is determined up to a braided tensor equivalence. This category admits a fiber functor if and only if $q$ is the diagonal of some bilinear form $r_0 : \Gamma \times \Gamma \to k^\times$, i.e., $q(g) = r_0(g,\, g)$ for all $g \in \Gamma$. Such a category is the corepresentation category of a co-quasitriangular pointed semisimple Hopf algebra $k[\Gamma]$ with the $r$-form $r_0$. We will use the following notation:

$$\mathcal{C}(\Gamma,\, r_0) := \mathsf{Corep}(k[\Gamma],\, r_0).$$

**2.5. The symmetric center.** Let $\mathcal{C}$ be a braided tensor category with braiding $\{c_{X,Y}\}_{X,Y \in \mathcal{C}}$. The *symmetric center* $\mathcal{Z}_{sym}(\mathcal{C})$ of $\mathcal{C}$ is the full tensor subcategory of $\mathcal{C}$ consisting of all objects $Y$ such that $c_{Y,X} \circ c_{X,Y} = \mathrm{id}_{X \otimes Y}$ for all $X \in \mathcal{C}$.

A braided tensor category $\mathcal{C}$ is called *symmetric* if $\mathcal{Z}_{sym}(\mathcal{C}) = \mathcal{C}$. Any symmetric fusion category $\mathcal{C}$ has a canonical (possibly trivial) $\mathbb{Z}/2\mathbb{Z}$-grading

$$\mathcal{C} = \mathcal{C}_+ \oplus \mathcal{C}_-, \tag{8}$$

where $\mathcal{C}_+$ is the maximal Tannakian subcategory of $\mathcal{C}$ [**De02**]. In terms of the canonical ribbon element $\theta$ of $\mathcal{C}$, one has $\theta_X = \pm \mathrm{id}_X$ when $X \in \mathcal{C}_\pm$.

A braided tensor category $\mathcal{C}$ is called *factorizable* if $\mathcal{Z}_{sym}(\mathcal{C})$ is trivial.

Let $\mathcal{C} = \mathsf{Corep}(H)$, where $H$ is a co-quasitriangular Hopf algebra with an $r$-form $r$ (so that the braiding of $\mathcal{C}$ is given by (5)). Then $\mathcal{Z}_{sym}(\mathcal{C}) = \mathsf{Corep}(H_{sym})$ for a Hopf subalgebra $H_{sym} \subset H$. The category $\mathcal{C}$ is symmetric if and only if $H_{sym} = H$ and it is factorizable if and only if $H_{sym} = k1$. This Hopf subalgebra $H_{sym}$ was described by Natale in [**Na06**] (note that in [**Na06**] quasitriangular Hopf algebras were considered while we deal with the dual situation). Below we reproduce this description using our terminology.

Consider the linear map $\Phi_r : H \to H^*$ given by

$$\Phi_r(x)(y) = r(y_{(1)},\, x_{(1)})\, r(x_{(2)},\, y_{(2)}), \qquad x, y \in H.$$

Its image $\Phi_r(H)$ is a normal coideal subalgebra of $H^*$. Hence, $H^*\Phi_r(H)^+$ is a Hopf ideal of $H^*$. We have $H_{sym} = (H^*\Phi_r(H)^+)^\perp$. Here for $I \subset H^*$ we denote $I^\perp \subset H$ its annihilator, i.e, $I^\perp = \{x \in H \mid f(x) = 0 \text{ for all } f \in I\}$. Explicitly,

$$H_{sym} = \{x \in H \mid x_{(1)} r(x_{(2)}, y_{(1)}) r(y_{(2)}, x_{(3)}) = \varepsilon(y)x, \text{ for all } y \in H\}, \tag{9}$$

Equivalently, $H_{sym}$ consists of all $x \in H$ such that the squared braiding $c^2_{H,H} : H \otimes H \to H \otimes H$ fixes $x \otimes y$ for all $y \in H$.

**2.6. The Drinfeld center of a tensor category and Yetter-Drinfeld modules.** An important example of a braided tensor category is the *Drinfeld center* $\mathcal{Z}(\mathcal{C})$ of a finite tensor category $\mathcal{C}$. The objects of $\mathcal{Z}(\mathcal{C})$ are pairs $(Z, \gamma)$ consisting of an object $Z$ of $\mathcal{C}$ and a natural isomorphism $\gamma : - \otimes Z \to Z \otimes -$ satisfying a hexagon axiom.

The braiding of $\mathcal{Z}(\mathcal{C})$ is

$$c_{(Z,\gamma),(Z',\gamma')} = \gamma'_Z : (Z, \gamma) \otimes (Z', \gamma') \to (Z', \gamma') \otimes (Z, \gamma).$$

EXAMPLE 2.2. If $H$ is a finite dimensional Hopf algebra then the Drinfeld center of $\mathsf{Rep}(H)$ is braided equivalent to $\mathsf{Rep}\,(D(H))$, where $D(H)$ is the *Drinfeld double* of $H$. As a coalgebra, $D(H) = H^{*\mathrm{cop}} \otimes H$, where $H^{*\mathrm{cop}}$ is the co-opposite dual of $H$, while the algebra structure is given by

$$(p \otimes h)(p' \otimes h') = p\left(h_{(1)} \rightharpoonup p' \leftharpoonup S^{-1}(h_{(3)})\right) \otimes h_{(2)}h', \quad p, p' \in H^*, \ h, h' \in H,$$

where $(h \rightharpoonup p \leftharpoonup g)(x) = p(gxh)$, for all $h$, $g$, $x \in H$, $p \in H^*$.

It is well-known that $\mathsf{Rep}\,(D(H))$ is braided equivalent to the category ${}^H_H\mathcal{YD}$ of *(left) Yetter-Drinfeld modules* over $H$. An object $V$ in this category has simultaneously a structure of a left $H$-module, $h \otimes v \mapsto h \cdot v$, and a structure of a left $H$-comodule $\delta : V \to H \otimes V$, $\delta(v) = v_{(-1)} \otimes v_{(0)}$, such that the following condition is satisfied:

$$\delta(h \cdot v) = h_{(1)}v_{(-1)}S(h_{(3)}) \otimes h_{(2)} \cdot v_{(0)}, \qquad h \in H, \, v \in V.$$

Morphisms between such objects are linear maps preserving both the action and the co-action of $H$. The braiding of ${}^H_H\mathcal{YD}$ is given by:

$$c_{U,V} : U \otimes V \to V \otimes U, \quad u \otimes v \mapsto u_{(-1)} \cdot v \otimes u_{(0)}$$

for all $u \in U$, $v \in V$ and $U$, $V \in {}^H_H\mathcal{YD}$.

For a finite group $\Gamma$ we denote ${}^{k[\Gamma]}_{k[\Gamma]}\mathcal{YD}$ by ${}^\Gamma_\Gamma\mathcal{YD}$. By definition, this category consists of $\Gamma$-equivariant vector bundles on $\Gamma$, where $\Gamma$ acts on itself by conjugation. If $\Gamma$ is abelian then

$${}^\Gamma_\Gamma\mathcal{YD} \simeq \mathcal{Z}(\mathsf{Vec}_\Gamma) \simeq \mathcal{C}(\Gamma \times \widehat{\Gamma}, \, h),$$

where $h : \Gamma \times \widehat{\Gamma} \to k^\times$ is the canonical quadratic form, $h(g, \chi) = \chi(g)$, $g \in \Gamma$, $\chi \in \widehat{\Gamma}$.

REMARK 2.3. For a braided tensor category $\mathcal{C}$ there exist canonical braided tensor embeddings

$$\mathcal{C} \hookrightarrow \mathcal{Z}(\mathcal{C}) : X \mapsto (X, c_{-,X}) \quad \text{and} \quad \mathcal{C}^{\mathrm{rev}} \hookrightarrow \mathcal{Z}(\mathcal{C}) : X \mapsto (X, c^{-1}_{X,-}). \tag{10}$$

The intersection of images of $\mathcal{C}$ and $\mathcal{C}^{\mathrm{rev}}$ in $\mathcal{Z}(\mathcal{C})$ is equivalent to $\mathcal{Z}_{sym}(\mathcal{C})$, the symmetric center of $\mathcal{C}$.

**2.7. 2-cocycles and deformations.** A *2-cocycle* on $H$ is a convolution invertible linear map $\sigma : H \otimes H \to k$ such that $\sigma(x, 1) = \varepsilon(x) = \sigma(1, x)$ and

$$\sigma(x_{(1)}, y_{(1)})\sigma(x_{(2)}y_{(2)}, z) = \sigma(y_{(1)}, z_{(1)})\sigma(x, y_{(2)}z_{(2)}) \tag{11}$$

for all $x$, $y$, $z \in H$. Two 2-cocycles $\sigma$ and $\sigma'$ are *gauge equivalent* if there exists a convolution invertible map $u : H \to k$ such that

$$\sigma'(x, y) = u^{-1}(x_{(1)})u^{-1}(y_{(1)})\sigma(x_{(2)}, y_{(2)})u(x_{(3)}y_{(3)}), \qquad x, y \in H.$$

for all $x$, $y \in H$.

The isomorphism classes of fiber functors on $\mathsf{Corep}(H)$ are in bijection with the set of gauge equivalence classes of 2-cocycles on $H$.

Twisting the multiplication of $H$ on both sides by a 2-cocycle $\sigma$, we obtain a new Hopf algebra, denoted $H^\sigma$ and called a *cocycle deformation* of $H$. We have $H^\sigma = H$ as a coalgebra and the multiplication of $H^\sigma$ is given by

$$x \cdot_\sigma y = \sigma(x_{(1)}, y_{(1)})x_{(2)}y_{(2)}\sigma^{-1}(x_{(3)}, y_{(3)}), \qquad x, y \in H. \tag{12}$$

If $H$ is a co-quasitriangular Hopf algebra with an $r$-form $r$ then $H^\sigma$ is also co-quasitriangular with $r$-form $r^\sigma$, given by

$$r^\sigma(x, y) = \sigma(y_{(1)}, x_{(1)})r(x_{(2)}, y_{(2)})\sigma^{-1}(x_{(3)}, y_{(3)}) \qquad x, y \in H. \tag{13}$$

For gauge equivalent 2-cocycles $\sigma$ and $\sigma'$ co-quasitriangular Hopf algebras $H^\sigma$ and $H^{\sigma'}$ are isomorphic.

A 2-cocycle $\sigma$ on $H$ is called *invariant* if

$$\sigma(x_{(1)}, y_{(1)})x_{(2)}y_{(2)} = x_{(1)}y_{(1)}\sigma(x_{(2)}, y_{(2)}) \tag{14}$$

for all $x, y \in H$. (Invariant cocycles are called *lazy* cocycles in [**BC06**]). Note that $\sigma$ is invariant if and only if $H^\sigma = H$ as Hopf algebras. The set of invariant 2-cocycles is a group under convolution product denoted by $\mathrm{Z}^2_{\mathrm{inv}}(H)$.

For a convolution invertible linear map $u : H \to k$ such that $u(1) = 1$ and $u(x_{(1)})x_{(2)} = x_{(1)}u(x_{(2)})$, for all $x \in H$, the map

$$\sigma_u : H \otimes H \to k, \quad \sigma_u(x, y) = u(x_{(1)})u(y_{(1)})u^{-1}(x_{(2)}y_{(2)}), \quad x, y \in H$$

is an invariant 2-cocycle. The set of all such 2-cocycles is a subgroup of $\mathrm{Z}^2_{\mathrm{inv}}(H)$ denoted by $\mathrm{B}^2_{\mathrm{inv}}(H)$.

The *second invariant (or lazy) cohomology group* of $H$ [**BC06**] is the quotient group

$$\mathrm{H}^2_{\mathrm{inv}}(H) = \mathrm{Z}^2_{\mathrm{inv}}(H)/\mathrm{B}^2_{\mathrm{inv}}(H).$$

For example, if $\Gamma$ is a group, then $\mathrm{H}^2_{\mathrm{inv}}(k[\Gamma]) = \mathrm{H}^2(\Gamma, k^\times)$, the second cohomology group of $\Gamma$ with coefficients in $k^\times$.

**2.8. Galois objects and 2-cocycles.** We recall here the connection between Galois objects and 2-cocycles. Let $H$ be a Hopf algebra.

A *left $H$-Galois object* is a non-zero left $H$-comodule algebra $A$ such that $A^{\mathrm{co}H} = k$ and the linear map $A \otimes A \to H \otimes A$, $a \otimes b \mapsto a_{(-1)} \otimes a_{(0)}b$, for all $a, b \in A$, is bijective.

If $\sigma$ is a 2-cocycle on $H$ then $H$, with the comodule structure given by $\Delta$ and multiplication

$$x \cdot y = x_{(1)}y_{(1)}\sigma^{-1}(x_{(2)}, y_{(2)}), \qquad x, y \in H, \tag{15}$$

is a left $H$-Galois object, denoted by $H_{\sigma^{-1}}$.

Conversely, if $A$ is a left $H$-Galois object, then there exists a left $H$-colinear isomorphism $\psi : H \to A$ such that $\psi(1) = 1$. The map $\kappa : H \otimes H \to k$, defined by

$$\kappa(x, y) = \varepsilon\Big(\psi^{-1}\big(\psi(x)\psi(y)\big)\Big), \qquad x, y \in H \tag{16}$$

is convolution invertible, $\sigma := \kappa^{-1}$ is a 2-cocycle and $\psi : H_{\sigma^{-1}} \to A$ is a left $H$-comodule algebra isomorphism.

## 3. Quantum linear spaces of symmetric type

**3.1. Quantum linear spaces.** An important class of pointed Hopf algebras with a given abelian group $\Gamma$ of group-like elements can be constructed as follows [**AS98**].

Let $g_1, \ldots, g_n$ be elements of $\Gamma$ and let $\chi_1, \ldots, \chi_n$ be elements of the dual group $\widehat{\Gamma}$ such that

$$\chi_i(g_i) \neq 1 \qquad \text{and} \qquad \chi_j(g_i)\chi_i(g_j) = 1$$

for all $i, j = 1, \ldots, n$, $i \neq j$.

DEFINITION 3.1. A *quantum linear space* associated to the above datum $(g_1, \ldots, g_n, \chi_1, \ldots, \chi_n)$ is the Yetter-Drinfeld module

$$V = \bigoplus_{i=1}^{n} kx_i \in {}^{\Gamma}_{\Gamma}\mathcal{YD}, \tag{17}$$

with $h \cdot x_i = \chi_i(h)x_i$, for all $h \in \Gamma$, and $\rho(x_i) = g_i \otimes x_i$, for all $i$.

Let $V_g^\chi$ denote the simple object in ${}^{\Gamma}_{\Gamma}\mathcal{YD}$ corresponding to $g \in \Gamma$ and $\chi \in \widehat{\Gamma}$. Then $x_i \in V_{g_i}^{\chi_i}$, $i = 1, \ldots, n$.

The braiding on $V \otimes V$ on the basic elements $x_i \otimes x_j$, $i, j = 1 \ldots, n$, is given by

$$c_{V \otimes V}(x_i \otimes x_j) = \chi_j(g_i)\, x_j \otimes x_i. \tag{18}$$

DEFINITION 3.2. We will say that a quantum linear space $V$ is of *symmetric type* if $c_{V,V}^2 = \mathrm{id}_{V \otimes V}$ (i.e., $\chi_i(g_i) = -1$ for all $i = 1, \ldots, n$).

Note that this definition does not depend on the choice of a "basis" $g_i, \chi_i$, $i = 1, \ldots, n$.

REMARK 3.3. Let $\theta$ denote the canonical ribbon element of ${}^{\Gamma}_{\Gamma}\mathcal{YD}$. Its value on the object $V_{g_i}^{\chi_i}$ is $\chi_i(g_i)$. Thus, a quantum linear space $V$ of symmetric type satisfies $\theta_V = -\mathrm{id}_V$.

Note that the transposition map

$$\tau_{V,V} : V \otimes V \to V \otimes V : v_1 \otimes v_2 \mapsto v_2 \otimes v_1, \qquad v_1, v_2 \in V, \tag{19}$$

is a morphism in ${}^{\Gamma}_{\Gamma}\mathcal{YD}$.

LEMMA 3.4. *Let $V \in {}^{\Gamma}_{\Gamma}\mathcal{YD}$ be a quantum linear space of symmetric type and $r : V \otimes V \to k$ be a morphism in ${}^{\Gamma}_{\Gamma}\mathcal{YD}$. Then $r \circ c_{V,V} = -r \circ \tau_{V,V}$.*

PROOF. It suffices to check that $c_{U,U^*} = -\tau_{U,U^*}$ for every simple object $U = V_{g_i}^{\chi_i} \subset V$ such that $U^*$ is also a subobject of $V$. In this case $(g_i, \chi_i) = (g_j^{-1}, x_j^{-1})$ for some $j$. Therefore,

$$c_{U,U^*}(y \otimes y') = \chi_j(g_i)y' \otimes y = \chi_i^{-1}(g_i)y' \otimes y = -y' \otimes y$$

for all $y \in U$, $y' \in U^*$, as required . □

Given a quantum linear space $V$ we associate to it the bosonization $\mathfrak{B}(V)\#k[\Gamma]$ of the Nichols algebra $\mathfrak{B}(V)$ by $k[\Gamma]$[1]. This Hopf algebra is generated by the group-like elements $h \in \Gamma$ and the $(g_i, 1)$-skew primitive elements $x_i$ (i.e., such that

[1] We will abuse the terminology and will also refer to $\mathfrak{B}(V)\#k[\Gamma]$ as a quantum linear space.

$\Delta(x_i) = g_i \otimes x_i + x_i \otimes 1)$, $i = 1, \ldots, n$, satisfying the following relations:

$$hx_i = \chi_i(h)x_ih, \qquad x_i^{h_i} = 0, \qquad h \in \Gamma,\ i = 1, \ldots, n,$$

$$x_ix_j = \chi_j(g_i)x_jx_i, \qquad i, j = 1, \ldots, n,$$

where $h_i$ is the order of the root of unity $\chi_i(g_i)$. The set

$$\{gx_1^{i_1} \cdots x_n^{i_n} \mid g \in \Gamma,\ 0 \le i_j < h_j,\ j = 1, \ldots, n\}$$

is a basis of $\mathfrak{B}(V)\#k[\Gamma]$.

If $V$ is a quantum linear space then the liftings of $\mathfrak{B}(V)\#k[\Gamma]$, i.e., the pointed Hopf algebras $H$ for which there exists a Hopf algebra isomorphism

$$\mathrm{gr}\, H \simeq \mathfrak{B}(V)\#k[\Gamma],$$

where $\mathrm{gr}\, H$ is the graded Hopf algebra associated to the coradical filtration of $H$, were classified in [**AS98**, Theorem 5.5]. Namely, for any such a lifting $H$, there exist scalars $\mu_i \in \{0, 1\}$ and $\lambda_{ij} \in k\,(1 \le i < j \le n)$, such that

$\mu_i$ is arbitrary if $g_i^{h_i} \neq 1$ or $\chi^{r_i} = 1$, and $\mu_i = 0$ otherwise,
$\lambda_{ij}$ is arbitrary if $g_ig_j \neq 1$ and $\chi_i\chi_j = 1$, and $\lambda_{ij} = 0$ otherwise.

The Hopf algebra $H$ is generated by the group-like elements $g \in \Gamma$ and the $(g_i, 1)$-skew-primitive elements $a_i$, $i = 1, \ldots, n$, such that

$$\begin{aligned} ga_i &= \chi_i(g)a_ig, \qquad a_i^{h_i} = \mu_i(1 - g_i^{h_i}), \qquad i = 1, \ldots, n,\\ a_ia_j &= \chi_j(g_i)a_ja_i + \lambda_{ij}(1 - g_ig_j), \qquad 1 \le i < j \le n. \end{aligned}$$

It was shown in [**Ma01**] that these liftings are cocycle deformations of $\mathfrak{B}(V)\#k[\Gamma]$ (see Section 2.7).

REMARK 3.5. Suppose that $V$ is a quantum linear space of symmetric type. Then $x_i^2 = 0$ for all $i = 1, \ldots, n$. For a subset $P = \{i_1, i_2, \ldots, i_s\} \subseteq \{1, \ldots, n\}$ such that $i_1 < i_2 < \cdots < i_s$ we denote the element $x_{i_1} \cdots x_{i_s}$ by $x_P$ and use the convention that $x_\emptyset = 1$. Clearly, the set $\{gx_P \mid g \in \Gamma,\ P \subseteq \{1, \ldots, n\}\}$ is a basis of $\mathfrak{B}(V)\#k[\Gamma]$.

Let $F \subseteq P$ be subsets of $\{1, \ldots, n\}$ and let $\psi(P, F)$ be the element of $k$ such that $x_P = \psi(P, F)x_Fx_{P\setminus F}$. Thus,

$$\psi(P, F) = \prod_{\substack{j \in F,\, i \in P\setminus F\\ i<j}} \chi_j(g_i). \tag{20}$$

It is easy to check that the comultiplication formula for $x_P$ is given by

$$\Delta(x_P) = \sum_{F \subseteq P} \psi(P, F)g_Fx_{P\setminus F} \otimes x_F, \tag{21}$$

where $g_F = \Pi_{i\in F}\, g_i$ and $g_\emptyset = 1$.

**3.2. The double of a quantum linear space.** We introduce here a construction which will appear in Section 8 where we discuss the adjoint subcategory of the center of a pointed braided finite tensor category.

Let $V \in {}^{\Gamma}_{\Gamma}\mathcal{YD}$ be the quantum linear space of symmetric type associated to a datum $(g_1, \ldots, g_n, \chi_1, \ldots, \chi_n)$. Let $\Sigma$ be the subgroup of $\Gamma \times \widehat{\Gamma}$ generated by $(g_i, \chi_i^{-1})$, $i = 1, \ldots, n$, and define characters $\varphi_i : \Sigma \to k^\times$ by

$$\varphi_i(g, \chi) = \chi_i(g), \quad \text{for all } (g, \chi) \in \Sigma, \quad i = 1, \ldots, n.$$

We have

$$\varphi_i(g_i, \chi_i^{-1}) = -1 \quad \text{and} \quad \varphi_j(g_i, \chi_i^{-1})\varphi_i(g_j, \chi_j^{-1}) = 1$$

for all $i, j = 1, \ldots, n$. Thus we can consider the quantum linear space of symmetric type $W \in {}^{\Sigma}_{\Sigma}\mathcal{YD}$ associated to the datum $\big((g_1, \chi_1^{-1}), \ldots, (g_n, \chi_n^{-1}), \varphi_1, \ldots, \varphi_n\big)$.

DEFINITION 3.6. We call the quantum linear space $D(V) := W \oplus W^* \in {}^{\Sigma}_{\Sigma}\mathcal{YD}$ the *Drinfeld double* of $V$.

Note that the quantum linear space $D(V)$ is of symmetric type.

There is a canonical bilinear form $r_{D(V)} : D(V) \otimes D(V) \to k$ given by

$$r_{D(V)}((w, f) \otimes (w', f')) := \mathrm{ev}_W(f \otimes w') + \mathrm{ev}_W c_{W,W^*}(w \otimes f') \tag{22}$$

for all $w, w' \in W$, $f, f' \in W^*$, where $\mathrm{ev}_W : W^* \otimes W \to k$ is the evaluation morphism and $c_{W,W^*} : W \otimes W^* \to W^* \otimes W$ is the braiding in ${}^{\Sigma}_{\Sigma}\mathcal{YD}$.

Note that $r_{D(V)}$ is a symplectic bilinear form on $W \oplus W^*$. Indeed, if $\{x_i\}$ is a basis of $W$ such that $x_i \in W^{\varphi_i}_{(g_i, \chi_i^{-1})}$, $i = 1, \ldots, n$, and $\{x_i^*\}$ is the dual basis of $W^*$, then the matrix of $r_{D(V)}$ with respect to the basis $(x_1, \ldots, x_n, x_1^*, \ldots, x_n^*)$ is

$$\begin{pmatrix} 0 & -I_n \\ I_n & 0 \end{pmatrix},$$

where $I_n$ denotes the $n$-by-$n$ identity matrix.

**3.3. Galois objects for quantum linear spaces.** Let $V \in {}^{\Gamma}_{\Gamma}\mathcal{YD}$ be a quantum linear space of symmetric type and let $H = \mathfrak{B}(V)\#k[\Gamma]$. In [**Mo11**] Mombelli classified equivalence classes of exact indecomposable $\mathsf{Rep}(H)$-module categories. In particular, he classified $H$-Galois objects and, hence, 2-cocycles on $H$. Here we recall this classification, see [**Mo11**, Section 4] for details. Galois objects for some quantum linear spaces were also classified by Bichon in [**Bi06**].

A typical $H$-Galois object is determined by a 2-cocycle $\psi \in Z^2(\Gamma, k^\times)$ and two families of scalars $\xi = (\xi_i)_{i=1,\ldots,n}$ and $\alpha = (\alpha_{ij})_{1\le i<j\le n}$, satisfying:

$$\xi_i = 0 \quad \text{if } \chi_i^2(g) \neq \frac{\psi(g, g_i^2)}{\psi(g_i^2, g)}, \tag{23}$$

$$\alpha_{ij} = 0 \quad \text{if } \chi_i\chi_j(g) \neq \frac{\psi(g, g_ig_j)}{\psi(g_ig_j, g)}, \tag{24}$$

for all $g \in \Gamma$. To this datum one assigns a left $H$-comodule algebra $\mathcal{A}(\psi, \xi, \alpha)$ generated as an algebra by $\{e_g\}_{g\in\Gamma}$ and $v_1, \ldots, v_n$ subject to the relations:

$$\begin{aligned} e_f e_g &= \psi(f, g) e_{fg}, && f, g \in \Gamma \\ e_f v_i &= \chi_i(f) v_i e_f, && f \in \Gamma,\ i = 1, \ldots, n \\ v_i v_j - \chi_j(g_i) v_j v_i &= \alpha_{ij} e_{g_i g_j}, && 1 \le i < j \le n \\ v_i^2 &= \xi_i e_{g_i^2}, && i = 1, \ldots, n \end{aligned}$$

The left $H$-comodule structure of $\mathcal{A}(\psi, \xi, \alpha)$ is $\lambda : \mathcal{A}(\psi, \xi, \alpha) \to H \otimes \mathcal{A}(\psi, \xi, \alpha)$,

$$\lambda(v_i) = g_i \otimes v_i + x_i \otimes 1 \quad \text{and} \quad \lambda(e_f) = f \otimes e_f$$

for all $i = 1, \ldots, n$ and $f \in \Gamma$. Two $H$-Galois objects $\mathcal{A}(\psi, \xi, \alpha)$ and $\mathcal{A}(\psi', \xi', \alpha')$ are isomorphic if and only $\psi$, $\psi'$ are cohomologous, $\xi = \xi'$, and $\alpha = \alpha'$.

REMARK 3.7. Suppose that $\psi = 1$. Then the 2-cocycle $\sigma$ corresponding to the $H$-Galois object $\mathcal{A}(1, \xi, \alpha)$ above satisfies

$$\sigma(x_i, x_j) - \chi_j(g_i)\sigma(x_j, x_i) = \alpha_{ij}, \quad \text{and} \quad \sigma(x_i, x_i) = \xi_i, \qquad 1 \le i < j \le n.$$

In this case the conditions (23) and (24) are equivalent to $\sigma - \sigma \circ c_{V,V} : V \otimes V \to k$ being a $\Gamma$-module map.

**3.4. 2-cocycles on quantum linear spaces of symmetric type.** Let $V \in {}^{\Gamma}_{\Gamma}\mathcal{YD}$ be a quantum linear space of symmetric type with braiding $c_{V,V} : V \otimes V \to V \otimes V$ and let $b : V \otimes V \to k$ be a bilinear form.

Let $\tau_{V,V} : V \otimes V \to V \otimes V$ denote the transposition map.

DEFINITION 3.8. We will say that $b$ is *symmetric* (respectively, *alternating*) if $b \circ \tau_{V,V} = b$ (respectively, $b \circ \tau_{V,V} = -b$).

There is a canonical decomposition of $b$ into the sum of symmetric and alternating parts:

$$b = b_{sym} + b_{alt}, \tag{25}$$

where $b_{sym} = \frac{1}{2}(b + b \circ \tau_{V,V})$ and $b_{alt} = \frac{1}{2}(b - b \circ \tau_{V,V})$.

We will denote $\mathsf{Sym}^2_{{}^{\Gamma}_{\Gamma}\mathcal{YD}}(V^*)$ (respectively, $\mathsf{Alt}^2_{{}^{\Gamma}_{\Gamma}\mathcal{YD}}(V^*)$) the spaces of symmetric (respectively, alternating) morphisms $V \otimes V \to k$ in ${}^{\Gamma}_{\Gamma}\mathcal{YD}$. We have

$$\mathsf{Sym}^2_{{}^{\Gamma}_{\Gamma}\mathcal{YD}}(V^*) \subset S^2(V^*), \quad \mathsf{Alt}^2_{{}^{\Gamma}_{\Gamma}\mathcal{YD}}(V^*) \subset \wedge^2(V^*),$$

where $S^2(V^*)$ and $\wedge^2(V^*)$ are spaces of usual symmetric and alternating bilinear forms on the vector space $V$.

Let $H = \mathfrak{B}(V)\#k[\Gamma]$.

Let $\sigma : H \otimes H \to k$ be a 2-cocycle on $H$ such that $\sigma|_{\Gamma\times\Gamma} = 1$ and set

$$\alpha = \frac{1}{2}\sigma|_{V\otimes V} \circ (\mathrm{id}_{V\otimes V} - c_{V\otimes V}). \tag{26}$$

Denote

$$\alpha_{ij} := \alpha(x_i, x_j) = \frac{1}{2}\left(\sigma(x_i, x_j) - \chi_j(g_i)\sigma(x_j, x_i)\right), \qquad i, j = 1, \ldots, n.$$

By Remark 3.7, $\alpha : V \otimes V \to k$ is a $\Gamma$-module map and $\alpha \circ c_{V,V} = -\alpha$.

We have

$$\alpha_{ij} = -\chi_j(g_i)\alpha_{ji} \tag{27}$$

and

$$\alpha_{ij} = 0 \text{ if } \chi_i\chi_j \neq \varepsilon, \qquad i, j = 1, \ldots, n. \tag{28}$$

Conversely, given $\alpha : V \otimes V \to k$ such that scalars $\{\alpha_{ij} = \alpha(x_i, x_j)\}_{i,j=1}^n$ satisfy conditions (27) and (28) it follows from Remark 3.7 that there is a 2-cocycle $\sigma$ on $H$ such that

$$\sigma|_{\Gamma\times\Gamma} = 1 \quad \text{and} \quad \sigma|_{V\otimes V} \circ (\mathrm{id}_{V\otimes V} - c_{V\otimes V}) = \alpha.$$

Such a 2-cocycle $\sigma$ is unique up to a gauge equivalence.

We summarize these results in the following statement.

PROPOSITION 3.9. *The map $\sigma \mapsto \sigma|_{V\otimes V} \circ (id_{V\otimes V} - c_{V\otimes V})$ is a bijection between the set of gauge equivalence classes of 2-cocycles on $H$ whose restriction on $\Gamma$ is trivial and the set of bilinear forms $r$ on $V$ that are $\Gamma$-module maps satisfying $r \circ c_{V,V} = -r$.*

Next, we discuss invariant 2-cocycles on $H$.

PROPOSITION 3.10. *Let $\sigma$ be a 2-cocycle on $H$ whose restriction on $\Gamma$ is trivial and let $\alpha$ be defined as in (26). Then $\sigma$ is invariant if and only if $\alpha(x_i, x_j) = 0$ whenever $g_i g_j \neq 1$, i.e., if and only if $\alpha$ is a morphism in ${}^{\Gamma}_{\Gamma}\mathcal{YD}$.*

PROOF. Suppose that $\sigma$ is invariant. Taking $x = x_i$, $y = g \in \Gamma$ in (14) we obtain $\sigma(x_i, g) = 1$ for all $i = 1, \dots, n$ and $g \in \Gamma$. Similarly, taking $x = g$ and $y = x_i$ we get $\sigma(g, x_i) = 1$. Next, taking $x = x_i$, $y = x_j$ in (14) we obtain $\sigma(x_i, x_j)(1 - g_i g_j) = 0$ for all $i, j = 1, \dots, n$. This means that $\sigma$ is a $\Gamma$-comodule map. Hence, $\alpha$ is a $\Gamma$-comodule map. Combining this with Proposition 3.9 we see that $\alpha$ is a morphism in ${}^{\Gamma}_{\Gamma}\mathcal{YD}$.

Conversely, suppose that a 2-cocycle $\sigma$ on $H$ is such that $\sigma|_{\Gamma\times\Gamma} = 1$ and $\alpha$ is a morphism in ${}^{\Gamma}_{\Gamma}\mathcal{YD}$. Then the multiplication in the corresponding twisted Hopf algebra $H^\sigma$ satisfies relations $g \cdot_\sigma x_i = \chi_i(g) x_i \cdot_\sigma g$ and

$$x_i \cdot_\sigma x_j - \chi_j(g_i) x_j \cdot_\sigma x_i = \alpha_{ij}(1 - g_i g_j), \quad i, j = 1, \dots, n.$$

But the right hand side of the last equality is equal to 0, so that $H^\sigma = H$, i.e., $\sigma$ is invariant. □

COROLLARY 3.11. *The map $\sigma \mapsto (\sigma|_{V\otimes V})_{sym}$ is an isomorphism between the space of gauge equivalence classes of invariant 2-cocycles on $H$ whose restriction on $\Gamma$ is trivial and the space $\mathsf{Sym}^2_{{}^{\Gamma}_{\Gamma}\mathcal{YD}}(V^*)$.*

PROOF. Note that $\mathrm{id}_{V\otimes V} - c_{V,V}$ is an invertible morphism, so by Propositions 3.9 and 3.10 for an invariant twist $\sigma$ its restriction $\sigma|_{V\otimes V}$ is a morphism in ${}^{\Gamma}_{\Gamma}\mathcal{YD}$. Using Lemma 3.4 we get

$$\sigma|_{V\otimes V} \circ \tau_{V,V} = -\sigma|_{V\otimes V} \circ c_{V,V} = \sigma,$$

i.e., $\sigma|_{V\otimes V} \circ (\mathrm{id}_{V\otimes V} - c_{V\otimes V}) = 2\,(\sigma|_{V\otimes V})_{sym}$. □

We now analyze the general situation when $\sigma|_{\Gamma\times\Gamma}$ is not necessarily trivial. Let $\Gamma_0$ denote the subgroup of $\Gamma$ generated by $g_i$, $i = 1, \dots, n$.

PROPOSITION 3.12. *Let $\sigma$ be an invariant 2-cocycle on $H$. There is $\rho \in Z^2(\Gamma/\Gamma_0, k^\times)$ such that $\sigma|_{\Gamma\times\Gamma}$ is cohomologous to $\rho \circ (\pi_{\Gamma_0} \times \pi_{\Gamma_0})$, where $\pi_{\Gamma_0} : \Gamma \to \Gamma/\Gamma_0$ is the quotient homomorphism.*

PROOF. It suffices to check that the alternating bilinear form $\mathrm{alt}(\sigma) : \Gamma \times \Gamma \to k^\times$ given by

$$\mathrm{alt}(\sigma)(g, h) = \frac{\sigma(g, h)}{\sigma(h, g)}, \qquad g, h \in \Gamma \tag{29}$$

vanishes on $\Gamma \times \Gamma_0$. But this follows from invariance of $\sigma$ since we must have $\sigma(g_i, g) = \sigma(g, g_i) = 1$ for all $i = 1, \dots, n$ and $g \in \Gamma$. □

PROPOSITION 3.13. $H^2_{inv}(H) \cong H^2(\Gamma/\Gamma_0, k^\times) \times \mathsf{Alt}^2_{{}^{\Gamma}_{\Gamma}\mathcal{YD}}(V^*)$.

PROOF. By Corollary 3.11 the group $\mathsf{Alt}^2_{{}^{\Gamma}_{\Gamma}\mathcal{YD}}(V^*)$ is identified with the normal subgroup of $\mathrm{H}^2_{\mathrm{inv}}(H)$ consisting of gauge equivalence classes of invariant 2-cocycles with trivial restriction on $\Gamma$.

Next, there is a surjective Hopf algebra homomorphism $p : H \to k[\Gamma/\Gamma_0]$ obtained by composing the canonical projection $H \to k[\Gamma]$ with $\pi_{\Gamma_0} : \Gamma \to \Gamma/\Gamma_0$. Thus, for any 2-cocycle $\rho \in Z^2(\Gamma/\Gamma_0, k^\times)$ its pullback $p^*(\rho)$ is a 2-cocycle on $H$.

Using the explicit formula (21) for the comultiplication on $H$ we check that this 2-cocycle satisfies

$$p^*(\rho)(x_{(1)}, y_{(1)})x_{(2)} \otimes y_{(2)} = x_{(1)} \otimes y_{(1)}p^*(\rho)(x_{(2)}, y_{(2)}), \qquad x, y \in H.$$

Indeed, for $x = hx_P$, $y = fx_Q$, where $h, f \in \Gamma$, both sides of this equality are equal to $\rho(\pi_{\Gamma_0}(h), \pi_{\Gamma_0}(f))hx_P \otimes fx_Q$.

In particular, $p^*(\rho)$ is an invariant 2-cocycle on $H$ and belongs to the center of $\mathrm{H}^2_{\mathrm{inv}}(H)$. Thus, there is a central embedding $H^2(\Gamma/\Gamma_0, k^\times) \subset \mathrm{H}^2_{\mathrm{inv}}(H)$. The statement follows from Proposition 3.12. □

## 4. Classification of co-quasitriangular structures on pointed Hopf algebras (Proof of Theorem 1.1)

Let $\Gamma$ be a finite abelian group with a bicharacter $r_0 : \Gamma \times \Gamma \to k^\times$. Recall that $\mathcal{C}(\Gamma, r_0)$ denotes the fusion category of corepresentations of cosemisimple co-quasitriangular Hopf algebra $(k[\Gamma], r_0)$.

Let $H$ be a co-quasitriangular pointed Hopf algebra with the $r$-form $r : H \otimes H \to k$ and let $\Gamma$ be the group of its group-likes. Clearly, $\Gamma$ must be abelian. By the result of Angiono [**An13**], the algebra $H$ is generated by $\Gamma$ and skew-primitive elements $x_i$, $i = 1, \dots, n$, with $\Delta(x_i) = g_i \otimes x_i + x_i \otimes 1$ for some $g_i \in \Gamma$, $i = 1, \dots, n$. Furthermore, there exist characters $\chi_i \in \widehat{\Gamma}$ such that $gx_ig^{-1} = \chi_i(g)x_i$.

Let $V$ be the span of $\{x_i\}_{i=1}^n$. Then $V$ is an object in ${}^{\Gamma}_{\Gamma}\mathcal{YD}$ with the above conjugation action and coaction $\rho(x_i) = g_i \otimes x_i$.

It is clear that $r$ restricts to a bicharacter $r_0$ on $\Gamma$. Observe that (2) for the pair $(x, y) = (x_i, g)$ yields $r_0(g, g_i) = \chi_i^{-1}(g)$, and the same condition for the pair $(x, y) = (g, x_i)$ yields $r_0(g_i, g) = \chi_i(g)$. Thus,

$$r_0(g_i, -) = \chi_i = r_0(-, g_i)^{-1}, \quad \text{for all } i = 1, \dots, n.$$

This condition is equivalent to $V \in \mathcal{Z}_{sym}(\mathcal{C}(\Gamma, r_0)) \subset {}^{\Gamma}_{\Gamma}\mathcal{YD}$. It follows that $\chi_i(g_j)\chi_j(g_i) = 1$ for all $i, j = 1, \dots, n$. Furthermore, $\chi_i(g_i) = -1$ for all $i = 1, \dots, n$ (indeed, if $\chi_i(g_i) = 1$ for some $i$ then the Hopf subalgebra of $H$ generated by $g_i$ and $x_i$ is non-semisimple commutative, and, hence, infinite dimensional).

Thus,

$$V \in \mathcal{Z}_{sym}(\mathcal{C}(\Gamma, r_0))_- \subset {}^{\Gamma}_{\Gamma}\mathcal{YD},$$

where the embedding $\mathcal{C}(\Gamma, r_0) \hookrightarrow {}^{\Gamma}_{\Gamma}\mathcal{YD}$ is given by (10). So $V$ is a quantum linear space of symmetric type.

REMARK 4.1. It follows from the classification of symmetric fusion categories [**De02**] that in this situation there is a character $\chi : \Gamma \to k^\times$ such that $\chi(g_i) = -1$ for all $i = 1, \dots, n$ (in other words, $\chi = -1$ on the support of $V$).

It follows from results of Andruskiewitsch and Schneider [**AS98**] and Masuoka [**Ma01**] that $H$ is a 2-cocycle twisting of $\mathfrak{B}(V)\#k[\Gamma]$. Since we are interested in the tensor category of corepresentations of $H$ (which does not change under 2-cocycle twisting) from now on we will assume that $H = \mathfrak{B}(V)\#k[\Gamma]$.

If $g \in \Gamma$ and $i \in \{1, \dots, n\}$, then it follows from (3) that $r(g, x_i^m) = r(g, x_i)^m$, for all $m \geq 1$. Since $x_i$ is nilpotent, we have $r(g, x_i) = 0$. Similarly, using (4), we deduce that $r(x_i, g) = 0$.

For a non-empty subset $P$ of $\{1, \dots, n\}$ and for $g, h \in \Gamma$ one can show, using again (3) and (4), that $r(gx_P, h) = 0$ and $r(h, gx_P) = 0$ (the notation $x_P$ was introduced in Remark 3.5).

Consider now subsets $P$ and $Q$ of $\{1, \dots, n\}$. Condition (2) for the pair $(x, y) = (x_P, x_Q)$ is

$$(x_P)_{(1)}(x_Q)_{(1)}r\big((x_Q)_{(2)}, (x_P)_{(2)}\big) = r\big((x_Q)_{(1)}, (x_P)_{(1)}\big)(x_Q)_{(2)}(x_P)_{(2)}.$$

The terms in the coradical of $H$ of each side of the equality are $g_P g_Q r(x_Q, x_P)$ and $r(x_Q, x_P)1$, respectively. Since these have to be equal, we have

$$r(x_Q, x_P)(1 - g_P g_Q) = 0. \tag{30}$$

This means that the restriction $r_1 := r|_{V\otimes V}$ is a morphism of $\Gamma$-comodules, i.e, a morphism in $\mathcal{C}(\Gamma, r_0) \subset {}^{\Gamma}_{\Gamma}\mathcal{YD}$.

Thus, we showed that if $H$ admits a co-quasitriangular structure $r$, then $V$ is a quantum linear space of symmetric type and the pair $(r_0, r_1) := (r|_{\Gamma\times\Gamma}, r|_{V\otimes V})$ satisfies the conditions of Theorem 1.1.

It remains to prove that, conversely, given an object $V \in \mathcal{Z}_{sym}(\mathcal{C}(\Gamma, r_0)_-$ and a pair $(r_0 : \Gamma \times \Gamma \to k^\times, r_1 : V \otimes V \to k)$ there is a unique $r$-form on $H$ such that $(r|_{\Gamma\times\Gamma}, r|_{V\otimes V}) = (r_0, r_1)$.

We need the following general facts.

LEMMA 4.2. *Let $H = \mathfrak{B}(V)\#k[\Gamma]$. If $r : H \otimes H \to k$ is a convolution invertible linear map satisfying conditions (3) and (4) then $r$ is a co-quasitriangular structure on $H$ if and only if condition (2) holds for all pairs $(x, y) \in H_1 \times H_1$, where $H_1$ is the second term in the coradical filtration of $H$.*

PROOF. We need only prove sufficiency. By induction on $m$ we show that condition (2) holds for all pairs $(x, y)$ for which either $x$ or $y$ is in $H_m$, the $m$-th term of the coradical filtration.

Assume first that $x \in H_1$. Using induction on $k \geq 1$ we show that condition (2) holds for all pairs $(x, y)$ and $(y, x)$ with $y \in H_k$. If $k = 1$ there is nothing to prove. Assume that the claim is true for $k \geq 1$ and consider $z \in H_1$. Then, using the induction hypothesis and (3), we have

$$\begin{aligned}
x_{(1)}(yz)_{(1)}r\big((yz)_{(2)}, x_{(2)}\big) &= x_{(1)}y_{(1)}z_{(1)}r\big(y_{(2)}z_{(2)}, x_{(2)}\big) \\
&= x_{(1)}y_{(1)}z_{(1)}r\big(y_{(2)}, x_{(2)}\big)r\big(z_{(2)}, x_{(3)}\big) \\
&= y_{(2)}x_{(2)}z_{(1)}r\big(y_{(1)}, x_{(1)}\big)r\big(z_{(2)}, x_{(3)}\big) \\
&= y_{(2)}z_{(2)}x_{(3)}r\big(y_{(1)}, x_{(1)}\big)r\big(z_{(1)}, x_{(2)}\big) \\
&= y_{(2)}z_{(2)}x_{(2)}r\big(y_{(1)}z_{(1)}, x_{(1)}\big) \\
&= (yz)_{(2)}x_{(2)}r\big((yz)_{(1)}, x_{(1)}\big)
\end{aligned}$$

and using (4) we have

$$\begin{aligned}
(yz)_{(1)}x_{(1)}r\big(x_{(2)}, (yz)_{(2)}\big) &= y_{(1)}z_{(1)}x_{(1)}r\big(x_{(2)}, y_{(2)}z_{(2)}\big) \\
&= y_{(1)}z_{(1)}x_{(1)}r\big(x_{(2)}, z_{(2)}\big)r\big(x_{(3)}, y_{(2)}\big) \\
&= y_{(1)}x_{(2)}z_{(2)}r\big(x_{(1)}, z_{(1)}\big)r\big(x_{(3)}, y_{(2)}\big) \\
&= x_{(3)}y_{(2)}z_{(2)}r\big(x_{(1)}, z_{(1)}\big)r\big(x_{(2)}, y_{(1)}\big) \\
&= x_{(2)}y_{(2)}z_{(2)}r\big(x_{(1)}, y_{(1)}z_{(1)}\big) \\
&= x_{(2)}(yz)_{(2)}r\big(x_{(1)}, (yz)_{(1)}\big).
\end{aligned}$$

Since $H_{k+1} = H_k H_1$ it follows that (2) holds for all pairs $(x, y)$ and $(y, x)$ with $y \in H_{k+1}$. Thus, (2) holds for all pairs $(x, y)$ with either $x$ or $y$ in $H_1$.

Suppose now that (2) holds for all pairs $(x, y)$ with either $x$ or $y$ in $H_m$. Then a similar argument as the one before shows that (2) holds for all pairs $(x, yz)$ and $(yz, x)$ with $y \in H_m$, $z \in H_1$ and arbitrary $x$. Since $H_{m+1} = H_m H_1$, it follows that (2) is satisfied for all pairs $(x, y)$ with either $x$ or $y$ in $H_{m+1}$. This proves the induction step and concludes the proof. □

LEMMA 4.3. *Let $H$ be a Hopf algebra generated as an algebra by $h_1, \dots, h_n$ and such that the vector space $V$ spanned by $h_1, \dots, h_n$ is a subcoalgebra. Then any co-quasitriangular structure on $H$ is uniquely determined by its restriction to $V \otimes V$.*

PROOF. Suppose $r'$ and $r''$ are two co-quasitriangular structures on $H$ such that $r'(h_i, h_j) = r''(h_i, h_j)$, for all $i$, $j \in \{1, \dots, n\}$. If $i$, $i_1$, …, $i_t \in \{1, \dots, n\}$ then, using (4), we have

$$\begin{aligned} r'(h_{i_1} \cdots h_{i_t}, h_i) &= r'\big(h_{i_1}, (h_i)_{(1)}\big) \cdots r'\big(h_{i_t}, (h_i)_{(t)}\big) \\ &= r''\big(h_{i_1}, (h_i)_{(1)}\big) \cdots r''\big(h_{i_t}, (h_i)_{(t)}\big) \\ &= r''(h_{i_1} \cdots h_{i_t}, h_i). \end{aligned}$$

Thus, $r'(h, h_i) = r''(h, h_i)$, for all $h \in H$ and $i \in \{1, \dots, n\}$. Let $h \in H$ and $i_1, \dots, i_t$ be in $\{1, \dots, n\}$. Then, using (3), we have

$$\begin{aligned} r'(h, h_{i_1} \cdots h_{i_t}) &= r'\big(h_{(1)}, h_{i_t}\big) \cdots r'\big(h_{(t)}, h_{i_1}\big) \\ &= r''\big(h_{(1)}, h_{i_t}\big) \cdots r''\big(h_{(t)}, h_{i_1}\big) \\ &= r''(h, h_{i_1} \cdots h_{i_t}). \end{aligned}$$

Since $h_1, \dots, h_n$ generate $H$ as an algebra, we conclude that $r' = r''$. □

We now proceed to complete the proof of Theorem 1.1.

For $g \in \Gamma$ and $i = 1, \dots, n$ let $\gamma_g$, $\xi_i \in H^*$ be defined by

$$\gamma_g(hx_P) = \delta_{P,\emptyset} r_0(g, h), \qquad \xi_i(hx_P) = \begin{cases} 0 & \text{if } |P| \neq 1 \\ r_1(x_i, x_j) & \text{if } P = \{j\} \end{cases}$$

for all $h \in \Gamma$ and $P \subseteq \{1, \dots, n\}$. We will show that $\varphi : H \to H^{*\mathrm{cop}}$ defined by

$$\varphi(g) = \gamma_g \text{ and } \varphi(x_i) = \xi_i, \quad \text{for all } g \in \Gamma,\ i = 1, \dots, n,$$

is a bialgebra map. Using Remark 2.1, this will prove that $r : H \otimes H \to k$, $r(x, y) = \varphi(x)(y)$, for all $x$, $y \in H$, is a linear map, satisfying (3) and (4), whose restriction to $\Gamma \times \Gamma$, respectively $V \otimes V$, is $r_0$, respectively $r_1$.

Notice first that, since $r_1 : V \otimes V \to k$ is a morphism of Yetter-Drinfeld modules, we have $\xi_i(x_j)(1 - \chi_i \chi_j) = 0$ and $\xi_i(x_j)(1 - g_i g_j) = 0$, for all $i$, $j \in \{1, \dots, n\}$. Using this, it is straightforward to check that $\gamma_g \gamma_h = \gamma_{gh}$, $\gamma_g \xi_i = \chi_i(g) \xi_i \gamma_g$, $\xi_i \xi_j = \chi_j(g_i) \xi_j \xi_i$ and $\xi_i^2 = 0$, for all $g$, $h \in \Gamma$ and $i$, $j = 1, \dots, n$. For example, $\xi_i \xi_j(hx_P) = 0 = \chi_j(g_i) \xi_j \xi_i(hx_P)$ when $|P| \neq 2$. If $1 \leq u < v \leq n$, then

$$\begin{aligned} \xi_i \xi_j(hx_u x_v) &= \xi_i(hg_u x_v) \xi_j(hx_u) + \chi_u^{-1}(g_v) \xi_i(hg_v x_u) \xi_j(hx_v) \\ &= \xi_i(x_v) \xi_j(x_u) + \chi_i(g_j)^{-1} \xi_i(x_u) \xi_j(x_v) \\ &= \chi_j(g_i) \big(\xi_i(x_u) \xi_j(x_v) + \chi_j(g_i)^{-1} \xi_i(x_v) \xi_j(x_u)\big) \\ &= \chi_j(g_i) \big(\xi_j(hg_u x_v) \xi_i(hx_u) + \chi_u^{-1}(g_v) \xi_j(hg_v x_u) \xi_i(hx_v)\big) \\ &= \chi_j(g_i) \xi_j \xi_i(hx_u x_v). \end{aligned}$$

Thus, $\varphi : H \to H^{*\mathrm{cop}}$, $\varphi(g) = \gamma_g$ and $\varphi(x_i) = \xi_i$, for all $g \in \Gamma$ and $i = 1, \ldots, n$, is an algebra map. It is straightforward to check that $\gamma_g$ is a group-like element and $x_i$ a $(\gamma_{g_i}, \varepsilon)$-skew primitive element of $H^{*\mathrm{cop}}$ for every $g \in \Gamma$ and $i = 1, \ldots, n$. Thus, $\varphi$ is a bialgebra map. It follows that the linear map $r : H \otimes H \to k$, $r(x, y) = \varphi(x)(y)$, for all $x$, $y \in H$, satisfies (3) and (4). Moreover, $r|_{\Gamma \times \Gamma} = r_0$ and $r|_{V \otimes V} = r_1$. Since $r|_{k[\Gamma] \otimes k[\Gamma]}$ is convolution invertible, it follows that $r$ is convolution invertible. Taking into account Lemma 4.2, in order to prove that $r$ is a co-quasitriangular structure on $H$, we need only check that (2) holds for every pair of elements in the second term of the coradical filtration of $H$. This is straightforward, as we next show for the pair $(gx_i, hx_j)$:

$$\begin{aligned} r\big((hx_j)_{(1)}, (gx_i)_{(1)}\big)(hx_j)_{(2)}(gx_i)_{(2)} &= r(hg_j, gg_i)hx_jgx_i + r(hx_j, gx_i)hg \\ &= r(h, g)\chi_i^{-1}(h)ghx_ix_j + r(hx_j, gx_i)gh \\ &= gx_ihx_jr(h, g) + gg_ihg_jr(hx_j, gx_i) \\ &= (gx_i)_{(1)}(hx_j)_{(1)}r\big((hx_j)_{(2)}, (gx_i)_{(2)}\big), \end{aligned}$$

where for the third equality we use the fact that $r(hx_j, gx_i)(1 - g_ig_j) = 0$. Thus, $r$ is a co-quasitriangular structure on $H$ which restricts on $\Gamma$ to $r_0$ and on $V$ to $r_1$. The uniqueness of $r$ follows from Lemma 4.3. This completes the proof of Theorem 1.1.

Let $\mathcal{C}(\Gamma, r_0, V, r_1)$ denote the braided tensor category constructed above.

EXAMPLE 4.4. Take $\Gamma = \mathbb{Z}/2\mathbb{Z}$ and let $r_0$ be the unique non-trivial bicharacter of $\mathbb{Z}/2\mathbb{Z}$. Let $V$ be a multiple of the non-identity simple object of $\mathsf{Corep}(k[\mathbb{Z}/2\mathbb{Z}])$. Then $V$ belongs to $\mathcal{Z}_{sym}(\mathcal{C}(\mathbb{Z}/2\mathbb{Z}], r_0))$, so the Nichols Hopf algebra [**Ni78**]

$$E(V) := \mathfrak{B}(V)\#k[\mathbb{Z}/2\mathbb{Z}] \tag{31}$$

admits co-quasitriangular structures. According to Theorem 1.1 such structures are in bijection with bilinear forms $r_1 : V \otimes V \to k$ or, equivalently, with $n$-by-$n$ square matrices, where $n = \dim_k(V)$. This fact was first established in [**PvO99**].

## 5. The symmetric center of $\mathcal{C}(\Gamma, r_0, V, r_1)$

The symmetric center $\mathcal{Z}_{sym}(\mathcal{C})$ of a braided tensor category $\mathcal{C}$ was defined in Section 2.5.

Let $\Gamma$, $r_0$, $V$, $r_1$ be as in Theorem 1.1. Let $H = \mathfrak{B}(V)\#k[\Gamma]$ and $r : H \otimes H \to k$ be the corresponding Hopf algebra and $r$-form.

We have

$$\mathcal{Z}_{sym}(\mathcal{C}(\Gamma, r_0, V, r_1)) \cong \mathsf{Corep}(H_{sym}, r|_{H_{sym} \otimes H_{sym}}),$$

where $H_{sym}$ is the Hopf subalgebra of $H$ defined by (9).

Let $b : \Gamma \times \Gamma \to k^\times$ be the symmetric bicharacter given by

$$b(g, h) = r(g, h)r(h, g), \qquad g, h \in \Gamma.$$

Let $\Gamma^\perp$ and $V^\perp$ denote the radicals of $b$ and $(r_1)_{alt}$, respectively, i.e.,

$$\begin{aligned} \Gamma^\perp &= \{g \in \Gamma \mid b(g, h) = 1 \text{ for all } h \in \Gamma\}, \\ V^\perp &= \{v \in V \mid (r_1)_{alt}(v, w) = 0 \text{ for all } w \in V\}. \end{aligned}$$

REMARK 5.1. If $V \neq 0$ then $\Gamma^\perp \neq 0$ since $V \in \mathcal{C}(\Gamma^\perp, r_0|_{\Gamma^\perp \times \Gamma^\perp})$.

LEMMA 5.2. *$H_{sym}$ is generated as an algebra by $\Gamma^\perp$ and $V^\perp$.*

PROOF. Since $H_{sym}$ is a Hopf subalgebra of a pointed Hopf algebra with abelian coradical, it is pointed with abelian coradical. By the result of Angiono [**An13**], $H_{sym}$ is generated by its group-like and skew-primitive elements.

It is easy to see that an element $g \in \Gamma$ is in $H_{sym}$ if and only if $r(g,h)r(h,g) = 1$, for all $h \in \Gamma$. Thus, the set of group-like elements of $H_{sym}$ is $\Gamma^{\perp}$.

Let now $g \in \Gamma^{\perp}$ and let $x$ be a $(g,1)$-primitive element. If $g \notin \{g_i \mid i = 1, \dots, n\}$ then $x$ is a scalar multiple of $1-g$, so it is contained in $H_{sym}$. If $g = g_i$ for some $i \in \{1, \dots, n\}$ then $x = a(1-g_i) + \sum_{j:g_j=g_i} a_j x_j$, for some $a$, $a_j \in k$. Let $y = \sum_{j:g_j=g_i} a_j x_j$. We claim that $x \in H_{sym}$ if and only if $y \in V^{\perp}$. It is clear that $x \in H_{sym}$ if and only if $y \in H_{sym}$. Next,

$$y_{(1)} r(y_{(2)}, z_{(1)}) r(z_{(2)}, y_{(3)}) = \sum_{j:g_j=g_i} a_j \Big( r(g_j, z_{(1)}) r(z_{(2)}, x_j) + r(x_j, z) \Big) g_i + \varepsilon(z) y,$$

for $z \in H$, so $y \in H_{sym}$ if and only if

$$\sum_{j:g_j=g_i} a_j \Big( r(g_j, z_{(1)}) r(z_{(2)}, x_j) + r(x_j, z) \Big) = 0 \tag{32}$$

for all $z \in \{hx_l \mid h \in \Gamma, l = 1, \dots, n\}$. For $z = hx_l$, the left-hand side of (32) becomes

$$\begin{aligned} \mathrm{LHS}(32) &= \sum_{j:g_j=g_i} a_j \Big( r(g_j, hg_l) r(hx_l, x_j) + r(x_j, hx_l) \Big) \\ &= \sum_{j:g_j=g_i} a_j \Big( r(g_j, g_l) r(x_l, x_j) + r(x_j, x_l) \Big) \\ &= r(g_l, g_i) r(x_l, y) + r(y, x_l) \\ &= (r + r \circ c_{V,V})(y, x_l) \\ &= (r - r \circ \tau_{V,V})(y, x_l) = 2(r_1)_{alt}(y, x_l), \end{aligned}$$

where we used the fact that $r|_{V\otimes V}$ is a morphism in ${}^{\Gamma}_{\Gamma}\mathcal{YD}$ and Lemma 3.4. Thus, $y \in H_{sym}$ if and only if $y \in V^{\perp}$. It follows that non-trivial skew-primitive elements of $H_{sym}$ generate $V^{\perp}$, so the claim follows. $\square$

COROLLARY 5.3. *$H_{sym} = \mathfrak{B}(V^{\perp}) \# k[\Gamma^{\perp}]$ and*

$$\mathcal{Z}_{sym}(\mathcal{C}(\Gamma, r_0, V, r_1)) \cong \mathcal{C}(\Gamma^{\perp}, r_0|_{\Gamma^{\perp}\times\Gamma^{\perp}}, V^{\perp}, r_1|_{V^{\perp}\otimes V^{\perp}}).$$

COROLLARY 5.4.
(i) *$\mathcal{C}(\Gamma, r_0, V, r_1)$ is symmetric if and only if $\mathcal{C}(\Gamma, r_0)$ is symmetric and $r_1$ is symmetric.*
(ii) *$\mathcal{Z}_{sym}(\mathcal{C}(\Gamma, r_0, V, r_1))$ is semisimple if and only if $(r_1)_{alt} : V \otimes V \to k$ is non-degenerate.*
(iii) *$\mathcal{C}(\Gamma, r_0, V, r_1)$ is factorizable if and only if $\mathcal{C}(\Gamma, r_0)$ is factorizable and $V = 0$.*

EXAMPLE 5.5. Let $E(V)$ be the Hopf algebra from Example 4.4 and let $r_1 : V \otimes V \to k$ be a bilinear form. In this case $c_{V,V} = -\tau_{V,V}$, so Corollary 5.4(i) says that the co-quasitriangular structure determined by $r_1$ is symmetric if and only if $r_1$ is symmetric (in the usual linear algebra sense). This was proved in [**CC04**].

## 6. Ribbon structures on $\mathcal{C}(\Gamma, r_0, V, r_1)$

Ribbon structures on braided tensor categories and ribbon elements of coquasitriangular Hopf algebras were defined in Section 2.3. We want to classify ribbon structures of $\mathcal{C}(\Gamma, r_0, V, r_1)$.

LEMMA 6.1. *Let $A$ be an abelian group and let $a_1, a_2, \ldots, a_n$ be elements of $A$. Then there exists a group homomorphism $\gamma : A \to \{\pm 1\}$ such that $\gamma(a_i) = -1$, for all $i = 1, \ldots, n$ if and only if there are no relations in $A$ of the form $a_{i_1} a_{i_2} \cdots a_{i_k} = x^2$, with $k$ odd.*

PROOF. Homomorphisms $A \to \{\pm 1\}$ are in bijection with homomorphisms $A/A^2 \to \{\pm 1\}$. Let $\bar{a}_1, \bar{a}_2 \ldots, \bar{a}_n$ be images of $a_1, a_2, \ldots, a_n$ in $A/A^2$. If there are no relations in $A$ of the form $a_{i_1} a_{i_2} \cdots a_{i_k} = x^2$ with $k$ odd then all relations in $A/A^2$ (viewed as a $\mathbb{Z}/2\mathbb{Z}$-vector space) are of the form $\bar{a}_{i_1} + \bar{a}_{i_2} \cdots + \bar{a}_{i_l} = 0$ with $l$ even. Therefore, there is a well-defined homomorphism $A/A^2 \to \{\pm 1\}$ sending each $a_i$ to $-1$. The converse implication is trivial. □

PROPOSITION 6.2. *The set of ribbon structures on $\mathcal{C}(\Gamma, r_0, V, r_1)$ is non-empty and is in bijection with the set of group homomorphisms $\gamma : \Gamma \to \{\pm 1\}$ such that $\gamma(g_i) = -1$, for all $i = 1, \ldots, n$.*

PROOF. Let $H = \mathfrak{B}(V)\#k[\Gamma]$. As explained in Section 2.3, ribbon structures on $\mathcal{C}(\Gamma, r_0, V, r_1)$ are in bijection with group-like elements $\gamma \in G(H^*)$ satisfying $\gamma^2 = (\eta \circ S) * \eta^{-1}$ and $S^2_{H^*}(p) = \gamma^{-1} * p * \gamma$, for all $p \in H^*$. Let $\gamma$ be such an element. We have

$$\gamma(g)^2 = \gamma^2(g) = (\eta \circ S) * \eta^{-1}(g) = r_0(g^{-1}, g) r_0(g, g) = 1$$

for all $g \in \Gamma$, so $\gamma(\Gamma) \subseteq \{\pm 1\}$. Now $S^2_{H^*}(p) = \gamma^{-1} * p * \gamma$ for all $p \in H^*$, if and only if $S^2_H = \gamma^{-1} * \mathrm{id}_H * \gamma$. Since both maps are algebra maps, they are equal if and only if they agree on algebra generators. We have

$$\begin{aligned} S^2_H(g) &= g, & (\gamma^{-1} * \mathrm{id}_H * \gamma)(g) &= g, \\ S^2_H(x_i) &= -x_i, & (\gamma^{-1} * \mathrm{id}_H * \gamma)(x_i) &= \gamma^{-1}(g_i) x_i \end{aligned}$$

for all $g \in \Gamma$ and $i = 1, \ldots, n$. Thus, $S^2_H = \gamma^{-1} * \mathrm{id}_H * \gamma$ if and only if $\gamma(g_i) = -1$ for all $i$.

It remains to show that there always exists a homomorphism $\gamma : \Gamma \to \{\pm 1\}$ such that $\gamma(g_i) = -1$, for all $i = 1, \ldots, n$. Since $k^\times$ is an injective $\mathbb{Z}$-module, it is enough to show that there is such a homomorphism on the subgroup $\Gamma_0 = \langle g_1, \ldots g_n \rangle \subset \Gamma$. Using Lemma 6.1, we have to show that there are no relations in $\Gamma_0$ of the form $g_{i_1} g_{i_2} \cdots g_{i_k} = x^2$ with $k$ odd. If $x = g_{i_1}^{e_1} g_{i_2}^{e_2} \cdots g_{i_t}^{e_t}$ is an element of $\Gamma_0$ then

$$r_0(x, x) = \prod_{r=1}^{t} r_0(g_{i_r}, g_{i_r})^{e_r^2} \prod_{1 \le r < s \le t} \left( r_0(g_{i_r}, g_{i_s}) r_0(g_{i_s}, g_{i_r}) \right)^{e_r e_s} = (-1)^{\sum_{r=1}^t e_r^2}.$$

In particular, $r_0(x, x)^2 = 1$ for all $x \in \Gamma_0$. On the other hand, if $g_{i_1} g_{i_2} \cdots g_{i_k} = x^2$ with $k$ odd, then $r_0(x, x)^4 = r_0(x^2, x^2) = (-1)^k = -1$, which is a contradiction. □

## 7. Metric data (Proof of Theorem 1.2)

It is well known that the 1-categorical truncation of the 2-category of pointed braided fusion categories is equivalent to the 2-category of *pre-metric groups*.

The objects of the former are pointed braided fusion categories and morphisms are natural isomorphism classes of braided tensor functors. The objects of the latter are pairs $(\Gamma, q)$, where $\Gamma$ is an abelian group and $q : \Gamma \to k^\times$ is a quadratic form, and morphisms are orthogonal homomorphisms (see [**JS93**] or [**EGNO15**, Section 8.4]).

The goal of this section is to extend this result to tensor categories that are not necessarily semisimple.

**7.1. The category of metric data.** Recall that a *groupoid* is a category in which all morphisms are isomorphisms.

Let $\mathcal{B}$ denote a groupoid whose objects are pointed braided tensor categories admitting a fiber functor and morphisms are natural isomorphism classes of equivalences of braided tensor categories. Thus, objects of $\mathcal{B}$ can be identified with the co-representation categories of co-quasitriangular pointed Hopf algebras.

Define a groupoid $\mathcal{M}$ as follows. The objects of $\mathcal{M}$ are data $(\mathcal{P}, V, r)$, where $\mathcal{P}$ is pointed braided fusion category admitting a fiber functor, $V$ is an object in $\mathcal{Z}_{sym}(\mathcal{P})_- \subset {}^{\Gamma}_{\Gamma}\mathcal{YD}$, and $r : V \otimes V \to k$ is an alternating bilinear form in $\mathcal{P}$. The set of morphisms

$$\mathsf{Hom}_{\mathcal{M}}((\mathcal{P}, V, r), (\mathcal{P}', V', r'))$$

is the set of pairs $(\alpha, f)$, where $\alpha : \mathcal{P} \to \mathcal{P}'$ is a braided equivalence and $f : \alpha(V) \to V'$ is an isomorphism in ${}^{\Gamma'}_{\Gamma'}\mathcal{YD}$ such that $r' \circ (f \otimes f) = \alpha(r)$, modulo the equivalence relation identifying pairs $(\alpha, f)$ and $(\alpha, -f)$.

DEFINITION 7.1. We will call $\mathcal{M}$ the groupoid of *metric data*.

Define a functor

$$F : \mathcal{M} \to \mathcal{B} \tag{33}$$

as follows. Given a datum $(\mathcal{P}, V, r)$ let $\Gamma$ be the underlying Abelian group of $\mathcal{P}$ and let $r_0 : \Gamma \times \Gamma \to k^\times$ be the corresponding $r$-form. Set

$$F(\mathcal{P}, V, r) = \mathcal{C}(\Gamma, r_0, V, r). \tag{34}$$

A morphism $(\alpha, f)$ between $(\mathcal{P}, V, r)$ and $(\mathcal{P}', V', r')$ (with $\mathcal{P} = \mathcal{C}(\Gamma, r_0)$ and $\mathcal{P} = \mathcal{C}(\Gamma', r_0'))$ gives rise to an isomorphism of Hopf algebras $\varphi_{(\alpha, f)} : \mathfrak{B}(V)\#k[\Gamma] \xrightarrow{\sim} \mathfrak{B}(V')\#k[\Gamma']$ given by

$$\varphi_{(\alpha, f)}(g) = \alpha(g), \quad \varphi_{(\alpha, f)}(x) = f(x) \tag{35}$$

for all $g \in \Gamma$ and $x \in V$. Here we slightly abuse notation and use $\alpha$ to denote the isomorphism between $\Gamma$ and $\Gamma'$ corresponding to the braided equivalence $\alpha : \mathcal{P} \xrightarrow{\sim} \mathcal{P}'$. The bicharacter $r_0' \circ (\alpha \times \alpha)/r_0$ is alternating and so is equal to $\mathrm{alt}(\mu)$ for some $\mu \in Z^2(\Gamma/\Gamma_0, k^\times)$, see (29). This means that the $r$-form $r_0' \circ (\alpha \times \alpha)$ is a twisting deformation of $r_0$ by means of the above 2-cocycle $\mu \in H^2(\Gamma/\Gamma_0, k^\times)$. But such $\mu$ defines an invariant 2-cocycle on $\mathfrak{B}(V)\#k[\Gamma]$ by Proposition 3.13. Thus, $\varphi_{(\alpha, f)}$ gives rise to a well defined braided tensor equivalence $F(\alpha, f)$ between $\mathcal{C}(\Gamma, r_0, V, r)$ and $\mathcal{C}(\Gamma', r_0', V', r')$.

**7.2. Proof of Theorem 1.2.** Our goal is to show that the functor $F$ defined in the previous Section is an equivalence.

Let $H := \mathfrak{B}(V)\#k[\Gamma]$ and $H' := \mathfrak{B}(V')\#k[\Gamma']$ with co-quasitriangular structures on $H$ and $H'$ defined by $(r_0, r_1)$ and $(r_0', r_1')$.

LEMMA 7.2. *Co-quasitriangular Hopf algebra isomorphisms $H \to H'$ are in bijection with pairs $(\alpha, f)$, where $\alpha : \Gamma \to \Gamma'$ is a group isomorphism, $f : V^\alpha \to V'$ is an isomorphism such that $r'_0 \circ (\alpha \times \alpha) = r_0$, and $r'_1 \circ (f \otimes f) = r_1^\alpha$. Here $V^\alpha$ is the object of ${}^{\Gamma'}_{\Gamma'}\mathcal{YD}$ induced from $V$ and $r_1^\alpha$ is the bilinear form induced from $r_1$ by means of $\alpha$.*

PROOF. Let $\phi : H \to H'$ be an isomorphism of co-quasitriangular Hopf algebras. Since $f$ takes group-like elements to group-like elements, it restricts to a group isomorphism $\alpha : \Gamma \to \Gamma'$. Let us show that $\phi$ induces an isomorphism $V^\alpha \to V'$.

Notice first that $\alpha(g_i) \in \{g'_j\}$. Indeed, if this is not the case, then $\phi(x_i) = a(1-\phi(g_i))$, for some $a \in k$. But $\phi(x_i)$ anti-commutes with $\phi(g_i)$, while $a(1-\phi(g_i))$ commutes with $\phi(g_i)$, so $\phi(x_i) = 0$. This, however, contradicts the injectivity of $\phi$. Thus, $\alpha(g_i) \in \{g'_j\}$, for all $i$. It follows that there exist scalars $a_i$, $b_{ij} \in k$ such that $b_{ji}\big(g'_j - \phi(g_i)\big) = 0$ and

$$\phi(x_i) = a_i(1 - \phi(g_i)) + \sum_j b_{ji}x'_j, \qquad i = 1, \ldots, n.$$

We must have $\phi(g)\phi(x_i) = \chi_i(g)\phi(x_i)\phi(g)$, for all $g \in \Gamma$. Now

$$\begin{aligned}\phi(g)\phi(x_i) &= a_i(\phi(g) - \phi(gg_i)) + \sum_j b_{ji}\phi(g)x'_j \\ &= a_i(\phi(g) - \phi(gg_i)) + \sum_j b_{ji}\chi'_j\big(\phi(g)\big)x'_j\phi(g)\end{aligned}$$

and

$$\chi_i(g)\phi(x_i)\phi(g) = a_i\chi_i(g)(\phi(g) - \phi(gg_i)) + \sum_j b_{ji}\chi_i(g)x'_j\phi(g).$$

Thus, $\phi(g)\phi(x_i) = \chi_i(g)\phi(x_i)\phi(g)$ if and only if $a_i = a_i\chi_i(g)$ and $\chi'_j\big(\phi(g)\big)b_{ji} = \chi_i(g)b_{ji}$, for all $j$ and $g \in \Gamma$. Taking $g = g_i$ in the first condition, we obtain $a_i = 0$. The second condition is equivalent to $b_{ji}(\chi_i - \chi'_j \circ \phi) = 0$. It follows that

$$\phi(x_i) = \sum_j b_{ji}x'_j$$

where $b_{ji}\big(g'_j - \alpha(g_i)\big) = 0$ and $b_{ji}(\chi_i\alpha^{-1} - \chi'_j) = 0$. This means precisely that the restriction $f$ of $\phi$ to $V$ is a morphism in $\mathcal{C}(\Gamma', r'_0)$ from $V^\alpha$ to $V'$. Since $\phi$ is an isomorphism, the restriction is also an isomorphism.

We have proven that if $\phi : H \to H'$ is a Hopf algebra isomorphism then it induces, by restriction, isomorphisms $\alpha : \Gamma \to \Gamma'$ and $f : V^\alpha \to V'$. We have $r' \circ (\phi \otimes \phi) = r$ if and only if $r'_0 \circ (\alpha \times \alpha) = r_0$ and $r'_1 \circ (f \otimes f) = r_1^\alpha$. It is easy to check that every such data comes from an isomorphism of co-quastriangular Hopf algebras $H \to H'$. This completes the proof. □

LEMMA 7.3. *Let $\sigma$ be a 2-cocycle on $H = \mathfrak{B}(V)\#k[\Gamma]$ such that $\sigma|_{\Gamma\times\Gamma} = 1$. Then the twisted $r$-form $r^\sigma$ (see (13)) on $H^\sigma$ satisfies*

$$r^\sigma|_{V\otimes V} = r|_{V\otimes V} + 2(\sigma|_{V\otimes V})_{sym}. \tag{36}$$

PROOF. Using formula (13) we compute

$$\begin{aligned} r^\sigma(x_i, x_j) &= r(g_i, g_j)\sigma^{-1}(x_i, x_j) + r(x_i, x_j) + \sigma(x_j, x_i) \\ &= r(x_i, x_j) + (\sigma(x_j, x_i) - \chi_i(g_j)\sigma(x_i, x_j)) \\ &= r(x_i, x_j) + (\sigma \circ \tau_{V,V} - \sigma \circ c_{V,V} \circ \tau_{V,V})(x_i, x_j). \end{aligned}$$

Since restrictions of $r$ and $r^\sigma$ on $V \otimes V$ are morphism in ${}^\Gamma_\Gamma\mathcal{YD}$ we conclude, using Lemma 3.4, that

$$\sigma|_{V\otimes V} \circ \tau_{V,V} - \sigma|_{V\otimes V} \circ c_{V,V} \circ \tau_{V,V} = 2(\sigma|_{V\otimes V})_{sym},$$

which implies the statement. □

We now proceed with the proof of Theorem 1.2. We need to show that the functor $F$ (33) is surjective and fully faithful.

(1) *$F$ is surjective.*

For this end it suffices to check that the co-quasitriangular structure on $H$ defined using $r : V \otimes V \to k$ is gauge equivalent, by means of an invariant 2-cocycle on $H$, to the one defined using an alternating morphism $V \otimes V \to k$ in $\mathcal{C}(\Gamma, r_0)$. Let $\sigma$ be an invariant 2-cocycle on $H$ such that $\sigma|_{\Gamma\times\Gamma} = 1$ (such 2-cocycles are classified in Corollary 3.11) and let $r^\sigma = \sigma_{21} * r * \sigma^{-1}$. By Lemma 7.3 we have

$$r^\sigma|_{V\otimes V} = r|_{V\otimes V} + 2(\sigma|_{V\otimes V})_{sym}.$$

Thus, we can take $\sigma$ such that $(\sigma|_{V\otimes V})_{sym} = -\frac{1}{2}(r \circ \tau)_{sym}$, so that $r^\sigma|_{V\otimes V} = (r|_{V\otimes V})_{alt}$, i.e., $r^\sigma|_{V\otimes V}$ is alternating.

Since $\mathcal{C}(\Gamma, r_0, V, r)$ and $\mathcal{C}(\Gamma, r_0, V, r^\sigma)$ are equivalent braided tensor categories, the surjectivity of $F$ follows.

(2) *$F$ is faithful.* We need to check that $F$ is injective on morphisms. It is clear from definitions that $F(\alpha, f) = F(\alpha', f')$ implies $\alpha = \alpha'$. Therefore, it remains to check that for an automorphism $(\mathrm{id}_\Gamma, f)$ of $(\mathcal{P}, V, r) \in \mathcal{M}$ one has $F(\mathrm{id}_\Gamma, f) \cong \mathrm{id}_{\mathcal{C}(\mathcal{P}, V, r)}$ as a tensor functor if and only if $f = \pm\mathrm{id}_V$. One implication is clear since $-\mathrm{id}_V$ preserves any bilinear form.

Note that the Hopf algebra automorphism $\varphi_{(\mathrm{id}_\Gamma, f)}$ of $H$ defined in (35) gives rise to a trivial tensor autoequivalence of $\mathsf{Corep}(H)$ if and only if it is given by

$$h \mapsto \chi \rightharpoonup h \leftharpoonup \chi^{-1}, \quad h \in H$$

for some character $\chi \in H^*$. The condition that it preserves $r$ is equivalent to $\chi(g_i)\chi(g_j) = 1$ for all $i, j = 1, \ldots, n$, i.e., to $\chi$ being identically equal to 1 or $-1$ on the support of $V$. By the Remark 4.1 there exists $\chi$ such that this value is $-1$, so that $f = \pm\mathrm{id}_V$.

(3) *$F$ is full.* We need to check that $F$ is surjective on morphisms. We claim that any braided tensor equivalence $\Phi$ between $\mathsf{Corep}(\mathfrak{B}(V)\#k[\Gamma], r)$ and $\mathsf{Corep}(\mathfrak{B}(V')\#k[\Gamma'], r')$ is isomorphic to one coming from a co-quasitriangular Hopf algebra isomorphism $\mathfrak{B}(V)\#k[\Gamma] \to \mathfrak{B}(V')\#k[\Gamma']$. By the result of Davydov [**Da10**]

$\Phi$ corresponds to a pair $(f, \sigma)$, where $\sigma$ is a 2-cocycle on $\mathfrak{B}(V)\#k[\Gamma]$ and $f : (\mathfrak{B}(V)\#k[\Gamma])^\sigma \to \mathfrak{B}(V')\#k[\Gamma']$ is a Hopf algebra isomorphism such that

$$r \circ (f \otimes f) = r^\sigma.$$

The last condition corresponds to the braided property of the equivalence.

We must have $\sigma|_{\Gamma\times\Gamma} = 1$ since a non-trivial twisting changes the braided equivalence class of $\mathsf{Corep}(k[\Gamma], r_0)$. By Lemma 7.3 we have

$$(r \circ (f \otimes f) - r)\,|_{V\otimes V} = 2(\sigma|_{V\otimes V})_{sym}.$$

The left hand side of the above equality is alternating, while the right hand side is symmetric. Hence, both sides are equal to 0 and so $\sigma$ is gauge equivalent to the trivial 2-cocyle by Proposition 3.9. This means that $\Phi$ is isomorphic to the equivalence induced by a co-quasitriangular Hopf algebra isomorphism, so the result follows from Lemma 7.2.

REMARK 7.4. We can give a conceptual explanation of the reason why $(r_1)_{alt}$ is an invariant of the braided tensor category $\mathcal{C} := \mathcal{C}(\Gamma, r_0, V, r_1)$.

Let $g \in \Gamma$. We will also use $g$ to denote the corresponding invertible $H$-comodule. Recall that $\mathsf{Ext}_\mathcal{C}(g, 1) \cong P_{1,g}(H)/k(1-g)$, where $P_{1,g}(H)$ denotes the space of $(1, g)$-skew primitive elements of $H$. Explicitly, elements of $\mathsf{Ext}_\mathcal{C}(g, 1)$ are in bijection with equivalence classes of short exact sequences

$$0 \to 1 \xrightarrow{\iota} V_x \xrightarrow{p} g \to 0,$$

where 1 denotes the trivial comodule $k$. The 2-dimensional comodule $V_x$ is a vector space with a basis $v_0$, $v_1$ and $H$-coaction given by

$$\rho(v_0) = v_0 \otimes 1, \qquad \rho(v_1) = v_0 \otimes x + v_1 \otimes g, \tag{37}$$

where $x \in P_{1,g}(H)$.

Let $x' \in P_{1,g'}(H)$, $g' \in \Gamma$, be another skew-primitive element of $H$, let

$$0 \to 1 \xrightarrow{\iota'} V_{x'} \xrightarrow{p'} g' \to 0$$

be the corresponding extension, and let $v_0'$, $v_1'$ be a basis of $V_{x'}$ defined analogously to (37).

Let $\beta_{x,x'} = c_{V_{x'},V_x} \circ c_{V_x,V_{x'}}$ denote the square of the braiding on $V_x \otimes V_{x'}$. Using formula (5) one computes

$$\begin{aligned}
\beta_{x,x'}(v_0 \otimes v_0') &= v_0 \otimes v_0', \\
\beta_{x,x'}(v_0 \otimes v_1') &= v_0 \otimes v_1', \\
\beta_{x,x'}(v_1 \otimes v_0') &= v_1 \otimes v_0', \\
\beta_{x,x'}(v_1 \otimes v_1') &= v_1 \otimes v_1' + (r_1(x, x') + r_0(g, g')r_1(x', x))v_0 \otimes v_0'.
\end{aligned}$$

Let $s := r_1 - r_1 \circ \tau$. Combining Lemma 3.4 with above computation we see that

$$\beta_{x,x'} = \mathrm{id}_{V_x\otimes V_{x'}} + s(x, x')(p \otimes p') \circ (\iota \otimes \iota') \tag{38}$$

for all $x, x' \in V = \mathsf{Ext}_\mathcal{C}(\Gamma, 1)$ (note that $(p\otimes p')\circ(\iota\otimes\iota') \in \mathrm{End}_\mathcal{C}(V_x \otimes V_{x'})$ whenever $s(x, x') \neq 0$).

It follows from (38) that $s = (r_1)_{alt}$ is an invariant of the braided equivalence class of $\mathcal{C}$ (a computation establishing this fact is straightforward and can be found in the proof of [**BN15**, Proposition 6.7]).

Let $\mathcal{C} = \mathcal{C}(\mathcal{P}, V, r)$. Theorem 1.2 allows to compute the group $\mathsf{Aut}^{br}(\mathcal{C})$ of isomorphism classes of braided autoequivalences of $\mathcal{C}$. Namely, let

$$\begin{aligned} \mathsf{Aut}(V, r) &:= \{f \in \mathsf{Aut}_{\mathcal{P}}(V) \mid r \circ (f \otimes f) = r\}/\{\pm \mathrm{id}_V\}, \\ O(\mathcal{P}, r) &:= \{\alpha \in \mathsf{Aut}(\Gamma) \mid q \circ \alpha = q \text{ and } r^{\alpha}, r : V \otimes V \to k \text{ are congruent in } \mathcal{P}\}. \end{aligned}$$

COROLLARY 7.5. *There is a short exact sequence*

$$1 \to \mathsf{Aut}(V, r) \to \mathsf{Aut}^{br}(\mathcal{C}) \to O(\mathcal{P}, r) \to 1. \tag{39}$$

EXAMPLE 7.6. Let us consider $\mathcal{C} = \mathsf{Corep}(E(V), r)$, where $E(V)$ is the Hopf algebra from Example 4.4 with the co-quasitriangular structure given by the zero bilinear form on $V$ (this structure is symmetric). In this case $\Gamma = \mathbb{Z}/2\mathbb{Z}$, so $O(\mathcal{P}, r) = 1$ and Corollary 7.5 implies that $\mathsf{Aut}^{br}(\mathcal{C}) = GL_n(V)/\{\pm \mathrm{id}_V\}$, cf. [**BN15**].

## 8. The Drinfeld center of a pointed braided tensor category

It is well known that the Drinfeld center of a pointed braided fusion category is pointed (see, e.g., [**DN13**, Proposition 5.8]). This is no longer true in the non-semisimple case. Indeed, the Drinfeld center is always factorizable, cf. Corollary 5.4(iii).

Let $\mathcal{C} = \mathcal{C}(\Gamma, r_0, V, r_1)$ be a pointed braided tensor category corresponding to a non-zero *self-dual* Yetter-Drinfeld module $V$. In this Section we show that the trivial component of the universal grading of $\mathcal{Z}(\mathcal{C})$ is pointed, i.e., $\mathcal{Z}(\mathcal{C})$ is nilpotent of nilpotency class 2 ([**GN08**]).

**8.1. The group-like elements of $D(H)^*$.** It is well known [**R93**, Proposition 10] that for a finite dimensional Hopf algebra $H$ the group of central group-like elements of $D(H)$ is

$$G\big(D(H)^*\big) \cong G\big(D(H)\big) \cap Z\big(D(H)\big).$$

More explicitly,

$$G\big(D(H)^*\big) = \{g \otimes \gamma \mid (g, \gamma) \in G(H) \times G(H^*), \\ \sum \gamma(h_{(1)})h_{(2)}g = \sum \gamma(h_{(2)})gh_{(1)}, \forall h \in H\}$$

Let $V$ be a quantum linear space of symmetric type and let $H = \mathfrak{B}(V)\#k[\Gamma]$. It is not hard to check that

$$G\big(D(H)^*\big) = \{g \otimes \gamma \mid (g, \gamma) \in \Gamma \times \widehat{\Gamma},\ \gamma(g_i) = \chi_i(g), \text{for all } i = 1, \dots, n\}.$$

Let $b : (\Gamma \times \widehat{\Gamma}) \times (\Gamma \times \widehat{\Gamma}) \to k^{\times}$ be the canonical non-degenerate bicharacter defined by

$$b\big((g, \chi), (g', \chi')\big) = \chi(g')\chi'(g), \qquad (g, \chi), (g', \chi') \in \Gamma \times \widehat{\Gamma}.$$

Consider the subgroup

$$\Sigma := \langle (g_i, \chi_i^{-1}) \mid i = 1, \dots, n\rangle \subset \Gamma \times \widehat{\Gamma}. \tag{40}$$

REMARK 8.1. The above $\Sigma$ is an isotropic subgroup of $\Gamma \times \widehat{\Gamma}$, i.e., $\Sigma \subset \Sigma^{\perp}$, where $\Sigma^{\perp}$ is the orthogonal complement of $\Sigma$ with respect to $b$.

It follows that

$$G(D(H)^*) \cong G(D(H)) \cap Z(D(H)) \cong \Sigma^{\perp}. \tag{41}$$

Let $H$ be a Hopf algebra. It was shown in [**GN08**] that for any Hopf subalgebra $K$ of $H$ contained in the center of $H$ tensor category $\mathsf{Rep}(H)$ is graded by $G(K^*)$. The trivial component of this grading is $\mathsf{Rep}(H/HK^+)$. The maximal central Hopf subalgebra of $H$ provides the universal grading of $\mathsf{Rep}(H)$. In this case, the trivial component is $\mathsf{Rep}(H)_{\mathrm{ad}}$, the adjoint subcategory of $\mathsf{Rep}(H)$.

If $H$ is quasitriangular then the maximal central Hopf subalgebra of $H$ is the group algebra of $G(H) \cap Z(H)$ and the universal grading group of $\mathsf{Rep}(H)$ is the group of characters of $G(H) \cap Z(H)$.

We will be interested in the adjoint subcategory of $\mathcal{Z}(\mathsf{Corep}(H))$, where $H = \mathfrak{B}(V)\#k[\Gamma]$ for a quantum linear space $V \in {}^{\Gamma}_{\Gamma}\mathcal{YD}$ of symmetric type. We have

$$\mathcal{Z}(\mathsf{Corep}(H)) \cong \mathsf{Rep}(D(H)^{\mathrm{cop}}) \cong \mathsf{Corep}(D(H)^{*\mathrm{op}}). \tag{42}$$

Since $D(H)^{\mathrm{cop}}$ is quasitriangular, the universal grading group of $\mathcal{Z}(\mathsf{Corep}(H))$ is isomorphic to $\widehat{\Sigma^{\perp}}$, according to (41).

Remark 8.2. The Drinfeld double $D(\mathfrak{B}(V)\#k[\Gamma])$ was studied by Beattie in [**Be03**].

**8.2. The adjoint subcategory of the center of $\mathcal{C}$.** If $(H, r)$ is a co-quasitriangular Hopf algebra then

$$\iota_r : H \to D(H)^{*\,\mathrm{op}}, \quad \iota_r(x) = x_{(1)} \otimes r(-, x_{(2)})$$

is a Hopf algebra homomorphism. This corresponds to the embedding $\mathcal{C} \hookrightarrow \mathcal{Z}(\mathcal{C})$ from Remark 2.3.

Let $V \in {}^{\Gamma}_{\Gamma}\mathcal{YD}$ be a quantum linear space of symmetric type associated to the datum $(g_1, \dots, g_n, \chi_1, \dots, \chi_n)$ and let $H = \mathfrak{B}(V)\#k[\Gamma]$. Suppose $H$ admits co-quasitriangular structures. Then, according to Theorem 1.1, $r$-forms on $H$ are in bijection with pairs $(r_0, r_1)$, where $r_0$ is a bicharacter of $\Gamma$ such that $V \in \mathcal{Z}_{sym}(\mathcal{C}(\Gamma, r_0))_-$ and $r_1 : V \otimes V \to k$ is a morphism in $\mathcal{C}(\Gamma, r_0)$.

Consider the $r$-form $r$ on $H$ corresponding to the pair $(r_0, 0)$. Then $\iota_r(g_i) = g_i \otimes \chi_i^{-1}$ and $\iota_r(x_i) = x_i \otimes \varepsilon$. Thus, $D(H)^*$ contains group-like elements $g_i \otimes \chi_i^{-1}$ and $(g_i \otimes \chi_i^{-1}, 1)$-skew primitive elements $x_i \otimes \varepsilon$, $i = 1, \dots, n$.

Assume now that $V$ is a self-dual object of ${}^{\Gamma}_{\Gamma}\mathcal{YD}$. In this case the set $\{(g_i, \chi_i) \mid i = 1, \dots, n\}$ is closed under taking inverses. Let $r_1 : V \otimes V \to k$ denote a non-degenerate evaluation morphism in ${}^{\Gamma}_{\Gamma}\mathcal{YD}$. Let $r'$ be the $r$-form on $H$ corresponding to the pair $(r_0, r_1)$. Then $\iota_{r'}(g_i) = g_i \otimes \chi_i^{-1}$ and $\iota_{r'}(x_i) = g_i \otimes r'(-, x_i) + x_i \otimes \varepsilon$. We see that in this case $D(H)^*$ contains, in addition to the above, $(g_i \otimes \chi_i^{-1}, 1)$-skew primitive elements $g_i \otimes r'(-, x_i)$.

Let $A$ be the Hopf subalgebra of $D(H)^*$ generated by group-like elements $g_i \otimes \chi_i^{-1}$ and skew-primitive elements $x_i \otimes \varepsilon$ and $g_i \otimes r'(-, x_i)$, $i = 1, \dots, n$.

Remark 8.3. By definition (40), the group of group-likes of $A$ is $\Sigma$. The above skew-primitive elements $x_i \otimes \varepsilon$ and $g_i \otimes r'(-, x_i)$, $i = 1, \dots, n$ constructed above are linearly independent and form a $2n$-dimensional quantum linear space of symmetric type in ${}^{\Sigma}_{\Sigma}\mathcal{YD}$. Therefore,

$$\dim_k(A) = |\Sigma|\, 2^{2n}.$$

PROPOSITION 8.4. *Let $K = k[G(D(H^*))]$. We have*

$$A^* \cong D(H)/D(H)K^+. \tag{43}$$

PROOF. Let $\pi : D(H) \to D(H)/D(H)K^+$ be the canonical projection. We claim that $A$ is the image of the dual Hopf algebra homomorphism

$$\pi^* : \big(D(H)/D(H)K^+\big)^* \to D(H)^*$$

According to (41), $K$ is the group algebra of $\{\gamma \otimes g \mid (g, \gamma) \in \Sigma^\perp\}$. Let

$$e = \frac{1}{|\Sigma^\perp|} \sum_{(g,\gamma)\in\Sigma^\perp} \gamma \otimes g.$$

Then $e$ is a central idempotent of $D(H)$, $ze = \varepsilon(z)e$, for all $z \in K$, and $K^+ = (1-e)K$. Thus, $D(H)K^+ = D(H)K(1-e) = (1-e)D(H)$. The image of $\pi^*$ is

$$\mathrm{Im}(\pi^*) = \{f \in \big(D(H)\big)^* \mid f(z) = f(ez), \text{ for all } z \in D(H)\}.$$

It is easy to check that any $f \in \{g_i \otimes \chi_i^{-1}, x_i \otimes \varepsilon, g_i \otimes r'(-, x_i)\}$ satisfies $f((\gamma \otimes g)z) = f(z)$, for all $(g, \gamma) \in \Sigma^\perp$ and $z \in D(H)$. For example, if $(g, \gamma) \in \Sigma^\perp$, then

$$\begin{aligned}(x_i \otimes \varepsilon)\big((\gamma \otimes g)z\big) &= (g_i \otimes \chi_i^{-1})(\gamma \otimes g)\,(x_i \otimes \varepsilon)(z) + (x_i \otimes \varepsilon)(\gamma \otimes g)\,(1 \otimes \varepsilon)(z)\\ &= \gamma(g_i)\chi_i^{-1}(g)\,(x_i \otimes \varepsilon)(z)\\ &= (x_i \otimes \varepsilon)(z).\end{aligned}$$

It follows that the generators of $A$ are contained in the image of $\pi^*$, so $A \subseteq \mathrm{Im}(\pi^*)$. Using Remark 8.3 we compute

$$\dim \mathrm{Im}(\pi^*) = \frac{\dim D(H)}{\dim K} = \frac{|\Gamma|^2 2^{2n}}{|\Sigma^\perp|} = |\Sigma|\, 2^{2n} = \dim A$$

and so $A = \mathrm{Im}(\pi^*)$. □

Recall the notion of the Drinfeld double of $V$ from Section 3.2. We have $D(V) = W \oplus W^* \in {}^{\Sigma}_{\Sigma}\mathcal{YD}$, where $W$ is the quantum linear space associated to the datum

$$\big((g_1, \chi_1^{-1}), \dots, (g_n, \chi_n^{-1}), \varphi_1, \dots, \varphi_n\big), \quad \text{with } \varphi_i \in \widehat{\Sigma},\ \varphi_i(g, \chi) = \chi_i(g),\ i = 1, \dots, n.$$

Define a bicharacter $r_\Sigma : \Sigma \times \Sigma \to k^\times$ by

$$r_\Sigma\left((g, \chi), (g', \chi')\right) = \chi'(g).$$

The diagonal of this bicharacter is a quadratic form $q_\Sigma : \Sigma \to k^\times$,

$$q_\Sigma(g, \chi) = \chi(g), \qquad (g, \gamma) \in \Sigma.$$

Then $D(V) \in \mathcal{Z}_{sym}(\mathcal{C}(\Sigma, r_\Sigma))_-$.

THEOREM 8.5. *Let $(\mathcal{P}, V, r)$ be a metric datum such that $V \in \mathcal{P}$ is self-dual. There is an equivalence of braided tensor categories:*

$$\mathcal{Z}\big(\mathcal{C}(\mathcal{P}, V, r)\big)_{\mathrm{ad}} \cong \mathcal{C}(\mathcal{C}(\Sigma, r_\Sigma), D(V), r_{D(V)}),$$

*where $r_{D(V)}$ is the canonical symplectic form on $D(V)$ defined in (22).*

PROOF. Let $H = \mathfrak{B}(V)\#k[\Gamma]$ and let $A$ be the Hopf subalgebra of $D(H)^*$ generated by the group-like elements $g_i \otimes \chi_i^{-1}$ and by the skew-primitive elements $x_i \otimes \varepsilon$ and $g_i \otimes r'(-, x_i)$, $i = 1, \ldots, n$, where $r'$ is an $r$-form on $H$ whose restriction to $V \otimes V$ is non-degenerate. Using Proposition 8.4, we have

$$\begin{aligned}\mathcal{Z}(\mathsf{Corep}(H))_{\mathrm{ad}} &= \big(\mathsf{Rep}(D(H)^{\mathrm{cop}})\big)_{\mathrm{ad}} \\ &= \mathsf{Rep}\left(A^{*\,\mathrm{cop}}\right) = \mathsf{Rep}\left(A^*\right)^{\mathrm{op}} \simeq \mathsf{Rep}(A^*) = \mathsf{Corep}(A).\end{aligned}$$

where the equivalence between $\mathsf{Rep}(A^*)$ and its opposite follows from the fact that $\mathsf{Rep}(A^*)$ is braided.

We claim that $A \cong \mathfrak{B}(D(V))\#k[\Sigma]$. Indeed, it is easy to check that for each $(g, \gamma) \in \Sigma$ we have

$$\begin{aligned}(g \otimes \gamma)(x_i \otimes \varepsilon) &= \chi_i^{-1}(g)(x_i \otimes \varepsilon)(g \otimes \gamma), \\ (g \otimes \gamma)(g_i \otimes r(-, x_i)) &= \chi_i^{-1}(g)(g_i \otimes r(-, x_i))(g \otimes \gamma), \\ (g_i \otimes r(-, x_i))(x_j \otimes \varepsilon) &= \chi_j^{-1}(g_i)(x_j \otimes \varepsilon)(g_i \otimes r(-, x_i)),\end{aligned}$$

for all $i$, $j = 1, \ldots, n$. Note that $W$ is self-dual because $V$ is self-dual. Thus, there exists a basis $\{y_i\}_{i=1}^{2n}$ of $D(V)$ such that $y_i$, $y_{n+i} \in D(V)^{\varphi_i}_{(g_i, \chi_i^{-1})}$. It follows from the above, that the map $A \to \mathfrak{B}(D(V))\#k[\Sigma]$, given by

$$g \otimes \gamma \mapsto (g, \gamma), \quad x_i \otimes \varepsilon \mapsto y_i, \quad g_i \otimes r(-, x_i) \mapsto y_{n+i}$$

for all $(g, \gamma) \in \Sigma$ and $i = 1, \ldots, n$, is a Hopf algebra isomorphism.

The braiding on $\mathcal{Z}\big(\mathcal{C}(\Gamma, q, V, r)\big)_{\mathrm{ad}}$ is obtained by restriction of the braiding of $\mathcal{Z}\big(\mathcal{C}(\Gamma, q, V, r)\big)$. It corresponds to the braiding on $\mathsf{Corep}(A)$ coming from the restriction to $A$ of the canonical $r$-form $r_{D(H)^*} : D(H)^* \otimes D(H)^* \to k$. on $D(H)^*$. The latter is given by

$$r_{D(H)^*}(\alpha, \beta) = \sum_{h \in \Gamma, P \subseteq \{1, \ldots, n\}} \alpha(\varepsilon \otimes hx_P)\beta((hx_P)^* \otimes 1), \quad \alpha, \beta \in D(H)^*.$$

We have

$$\begin{aligned}r_{D(H)^*}(g \otimes \gamma, g' \otimes \gamma') &= \sum_{h,P} \varepsilon(g)\gamma(hx_P)(hx_P)^*(g')\gamma'(1) = \gamma(g'), \\ r_{D(H)^*}(x_i \otimes \varepsilon, x_j \otimes \varepsilon) &= \sum_{h,P} \varepsilon(x_i)\varepsilon(hx_P)(hx_P)^*(x_j\varepsilon(1)) = 0, \\ r_{D(H)^*}\big(x_i \otimes \varepsilon, g_j \otimes r(-, x_j)\big) &= \sum_{h,P} \varepsilon(x_i)\varepsilon(hx_P)(hx_P)^*(g_j)r(1, x_j) = 0, \\ r_{D(H)^*}\big(g_j \otimes r(-, x_j), x_i \otimes \varepsilon\big) &= \sum_{h,P} \varepsilon(g_j)r(hx_P, x_j)(hx_P)^*(x_i)\varepsilon(1) = r(x_i, x_j), \\ r_{D(H)^*}\big(g_i \otimes r(-, x_i), g_j \otimes r(-, x_j)\big) &= \sum_{h,P} \varepsilon(g_i)r(hx_P, x_i)(hx_P)^*(g_j)r(1, x_j) = 0.\end{aligned}$$

Thus, $\mathsf{Corep}(A, r_{D(H)^*}|_{A \otimes A}) \simeq \mathcal{C}(\Sigma, q_\Sigma, D(V), s)$, where the quadratic form $q_\Sigma : \Sigma \to k^\times$ is given by $q_\Sigma(g, \gamma) = \gamma(g)$ for all $(g, \gamma) \in \Sigma$ and the matrix of $s : D(V) \otimes D(V) \to k$ with respect to the basis $\{y_i\}$ is the block matrix

$$\begin{pmatrix} 0 & 0 \\ X^t & 0 \end{pmatrix}.$$

Here $X^t$ is the transpose of the matrix $X = \big(s(x_i, x_j)\big)_{i,j}$. Changing $s$ by a cocycle deformation $s^\sigma$ will not change the braided equivalence class of $\mathcal{C}(\Sigma, q_\Sigma, D(V), s)$. As explained in Section 7.2, we can choose invariant $\sigma$ such that $s^\sigma$ is alternating. The matrix of $s^\sigma$ with respect to the basis $\{y_i\}$ is then

$$\frac{1}{2}\begin{pmatrix} 0 & -X \\ X^t & 0 \end{pmatrix}.$$

This matrix is easily seen to be congruent to $\begin{pmatrix} 0 & -I_n \\ I_n & 0 \end{pmatrix}$. So after a change of basis the matrices of $s^\sigma$ and $r_{D(V)}$ coincide. Therefore, $\mathsf{Corep}(A, r_{D(H)^*}|_{A\otimes A}) \simeq \mathcal{C}(\Sigma, q_\Sigma, D(V), r_{D(V)})$. □

Let $G$ be a finite group. Recall that a tensor category $\mathcal{C}$ is called a *$G$-extension* of a tensor category $\mathcal{A}$ if there is a faithful grading $\mathcal{C} = \bigoplus_{g\in G} \mathcal{C}_g$ such that $\mathcal{C}_e \cong \mathcal{A}$.

COROLLARY 8.6. *$\mathcal{Z}(\mathcal{C}(\mathcal{P}, V, r))$ is a $\widehat{\Sigma^\perp}$-extension of $\mathcal{C}(\mathcal{C}(\Sigma, r_\Sigma), D(V), r_{D(V)})$.*


## Acknowledgments.

The authors are grateful to Adriana Nenciu for helpful discussions.


## References


[A14] Nicolás Andruskiewitsch, *On finite-dimensional Hopf algebras*, Proceedings of the International Congress of Mathematicians—Seoul 2014. Vol. II, Kyung Moon Sa, Seoul, 2014, pp. 117–141. MR3728608

[AS98] N. Andruskiewitsch and H.-J. Schneider, *Lifting of quantum linear spaces and pointed Hopf algebras of order $p^3$*, J. Algebra **209** (1998), no. 2, 658–691, DOI 10.1006/jabr.1998.7643. MR1659895

[AS10] Nicolás Andruskiewitsch and Hans-Jürgen Schneider, *On the classification of finite-dimensional pointed Hopf algebras*, Ann. of Math. (2) **171** (2010), no. 1, 375–417, DOI 10.4007/annals.2010.171.375. MR2630042

[An13] Iván Angiono, *On Nichols algebras of diagonal type*, J. Reine Angew. Math. **683** (2013), 189–251. MR3181554

[Be03] M. Beattie, *Duals of pointed Hopf algebras*, J. Algebra **262** (2003), no. 1, 54–76, DOI 10.1016/S0021-8693(03)00034-6. MR1970802

[Bi06] Julien Bichon, *Galois and bigalois objects over monomial non-semisimple Hopf algebras*, J. Algebra Appl. **5** (2006), no. 5, 653–680, DOI 10.1142/S0219498806001934. MR2269410

[BC06] Julien Bichon and Giovanna Carnovale, *Lazy cohomology: an analogue of the Schur multiplier for arbitrary Hopf algebras*, J. Pure Appl. Algebra **204** (2006), no. 3, 627–665, DOI 10.1016/j.jpaa.2005.06.002. MR2185622

[BN15] Costel-Gabriel Bontea and Dmitri Nikshych, *On the Brauer-Picard group of a finite symmetric tensor category*, J. Algebra **440** (2015), 187–218, DOI 10.1016/j.jalgebra.2015.06.006. MR3373393

[CC04] G. Carnovale and J. Cuadra, *Cocycle twisting of $E(n)$-module algebras and applications to the Brauer group*, $K$-Theory **33** (2004), no. 3, 251–276, DOI 10.1007/s10977-004-5115-2. MR2138543

[Da10] A. Davydov, *Twisted automorphisms of Hopf algebras*, Noncommutative structures in mathematics and physics, K. Vlaam. Acad. Belgie Wet. Kunsten (KVAB), Brussels, 2010, pp. 103–130. MR2742734

[DN13] Alexei Davydov and Dmitri Nikshych, *The Picard crossed module of a braided tensor category*, Algebra Number Theory **7** (2013), no. 6, 1365–1403, DOI 10.2140/ant.2013.7.1365. MR3107567

[De02] P. Deligne, *Catégories tensorielles* (French, with English summary), Mosc. Math. J. **2** (2002), no. 2, 227–248. Dedicated to Yuri I. Manin on the occasion of his 65th birthday. MR1944506

[DGNO10] Vladimir Drinfeld, Shlomo Gelaki, Dmitri Nikshych, and Victor Ostrik, *On braided fusion categories. I*, Selecta Math. (N.S.) **16** (2010), no. 1, 1–119, DOI 10.1007/s00029-010-0017-z. MR2609644

[EGNO15] Pavel Etingof, Shlomo Gelaki, Dmitri Nikshych, and Victor Ostrik, *Tensor categories*, Mathematical Surveys and Monographs, vol. 205, American Mathematical Society, Providence, RI, 2015. MR3242743

[GN08] Shlomo Gelaki and Dmitri Nikshych, *Nilpotent fusion categories*, Adv. Math. **217** (2008), no. 3, 1053–1071, DOI 10.1016/j.aim.2007.08.001. MR2383894

[JS93] André Joyal and Ross Street, *Braided tensor categories*, Adv. Math. **102** (1993), no. 1, 20–78, DOI 10.1006/aima.1993.1055. MR1250465

[Ma01] Akira Masuoka, *Defending the negated Kaplansky conjecture*, Proc. Amer. Math. Soc. **129** (2001), no. 11, 3185–3192, DOI 10.1090/S0002-9939-01-06005-1. MR1844991

[Mo11] Martín Mombelli, *Representations of tensor categories coming from quantum linear spaces*, J. Lond. Math. Soc. (2) **83** (2011), no. 1, 19–35, DOI 10.1112/jlms/jdq059. MR2763942

[M93] Susan Montgomery, *Hopf algebras and their actions on rings*, CBMS Regional Conference Series in Mathematics, vol. 82, Published for the Conference Board of the Mathematical Sciences, Washington, DC; by the American Mathematical Society, Providence, RI, 1993. MR1243637

[Na06] Sonia Natale, *R-matrices and Hopf algebra quotients*, Int. Math. Res. Not., posted on 2006, Art. ID 47182, 18, DOI 10.1155/IMRN/2006/47182. MR2250024

[Ne04] Adriana Nenciu, *Quasitriangular pointed Hopf algebras constructed by Ore extensions*, Algebr. Represent. Theory **7** (2004), no. 2, 159–172, DOI 10.1023/B:ALGE.0000026785.03997.60. MR2063007

[Ni78] Warren D. Nichols, *Bialgebras of type one*, Comm. Algebra **6** (1978), no. 15, 1521–1552, DOI 10.1080/00927877808822306. MR0506406

[PvO99] Florin Panaite and Freddy Van Oystaeyen, *Quasitriangular structures for some pointed Hopf algebras of dimension* $2^n$, Comm. Algebra **27** (1999), no. 10, 4929–4942, DOI 10.1080/00927879908826739. MR1701714

[R93] David E. Radford, *Minimal quasitriangular Hopf algebras*, J. Algebra **157** (1993), no. 2, 285–315, DOI 10.1006/jabr.1993.1102. MR1220770

[R94] David E. Radford, *On Kauffman's knot invariants arising from finite-dimensional Hopf algebras*, Advances in Hopf algebras (Chicago, IL, 1992), Lecture Notes in Pure and Appl. Math., vol. 158, Dekker, New York, 1994, pp. 205–266. MR1289427

[R11] David E. Radford, *Hopf algebras*, Series on Knots and Everything, vol. 49, World Scientific Publishing Co. Pte. Ltd., Hackensack, NJ, 2012. MR2894855



Department of Mathematics and Statistics, University of New Hampshire, Durham, NH 03824, USA

*Email address*: costel.bontea@gmail.com

Department of Mathematics and Statistics, University of New Hampshire, Durham, NH 03824, USA

*Email address*: dmitri.nikshych@unh.edu

Contemporary Mathematics
Volume **728**, 2019
https://doi.org/10.1090/conm/728/14657

# Eigenvalues of the squared antipode in finite dimensional weak Hopf algebras

Pavel Etingof

**With an appendix by Gregor Schaumann**

ABSTRACT. We extend Schaumann's theory of pivotal structures on fusion categories matched to a module category and of module traces to the case of non-semisimple tensor categories, and use it to study eigenvalues of the squared antipode $S^2$ in weak Hopf algebras. In particular, we diagonalize $S^2$ for semisimple weak Hopf algebras in characteristic zero, generalizing the result of Nikshych in the pseudounitary case. We show that the answer depends only on the Grothendieck group data of the pivotalizations of the categories involved and the global dimension of the fusion category (thus, all eigenvalues belong to the corresponding number field). On the other hand, we study the eigenvalues of $S^2$ on the non-semisimple weak Hopf algebras attached to dynamical quantum groups at roots of 1 defined by D. Nikshych and the author, and show that they depend nontrivially on the continuous parameters of the corresponding module category. We then compute these eigenvalues as rational functions of these parameters. The paper also contains an appendix by G. Schaumann discussing the connection between our generalization of module traces and the notion of an inner product module category introduced in *Pivotal tricategories and a categorification of inner-product modules* in a 2015 issue of Algebras and Representation Theory (see [Sch2]).

## 1. Introduction

Let $\mathbf{k}$ be an algebraically closed field. Let $\mathcal{C}$ be a finite tensor category over $\mathbf{k}$, and $\mathcal{M}$ a semisimple indecomposable (left) module category over $\mathcal{C}$. Let $A$ be a commutative semisimple algebra such that $\mathcal{M} = A-\text{mod}$, i.e., $A = \oplus_{i\in I}\mathbf{k}$, where $I$ is the set labeling the simple objects $M_i$ of $\mathcal{M}$. Then the action of $\mathcal{C}$ on $\mathcal{M}$ gives rise to a tensor functor $F : \mathcal{M} \to A-\text{bimod}$, and the algebra $H := \text{End}_{\mathbf{k}}F$ is a regular biconnected weak Hopf algebra with base $A$ (where $\text{End}_{\mathbf{k}}F = \text{End}\overline{F}$ is the endomorphism algebra of the composition $\overline{F}$ of $F$ with the forgetful functor from $A$-bimodules to vector spaces). Conversely, if $H$ is a regular biconnected weak Hopf algebra with commutative base $A$ then $\mathcal{C} = \text{Rep}\, H$ is a finite tensor category, and $\mathcal{M} = A-\text{mod}$ is an indecomposable semisimple $\mathcal{C}$-module category, and these assignments are inverse to each other (see [**N1**] and [**EGNO**], Subsection 7.23).

This work of the author of the main paper was partially supported by the NSF grant DMS-1502244. The work of the author of the appendix was partially supported by the stand-alone project P 27513-N27 of the Austrian Science Fund.

Recall that $H$ has a natural automorphism – the squared antipode $S^2$, which is semisimple in characteristic zero ([**N1**], Corollary 5.15). Therefore, the characteristic polynomial of $S^2$ is an invariant of the pair $(\mathcal{C}, \mathcal{M})$. The goal of this paper is to study this invariant.

If $\mathcal{M} = \mathrm{Vec}$ is the category of vector spaces, then $H$ is a finite dimensional Hopf algebra, and by Radford's theorem $S^2$ has finite order dividing $2 \dim H$. In particular, the eigenvalues of $S^2$ are roots of unity of order dividing $2 \dim H$. A natural question is whether this statement extends to the case of weak Hopf algebras.

Consider first the case when $H$ is semisimple (i.e., $\mathcal{C}$ a fusion category) and $\mathrm{char}(\mathbf{k}) = 0$. In this case, it follows from Ocneanu rigidity ([**EGNO**], Subsection 9.1) that the eigenvalues of $S^2$ are algebraic numbers, but getting more detailed information about them requires further study. When $\mathcal{C}$ is pseudounitary, this problem was solved by D. Nikshych in [**N2**]. Namely, it is shown in [**N2**] that in this case the eigenvalues of $S^2$, while in general not roots of unity, can be computed explicitly, and in fact depend only on the combinatorial information – the $\mathbb{Z}_+$-module $\mathrm{Gr}(\mathcal{M})$ over the Grothendieck ring $\mathrm{Gr}(\mathcal{C})$ of $\mathcal{C}$. More precisely, the characteristic polynomial of $S^2$ has the form

$$\chi(z) = \prod_{i,j,k,l \in I} (z - \lambda_{ijkl})^{n_{ijkl}}, \tag{1}$$

with

$$\lambda_{ijkl} := \frac{m_j m_l}{m_i m_k}, \quad n_{ijkl} := \sum_{r \in J} N_{ri}^j N_{rl}^k,$$

where $m_i$ are the Frobenius-Perron dimensions of $M_i$, $J$ labels the simple objects $X_r$ of $\mathcal{C}$, and

$$N_{ri}^j := \dim \mathrm{Hom}(X_r \otimes M_i, M_j).$$

If a fusion category $\mathcal{C}$ is not pseudounitary, the situation is more complicated. So let us first assume that the squared antipode is the conjugation by a trivial group-like element, i.e., $S^2(x) = gxg^{-1}$, where $g = yS(y)^{-1}$ and $y \in A_s \cong A$ is an invertible element of the source base of $H$ (this holds in all the examples we know). In this case, following Schaumann ([**Sch1**]), we will say that the pivotal structure on $\mathcal{C}$ defined by $g$ is *matched* to the module category $\mathcal{M}$; we show that our definition is equivalent to that of [**Sch1**]. In the pseudounitary case it is shown in [**N2**] that this is automatic. In general, in the matched case it follows from [**Sch1**] that the characteristic polynomial of $S^2$ is still given by formula (1), but with $m_i$ now being nonzero numbers such that for all $r, i$ we have

$$\sum_{j \in I} N_{ri}^j m_j = \dim(X_r) m_i,$$

where $\dim(X_i)$ is the dimension of $X_i$ in the pivotal structure defined by $g$ (i.e., $N_r \mathbf{m} = \dim(X_r)\mathbf{m}$, where $N_r := (N_{ri}^j)$ and $\mathbf{m} = (m_i)$). More precisely, the numbers $m_i$ are uniquely determined up to scaling (as the dimension character $X \mapsto \dim(X)$ occurs with multiplicity one in the module $\mathrm{Gr}(\mathcal{M}) \otimes \mathbf{k}$ over $\mathrm{Gr}(\mathcal{C}) \otimes \mathbf{k}$), and are all nonzero. We show that this answer depends only on the global dimension $\dim(\mathcal{C})$ of $\mathcal{C}$ and $\mathrm{Gr}(\mathcal{M})$ as a $\mathbb{Z}_+$-module over $\mathrm{Gr}(\mathcal{C})$.

For a general fusion category $\mathcal{C}$ (i.e., without assuming the existence of a matched pivotal structure) we obtain a slightly more complicated but similar answer (see Theorem 3.6). This answer depends only on $\dim(\mathcal{C})$ and $\mathrm{Gr}(\widetilde{\mathcal{M}})$ as a $\mathbb{Z}_+$-module over $\mathrm{Gr}(\widetilde{\mathcal{C}})$, where $\widetilde{\mathcal{C}}$ and $\widetilde{\mathcal{M}}$ are the pivotalizations of $\mathcal{C}, \mathcal{M}$.

We also extend these results to the non-semisimple case (for $\mathrm{char}(\mathbf{k}) = 0$). In this case, Ocneanu rigidity is no longer valid, and module categories over $\mathcal{C}$ (as well as $\mathcal{C}$ itself) may depend on continuous parameters. So one may wonder if the eigenvalues of $S^2$ may also depend on these parameters (and hence fail to be algebraic numbers, in general). We again consider the setting when $\mathcal{M}$ is matched to the pivotal structure on $\mathcal{C}$, and show that if the dimension character occurs with multiplicity (at most) one in the module $\mathrm{Gr}(\mathcal{M}) \otimes \mathbf{k}$ over $\mathrm{Gr}(\mathcal{C}) \otimes \mathbf{k}$ (e.g., if $\mathrm{Gr}(\mathcal{M}) \otimes \mathbf{k}$ is a cyclic module over $\mathrm{Gr}(\mathcal{C}) \otimes \mathbf{k}$) then this does not happen and the eigenvalues of $S^2$ are algebraic numbers which we compute explicitly. Namely, we show that the characteristic polynomial of $S^2$ is still given by (1), but with

$$n_{ijkl} := \sum_{q,r \in J} N_{qi}^{j} C_{qr} N_{rl}^{k},$$

where $C = (C_{qr})$ is the Cartan matrix of $\mathcal{C}$. A similar result holds in positive characteristic.

On the other hand, we consider the example of dynamical quantum groups at roots of 1 introduced in [**EN**]. In this case, the pivotal structure is matched to the corresponding module category, but the dimension character turns out to occur more than once in the module $\mathrm{Gr}(\mathcal{M}) \otimes \mathbf{k}$. As a result, the particular choice of the eigenvector $\mathbf{m} = (m_i)$ of $X_r$ with eigenvalues given by this character actually depends on the continuous parameter of the module category (an element $\Lambda$ of the maximal torus of the corresponding simple algebraic group), and we obtain a formula for the eigenvalues of $S^2$ as rational functions of $\Lambda$.

The organization of the paper is as follows. In Section 2 we recall the results of Schaumann on pivotal structures matched to module categories and extend them to the non-semisimple case. We also give two other equivalent definitions of this notion. In Section 3 we prove the formula for the characteristic polynomial of $S^2$ in the case when the dimension character occurs at most once in $\mathrm{Gr}(\mathcal{M}) \otimes \mathbf{k}$, and use it to obtain a formula for the characteristic polynomial of $S^2$ for arbitrary semisimple $\mathcal{C}, \mathcal{M}$ when $\mathrm{char}(\mathbf{k}) = 0$. In Section 4 we consider the example of dynamical quantum groups and compute the eigenvalues of $S^2$ in this example as rational functions of the parameters.

We note that the notion of a module trace has been generalized by G. Schaumann to the case when $\mathcal{C}$ is a finite tensor category and $\mathcal{M}$ an exact (not necessarily semisimple) $\mathcal{C}$-module category, [**Sch2**]; namely, he introduced the notion of an inner product module category ([**Sch2**], Definition 5.2). Specifically, $\mathcal{M}$ is an inner product module category if and only if the pivotal structure on $\mathcal{C}$ extends to a pivotal structure on the category $\mathcal{C}^{*\mathrm{op}}_{\mathcal{C} \oplus \mathcal{M}}$. G. Schaumann showed that this generalization coincides with ours when $\mathcal{C}$ is unimodular (i.e., the distinguished invertible object of $\mathcal{C}$ is $\mathbf{1}$), up to switching the pivotal structure to the conjugate one, and also when $\mathcal{C}^*_{\mathcal{M}}$ is unimodular, but they differ in general. The paper contains an appendix by G. Schaumann explaining the exact relationship between these two generalizations.

## 2. Pivotal structures matched to a module category

We start with giving a straightforward generalization of the results of [**Sch1**] to the setting of non-semisimple tensor categories in any characteristic. So this subsection is essentially an exposition of the results of [**Sch1**] in a slightly more general setting using a somewhat different approach.

**2.1. Definitions and examples of matched pivotal structures.** Let $\mathcal{C}$ be a finite tensor category over an algebraically closed field $\mathbf{k}$. Let $\mathcal{M}$ be a finite semisimple indecomposable module category over $\mathcal{C}$. In other words, $\mathcal{M}$ is a finite semisimple category and we are given a tensor functor $F : \mathcal{C} \to \mathrm{End}\mathcal{M}$. Let $\mathcal{C}^*_{\mathcal{M}} = \mathrm{Fun}_{\mathcal{C}}(\mathcal{M}, \mathcal{M})$ be the dual category. (For basics on tensor categories, module categories and dual categories, see [**EGNO**]).

Let $a : \mathrm{Id} \to **$ be a tensor isomorphism, i.e., a pivotal structure on $\mathcal{C}$. Then we can canonically define an invertible object $\chi(a, \mathcal{M}) \in \mathcal{C}^*_{\mathcal{M}}$ as follows. For $V \in \mathrm{End}\mathcal{M}$, we may naturally identify $V$ with $V^{**}$, so for $X \in \mathcal{C}$ we have a natural isomorphism $F(X^{**}) \cong F(X)^{**} \cong F(X)$. Hence we have a canonical identification $X^{**} \otimes M \cong X \otimes M$, $M \in \mathcal{M}$. Thus the pivotal structure $a$ defines a functorial automorphism

$$a \otimes 1 : X \otimes M \to X^{**} \otimes M = X \otimes M,$$

i.e., an invertible module functor $\mathcal{M} \to \mathcal{M}$ (with the underlying additive functor being the identity). But such a functor is nothing but an invertible object of the dual category $\mathcal{C}^*_{\mathcal{M}}$, which we denote by $\chi(a, \mathcal{M})$.

DEFINITION 2.1. We say that $a$ is matched to $\mathcal{M}$ if $\chi(a, \mathcal{M}) = \mathbf{1}$.

This definition can be extended to not necessarily indecomposable categories in an obvious way. Namely, if $\mathcal{M} = \oplus_i \mathcal{M}_i$ and $\mathcal{M}_i$ are indecomposable then $a$ is matched to $\mathcal{M}$ if and only if it is matched to each $\mathcal{M}_i$.

One can reformulate this definition in terms of weak Hopf algebras. (For basics on weak Hopf algebras, see [**N1**] and references therein). Namely, let $A$ be a commutative semisimple algebra such that $\mathcal{M} = A - \mathrm{mod}$, so that $\mathrm{End}\mathcal{M} = A -$ bimod. Let $F : \mathcal{C} \to A -$ bimod be the corresponding tensor functor. Recall that to $\mathcal{C}, \mathcal{M}$ we can associate the biconnected regular Frobenius weak Hopf algebra $H := \mathrm{End}_{\mathbf{k}} F = \mathrm{End}\overline{F}$ with bases $A_s, A_t \cong A$, and $\mathcal{C} = \mathrm{Rep}\, H$ (here $\overline{F}$ is the composition of $F$ with the forgetful functor $A -$bimod $\to$ Vec). A pivotal structure $a$ on $\mathcal{C}$ then gives rise to a grouplike element $g_a \in H$ such that $g_a x g_a^{-1} = S^2(x)$, $x \in H$, called the *pivotal* element. Such an element defines an invertible $H^*$-module ([**N1**], Section 4), which is exactly the object $\chi(a, \mathcal{M}) \in \mathrm{Rep}\, H^*_{\mathrm{op}} = \mathcal{C}^*_{\mathcal{M}}$ ([**ENO1**], p.590; [**O**], Theorem 4.2; [**EGNO**], Example 7.12.26). [1] Thus, we have the following proposition, giving an equivalent definition of a pivotal structure matched to $\mathcal{M}$:

PROPOSITION 2.2. *The pivotal structure $a$ is matched to $\mathcal{M}$ if and only if the corresponding pivotal element $g_a$ is a trivial grouplike element of $H$, i.e., $g_a = yS(y)^{-1}$ where $y \in A_s$ is invertible.*

Another definition of a matched pivotal structure (in fact, several equivalent definitions) is given by G. Schaumann in [**Sch1**]. Namely, fix a collection of numbers

[1]Note that in Example 7.12.26 of [**EGNO**], there is a misprint: $H^*$ should be $H^*_{\mathrm{op}}$ or $H^{*\mathrm{cop}}$.

$\mathbf{m} = (m_i, i \in I)$ in $\mathbf{k}$, and for any $M \in \mathcal{M}$ and endomorphism $f : M \to M$, define its trace

$$\mathrm{Tr}_M^{\mathbf{m}}(f) := \sum_{i \in I} m_i \mathrm{Tr}(f|_{\mathrm{Hom}(M_i, M)}).$$

DEFINITION 2.3. ([**Sch1**]) The trace $\mathrm{Tr}^{\mathbf{m}}$ is said to be a *module trace* on $\mathcal{M}$ with respect to a pivotal structure $a$ on $\mathcal{C}$ if for any $X \in \mathcal{C}$, $M \in M$ and morphism $f : X \otimes M \to X \otimes M$ one has

$$\mathrm{Tr}_{X \otimes M}^{\mathbf{m}}(f) = \mathrm{Tr}_M^{\mathbf{m}}(\mathrm{Tr}_X(f)),$$

where $\mathrm{Tr}_X(f) : M \to M$ is the composition

$$M \xrightarrow{\mathrm{coev}_X \otimes 1} X \otimes X^* \otimes M \xrightarrow{a_X \otimes f} X^{**} \otimes X^* \otimes M \xrightarrow{\mathrm{ev}_{X^*} \otimes 1} M.$$

PROPOSITION 2.4. *The pivotal structure $a$ is matched to $\mathcal{M}$ if and only if $\mathcal{M}$ admits a nonzero module trace with respect to $a$. In this case, the module trace is unique up to scaling.*

PROOF. Suppose that $a$ is matched to $M$, and let $y \in A = A_s$ be such that $g_a = yS(y)^{-1}$ in the corresponding weak Hopf algebra $H$ (it exists by Proposition 2.2). Recall that $A = \oplus_{i \in I} \mathbf{k}$, and let $m_i \in \mathbf{k}$ be the projection of $y$ onto the $i$-th summand. We claim that $\mathrm{Tr}^{\mathbf{m}}$ is a module trace. Indeed, let $X \in \mathcal{C}$ and $f : X \otimes M_i \to X \otimes M_i$ be a morphism. Then we have

$$\mathrm{Tr}^{\mathbf{m}}(f) = \sum_{j \in I} \beta_{ij} m_j,$$

where

$$\beta_{ij} := \mathrm{Tr}(f|_{\mathrm{Hom}(M_j, X \otimes M_i)}).$$

It is clear that it suffices to check the module trace property for simple $M$, so it is enough to check that

$$\mathrm{Tr}_X(f) = m_i^{-1} \sum_{j \in I} \beta_{ij} m_j.$$

Let $F(X)$ denote the $A$-bimodule corresponding to $X$. Then

$$F(X) = \oplus_{i,j \in I} \mathrm{Hom}(M_j, X \otimes M_i).$$

Thus

$$X \otimes M_i = \oplus_{j \in I} \mathrm{Hom}(M_j, X \otimes M_i) \otimes M_j$$

and $g_a = yS(y)^{-1}$ acts on the summand $\mathrm{Hom}(M_j, X \otimes M_i) \otimes M_j$ by multiplication by $m_j m_i^{-1}$. Thus,

$$\mathrm{Tr}_X(f) = \mathrm{Tr}(f \circ (g \otimes 1)) = m_i^{-1} \sum_{j \in I} \beta_{ij} m_j,$$

as desired.

Conversely, suppose that $\mathrm{Tr}^{\mathbf{m}}$ is a nonzero module trace on $\mathcal{M}$. We claim that $m_i \neq 0$ for all $i$. Indeed, if $m_i = 0$ for some $i$ then for any $X \in \mathcal{C}$, the trace of any endomorphism of $X \otimes M_i$ is zero. But any object of $\mathcal{M}$ is a direct summand of $X \otimes M_i$ for some $X$, which implies that $m_j = 0$ for all $j$.

Thus, we have an invertible element $y \in A = A_s$ whose projection to the $i$-th summand of $A$ is $m_i$. Let $g = yS(y)^{-1} \in H$. It is clear that $g_a$ and $g$ act naturally on $\mathrm{Hom}(M_j, X \otimes M_i)$ for each $i, j$, and the module trace property implies that

$$\mathrm{Tr}((g_a - g) \circ f|_{\mathrm{Hom}(M_j, X \otimes M_i)}) = 0$$

for any $X \in \mathcal{C}$ and $f : X \otimes M_i \to X \otimes M_i$. This implies that $g_a - g$ acts by zero on $\mathrm{Hom}(M_j, X \otimes M_i)$ for all $X, i, j$, i.e., $g_a = g = yS(y)^{-1}$, as desired.

The uniqueness of the module trace up to scaling (when it exists) follows from the fact proved above that for a module trace, $m_i \neq 0$ for all $i$ (as in [**Sch1**]). □

Now observe that any collection of numbers $m_i \neq 0$ gives rise to a functorial isomorphism $\phi^{\mathbf{m}}_{MN} : \mathrm{Hom}(M,N) \cong \mathrm{Hom}(N,M)^*$ for $M, N \in \mathcal{M}$ defined by the pairing $\mathrm{Hom}(M,N) \otimes \mathrm{Hom}(N,M) \to \mathbf{k}$ given by

$$\langle f, h \rangle := \mathrm{Tr}^{\mathbf{m}}_N(f \circ h)$$

(note that $\mathrm{Tr}^{\mathbf{m}}_N(f \circ h) = \mathrm{Tr}^{\mathbf{m}}_M(h \circ f)$). Moreover, any functorial isomorphism $\mathrm{Hom}(M,N) \cong \mathrm{Hom}(N,M)^*$ is uniquely obtained in this way.

Definition 2.5. A functorial isomorphism

$$\phi_{MN} : \mathrm{Hom}(M,N) \to \mathrm{Hom}(N,M)^*$$

is said to be $\mathcal{C}$-balanced (for a given pivotal structure $a : X \to X^{**}$ on $\mathcal{C}$) if for each $X \in \mathcal{C}$ the composite isomorphism

$$\mathrm{Hom}(X \otimes M, N) \xrightarrow{\phi_{X\otimes M,N}} \mathrm{Hom}(N, X \otimes M)^* \cong \mathrm{Hom}(X^* \otimes N, M)^*$$

$$\xrightarrow{\phi^{-1}_{M,X^*\otimes N}} \mathrm{Hom}(M, X^* \otimes N) \cong \mathrm{Hom}(X^{**} \otimes M, N) \xrightarrow{\circ(a_X \otimes 1)} \mathrm{Hom}(X \otimes M, N)$$

is the identity.

Proposition 2.6. *([**Sch1**], Theorem 4.2) The isomorphism $\phi^{\mathbf{m}}$ is $\mathcal{C}$-balanced if and only if $Tr^{\mathbf{m}}$ is a module trace.*

Proof. The proof is the same as the proof of Theorem 4.2 in [**Sch1**]. □

Thus, thanks to Propositions 2.2, 2.4, 2.6 we have four equivalent definitions of a pivotal structure matched to a module category.

Example 2.7. 1. Let $\mathcal{C} = \mathrm{Rep}\, H$, where $H$ is a finite dimensional Hopf algebra over $\mathbf{k}$, and $\mathcal{M} = \mathrm{Vec}$, so that $F$ is the corresponding fiber functor. Then a pivotal structure $a$ is defined by a grouplike element $g_a \in H$ such that $g_a x g_a^{-1} = S^2(x)$, and $\mathcal{C}^*_{\mathcal{M}} = \mathrm{Rep}\, H^*_{\mathrm{op}}$, so $\chi(a, \mathcal{M}) \in H$ is a grouplike element. Recall that an $H^*_{\mathrm{op}}$-module $V$ gives rise to the $\mathcal{C}$-linear functor $\mathrm{Vec} \to \mathrm{Vec}$ given by $F_V(\mathbf{k}) = V$ with the map $X \otimes V \to X \otimes V$ for $X \in \mathcal{C}$ defined by $x \otimes v \mapsto \rho(v)(x \otimes 1)$, where $\rho : V \to H \otimes V$ is the left $H$-comodule structure attached to the right $H^*$-module structure on $V$. Therefore, we have $\chi(a, \mathcal{M}) = g_a$. Thus, $a$ is matched to $\mathcal{M}$ if and only if $g_a = 1$. In this case $S^2 = \mathrm{Id}$, so $a$ is the canonical spherical structure, and $\mathcal{C}$ is semisimple if $\mathrm{char}(\mathbf{k}) = 0$ by the Larson-Radford theorem. In positive characteristic $\mathcal{C}$ need not be semisimple (e.g., $H$ can be a group algebra).

2. If $\mathcal{M} = \mathcal{C}$ for $\mathcal{C}$ semisimple then it is easy to see that $\chi(a, \mathcal{M}) = \mathbf{1}$ for any pivotal structure $a$, so all pivotal structures are automatically matched to $\mathcal{M}$ (and the corresponding module trace is just the left trace in $\mathcal{C}$).

3. If $\mathcal{C}$ is semisimple and pseudounitary over $\mathbf{k} = \mathbb{C}$ and $a$ is the canonical spherical structure on $\mathcal{C}$, then $a$ is matched to $\mathcal{M}$, since $S^2$ is given by conjugation by a trivial grouplike element, as shown in [**N2**].

4. ([**Sch1**], Example 3.13). Let $\mathrm{Vec}^{\omega}_G$ be the category of $G$-graded finite dimensional vector spaces over $\mathbf{k}$ with associativity defined by a 3-cocycle $\omega$ on $G$.

Then pivotal structures on $\mathcal{C}$ correspond to characters $\kappa : G \to \mathbf{k}^\times$. Also, indecomposable module categories over $\mathcal{C}$ are $\mathcal{M}(H, \psi)$, where $H \subset G$ is a subgroup and $\psi$ a 2-cochain on $H$ such that $d\psi = \omega|_H$ ([**EGNO**], Example 9.7.2). It is not hard to show that $a$ is matched to $\mathcal{M}(H, \psi)$ if and only if $\kappa|_H = 1$ (as shown in [**Sch1**]).

5. ([**Sch1**], Example 3.14). Let $\mathcal{C}$ be a fusion category, and $\widetilde{\mathcal{C}}$ be the pivotalization of $\mathcal{C}$, [**EGNO**], Definition 7.21.9. Thus we have a forgetful tensor functor $\widetilde{\mathcal{C}} \to \mathcal{C}$, hence $\mathcal{C}$ becomes a $\widetilde{\mathcal{C}}$-module category, and the dual category is the semidirect product $\mathrm{Vec}_{\mathbb{Z}/2} \ltimes \mathcal{C}$. Let $a$ be the canonical pivotal structure on $\widetilde{\mathcal{C}}$. Then $\chi(a, \mathcal{C})$ is the invertible object of $\mathrm{Vec}_{\mathbb{Z}/2}$ corresponding to the generator $1 \in \mathbb{Z}/2$. Thus, $a$ is not matched to $\mathcal{C}$.

6. If $F : \mathcal{C} \to \mathcal{D}$ is a pivotal tensor functor between pivotal finite tensor categories and the pivotal structure of $\mathcal{D}$ is matched to $\mathcal{M}$ then the pivotal structure of $\mathcal{C}$ is also matched to $\mathcal{M}$.

**2.2. Properties of $m_i$.** Let $X_r$ be the simple objects of $\mathcal{C}$ and $N_r$ be the matrix of multiplication by $X_r$ in the basis $(M_i)$. Assume that $\mathcal{C}$ is equipped with a pivotal structure $a$ matched to $\mathcal{M}$, and let $\mathbf{m} = (m_i)$ be the corresponding vector.

PROPOSITION 2.8. *([**Sch1**]) One has*

$$N_r\mathbf{m} = \dim(X_r)\mathbf{m}.$$

PROOF. Consider the module trace condition for the identity morphism:

$$\mathrm{Tr}^{\mathbf{m}}_{X_r \otimes M_i}(\mathrm{Id}) = \mathrm{Tr}^{\mathbf{m}}_{M_i}(\mathrm{Tr}_{X_r}(\mathrm{Id}))$$

for $X \in \mathcal{C}, M \in \mathcal{M}$. This yields

$$\sum_{j \in I} N_{ri}^j m_j = \dim(X_r) m_i,$$

or in the matrix form $N_r\mathbf{m} = \dim(X_r)\mathbf{m}$, as desired. □

Consider now the Grothendieck group $\mathrm{Gr}(\mathcal{M}) \otimes \mathbf{k}$ as a module over $\mathrm{Gr}(\mathcal{C}) \otimes \mathbf{k}$. Note that in characteristic zero this is a semisimple module (even for non-semisimple $\mathcal{C}$), since for $\mathbf{k} = \mathbb{C}$ it has a positive definite invariant Hermitian form in which $M_i$ are orthonormal. Using Proposition 2.8, we obtain the following proposition.

PROPOSITION 2.9. *If the pivotal structure $a$ is matched to $\mathcal{M}$ then the dimension character $X \mapsto \dim(X)$ occurs in the decomposition of the $\mathrm{Gr}(\mathcal{C}) \otimes \mathbf{k}$-module $\mathrm{Gr}(\mathcal{M}) \otimes \mathbf{k}$, and $\mathbf{m}$ is an eigenvector corresponding to this character. Moreover, if this character occurs with multiplicity 1, then this determines $\mathbf{m}$ up to a scalar.*

EXAMPLE 2.10. If $\mathcal{C}$ is pseudounitary then the dimension character occurs in $\mathrm{Gr}(\mathcal{M}) \otimes \mathbf{k}$ with multiplicity 1 by the Frobenius-Perron theorem, so $m_i = \mathrm{FPdim}(M_i)$, as shown in [**N2**].

**2.3. Tensor isomorphisms matched to a module category.** Let us generalize Definition 2.1 to more general tensor isomorphisms. Let $r \in \mathbb{Z}$ and $b_X : X \to X^{(2r)}$ be a functorial tensor isomorphism (where $X^{(2r)}$ denotes the $2r$-th dual of $X$). Then we can define an invertible object $\chi(b, \mathcal{M}) \in \mathcal{C}^*_{\mathcal{M}}$ similarly to the above.

DEFINITION 2.11. We say that $b$ is matched to $\mathcal{M}$ if $\chi(b, \mathcal{M}) = \mathbf{1}$.

In the language of weak Hopf algebras, this means that $S^{2r}(x) = h_b x h_b^{-1}$, where $h_b \in H$ is a grouplike element, and the matched condition means that $h_b = zS(z)^{-1}$ is a trivial grouplike element (for some invertible $z \in A_s$).

It is clear that

$$\chi(b_1 \circ b_2, \mathcal{M}) = \chi(b_1, \mathcal{M}) \otimes \chi(b_2, \mathcal{M}) = \chi(b_2, \mathcal{M}) \otimes \chi(b_1, \mathcal{M})$$

(as $b_1 \circ b_2 = b_2 \circ b_1$) so if $b_1, b_2$ are matched to $\mathcal{M}$ then so is $b_1 \circ b_2$.

EXAMPLE 2.12. 1. (V. Ostrik) Let $b_X : X \to X$ be a tensor automorphism of the identity functor of $\mathcal{C}$ (i.e., $r = 0$). Then $b$ canonically defines an invertible object $Z_b$ in the Drinfeld center $\mathcal{Z}(\mathcal{C})$ of $\mathcal{C}$ which is $\mathbf{1}$ as an object of $\mathcal{C}$ (so $b_X$ is the corresponding isomorphism between $Z_b \otimes X$ and $X \otimes Z_b$). It follows from the definition that $\chi(b, \mathcal{M})$ is the image of $Z_b$ in $\mathcal{C}^*_{\mathcal{M}}$.

2. It is easy to show that $b$ is matched to $\mathcal{M}$ if and only if there is an automorphism $b_{\mathcal{M}}$ of the identity functor of $\mathcal{M}$ compatible with $b$ (which is necessarily unique and also has order $d$; in particular, $d \neq 0$ in $\mathbf{k}$). In other words, $b$ defines a faithful $\mathbb{Z}/d$-grading on $\mathcal{C}$, and $b_{\mathcal{M}}$ defines a compatible faithful $\mathbb{Z}/d$-grading on $\mathcal{M}$ (where $d$ is the order of $b$).

Now let $a, a'$ be two pivotal structures on $\mathcal{C}$ matched to $\mathcal{M}$. Then $a' = a \circ b$, where $b$ is a tensor automorphism of the identity fiunctor of $\mathcal{C}$ matched to $\mathcal{M}$. Thus there is a compatible automorphism $b_{\mathcal{M}}$ of the identity functor of $\mathcal{M}$. Let $b_i$ be the scalar by which $b_{\mathcal{M}}$ acts on $M_i$ (a root of unity of order dividing $d$). Then the numbers $m_i'$ corresponding to the pivotal structure $a'$ are related to the numbers $m_i$ corresponding to $a$ by the formula $m_i' = m_i b_i$.

EXAMPLE 2.13. Let $\mathcal{C}$ be a finite tensor category and $\mathcal{M}$ be a semisimple indecomposable $\mathcal{C}$-module category. Let $D$ be the dual distinguished invertible object of $\mathcal{C}$ and $\iota : X \to D^{-1} \otimes X^{****} \otimes D$ be the categorical Radford isomorphism (see [**EGNO**], Theorem 7.19.1). Let $d$ be the least common multiple of the order order of $D$ and the order of the (dual) distinguished invertible object of $\mathcal{C}^*_{\mathcal{M}}$. Then the tensor isomorphism $b := \iota^d : X \to X^{(4d)}$ is matched to $\mathcal{M}$. Indeed, by the Radford $S^4$ formula for weak Hopf algebras ([**N1**]), the operator $S^{4d}$ is given by conjugation by a trivial grouplike element.

In particular, if $\mathcal{C}$ is semisimple then $d = 1$ ([**N1**]), so $\iota$ is matched to $\mathcal{M}$. In other words, in a semisimple weak Hopf algebra $S^4(x) = hxh^{-1}$ where $h$ is a trivial grouplike element, as shown in [**N2**]. Thus, if $a$ is a spherical structure then $\chi(a, M)$ has order $\leq 2$ (as $a^2 = \iota$).

**2.4. The dual pivotal structure on the dual category.** It is shown in [**Sch1**] that if $\mathcal{C}$ is a finite tensor category with pivotal structure $a$ and $\mathcal{M}$ a semisimple indecomposable $\mathcal{C}$-module category such that $a$ is matched to $\mathcal{M}$ then the dual category $\mathcal{C}^*_{\mathcal{M}}$ acquires a canonical "dual" pivotal structure $\overline{a}$, also matched to $\mathcal{M}$, with the same module trace. This gives rise to a pivotal structure on $\mathcal{C}^{*\mathrm{op}}_{\mathcal{M}}$ which we will also denote by $\overline{a}$. An easy way to see this is to note that since in the corresponding weak Hopf algebra $H$, one has $S^2(x) = g_a x g_a^{-1}$ where $g_a = yS(y)^{-1}$, a similar formula holds in $H^*$ ([**N1**], Proposition 4.8).

For example, if $\mathcal{M} = \mathcal{C}$ for $\mathcal{C}$ semisimple, we get a dual pivotal structure $\overline{a}$ on $\mathcal{C}$ such that the left dimensions of objects with respect to $a$ are equal to their right dimensions with respect to $\overline{a}$, and vice versa. Namely, one has $\overline{a} = \iota \circ a^{-1}$. Thus, if $a$ is matched to a module category $\mathcal{N}$ then so is $\overline{a}$, and $a$ is a spherical structure if and only if $a = \overline{a}$. In particular, for $\mathbf{k} = \mathbb{C}$, since the left and right dimensions are complex conjugate, we get that the left dimensions with respect to $a$ are complex conjugate to the left dimensions with respect to $\overline{a}$ (as shown in [**Sch1**]).

EXAMPLE 2.14. Let $\mathcal{C} = \mathcal{E} \boxtimes \mathcal{E}^{\mathrm{op}}$ and $\mathcal{M} = \mathcal{E}$ for a fusion category $\mathcal{E}$. For a pair of pivotal structures $a, a'$ on $\mathcal{E}$, denote by $a \boxtimes a'$ the corresponding pivotal structure on $\mathcal{C}$. Then it is easy to show that $\chi(a \boxtimes a', \mathcal{M}) = Z_b$ (see Example 2.12), where $b = 1$ iff $\overline{a} = a'$. Thus $a \boxtimes a'$ is matched to $\mathcal{M}$ if and only if $a' = \overline{a}$.

More generally, for any finite tensor category $\mathcal{C}$ and indecomposable semisimple $\mathcal{C}$-module $\mathcal{M}$, regard $\mathcal{M}$ as a module over $\mathcal{C} \boxtimes \mathcal{C}^*_{\mathcal{M}}$. Let $a, a'$ be pivotal structures on $\mathcal{C}$ and $\mathcal{C}^*_{\mathcal{M}}$, such that $a$ is matched to $\mathcal{M}$, and let $a \boxtimes a'$ be their product. Then $\chi(a \boxtimes a', \mathcal{M})$ is an object of the Drinfeld center $\mathcal{Z}(\mathcal{C})$ which projects to $\mathbf{1} = \chi(a, \mathcal{M}) \in \mathcal{C}^*_{\mathcal{M}}$. Hence $\chi(a \boxtimes a', \mathcal{M}) = Z_b$ for some $b \in \mathrm{Aut}_{\otimes}(\mathrm{Id}_{\mathcal{C}^*_{\mathcal{M}}})$. It is easy to show that $b = 1$ iff $a' = \overline{a}$. Hence $a \boxtimes a'$ is matched to $\mathcal{M}$ if and only if $a' = \overline{a}$.

**2.5. Matched pivotal structures for fusion categories over fields of characteristic zero.** Let $\mathcal{C}$ be a fusion category over a field $\mathbf{k}$ of characteristic zero, and $\mathcal{M}$ an indecomposable semisimple $\mathcal{C}$-module category. Recall that $X_r$, $r \in J$ denote the simple objects of $\mathcal{C}$. Let $a$ be a pivotal structure on $\mathcal{C}$ and $\dim(X_r)$ be the dimension of $X_r$ in this pivotal structure. Let

$$Q := \sum_{r \in J} \dim(X_r^*) X_r \in \mathrm{Gr}(\mathcal{C}) \otimes \mathbf{k}.$$

It is easy to check that $Q^2 = \dim(\mathcal{C}) Q$, where

$$\dim(\mathcal{C}) = \sum_{r \in J} \dim(X_r) \dim(X_r^*)$$

is the global dimension of $\mathcal{C}$, which is independent on the pivotal structure $a$ and is nonzero by [**EGNO**], Theorem 7.21.12.

LEMMA 2.15. *([**Sch1**], Proposition 5.7) If $a$ is matched to $\mathcal{M}$ then one has*

$$Tr(Q|_{\mathrm{Gr}(\mathcal{M})}) = \dim(\mathcal{C}).$$

*Thus, $Q|_{\mathrm{Gr}(\mathcal{M})}$ has rank* 1.

PROOF. By [**EGNO**], Proposition 7.12.28, for any $M, N, P \in \mathcal{M}$ we have a canonical isomorphism

$$\underline{\mathrm{Hom}}_{\mathcal{C}}(N, P) \otimes M \cong {}^*\underline{\mathrm{Hom}}_{\mathcal{C}^*_{\mathcal{M}}}(M, N) \otimes P.$$

Recall that

$$\underline{\mathrm{Hom}}_{\mathcal{C}}(N, P) = \oplus_{r \in J} \mathrm{Hom}(X_r \otimes N, P) \otimes X_r$$

Thus the above isomorphism can be rewritten as

$$\oplus_{r \in J} \mathrm{Hom}(X_r \otimes N, P) \otimes X_r \otimes M \cong \oplus_{r \in J} \mathrm{Hom}(X_r \otimes M, N)^* \otimes X_r^* \otimes P.$$

In particular, setting $N = P = M_i$, we get

$$\oplus_{r \in J} \mathrm{Hom}(X_r \otimes M_i, M_i) \otimes X_r \otimes M \cong \oplus_{r \in J} \mathrm{Hom}(X_r \otimes M, M_i)^* \otimes X_r^* \otimes M_i.$$

Thus, summing over $i$, we get

$$\oplus_{r \in J} \mathrm{Tr}(X_r|_{\mathrm{Gr}(\mathcal{M})}) X_r \otimes M \cong \oplus_{r \in J, i \in I} \mathrm{Hom}(X_r \otimes M, M_i)^* \otimes X_r^* \otimes M_i.$$

Let us regard this as an identity in $\mathrm{Gr}(\mathcal{M})\otimes\mathbf{k}$, and let $M=\sum_{i\in I} m_iM_i$, where $\mathbf{m}=(m_i)$ is the vector defining the module trace on $\mathcal{M}$. Then the above identity takes the form

$$\sum_{r\in J}\mathrm{Tr}(X_r|_{\mathrm{Gr}(\mathcal{M})})\dim(X_r)M=\sum_{r\in J,i\in I}\dim(X_r)\dim\mathrm{Hom}(M,M_i)X_r^*M_i,$$

i.e., since $[X_r^*]=[X_r]^T$ in the standard basis of $\mathrm{Gr}(\mathcal{M})$, we have

$$\mathrm{Tr}(Q|_{\mathrm{Gr}(\mathcal{M})})M=\dim(\mathcal{C})M,$$

which yields

$$\mathrm{Tr}(Q|_{\mathrm{Gr}(\mathcal{M})})=\dim(\mathcal{C}),$$

as desired. □

COROLLARY 2.16. *(***[Sch1]***) The dimension character occurs with multiplicity* 1 *in the* $\mathrm{Gr}(\mathcal{C})\otimes\mathbf{k}$*-module* $\mathrm{Gr}(\mathcal{M})\otimes\mathbf{k}$*, so the vector* $\mathbf{m}$ *is uniquely determined by the equation* $N_r\mathbf{m}=\dim(X_r)\mathbf{m}$.

PROOF. This follows directly from Lemma 2.15. □

Let $\overline{\mathbf{m}}=(\overline{m}_i)$ be a vector defining a module trace on $\mathcal{M}$ for the pivotal structure $\overline{a}$ on $\mathcal{C}$ (the dual pivotal structure to $a$ with respect to $\mathcal{C}$ as a $\mathcal{C}$-module) defined in Subsection 2.4.

PROPOSITION 2.17. *We have* $\sum_{i\in I} m_i\overline{m}_i\neq 0$.

PROOF. Denote by $(,)$ the inner product on $\mathrm{Gr}(\mathcal{M})$ in which $M_i$ are orthonormal. Let $M=\sum_{i\in I} m_iM_i$, $\overline{M}=\sum_{i\in I}\overline{m}_iM_i$. Then we have $Q^*M=\dim(\mathcal{C})M$, hence

$$(M,QM_j)=(Q^*M,M_j)=\dim(\mathcal{C})(M,M_j)=\dim(\mathcal{C})m_j,$$

which implies that $QM_j=\beta_j\overline{M}$, where

$$(\sum_{i\in I} m_i\overline{m}_i)\beta_j=\dim(\mathcal{C})m_j,$$

implying that $\sum_{i\in I} m_i\overline{m_i}\neq 0$. □

Proposition 2.17 shows that we can normalize $m_i$ and $\overline{m}_i$ so that $\sum_{i\in I} m_i\overline{m}_i=\dim(\mathcal{C})$, i.e., $\beta_j=m_j$. This normalization is unique up to the action of $\mathbf{k}^\times$ by $\mathbf{m}\to\lambda\mathbf{m}$ and $\overline{\mathbf{m}}\mapsto\lambda^{-1}\overline{\mathbf{m}}$. Let us make such a normalization from now on, and call it a *pivotal normalization*.

Note that if $\mathbf{k}=\mathbb{C}$ then a pivotal normalization can be fixed so that $\overline{m}_i$ are complex conjugate to $m_i$, which justifies the notation. Note also that if $a$ is a spherical structure then there is a unique up to sign pivotal normalization such that $m_i=\overline{m_i}$ (so that $m_i\in\mathbb{R}$ if $\mathbf{k}=\mathbb{C}$). Let us call such a pivotal normalization *spherical*.

PROPOSITION 2.18. *For a pivotal normalization, one has*

$$\dim\underline{\mathrm{Hom}}_{\mathcal{C}}(M_i,M_j)=\overline{m}_im_j.$$

*Moreover, this property characterizes pivotal normalizations.*

PROOF. It is easy to see that $\dim\underline{\mathrm{Hom}}_{\mathcal{C}}(M_i,M_j)=(M_i,QM_j)$, but $QM_j=m_j\overline{M}$, so $(M_i,QM_j)=\overline{m}_im_j$. The second statement is clear. □

In the pseudounitary case this recovers the theory of Frobenius-Perron dimensions in module categories given in [**ENO2**], Subsection 2.5.

**2.6. Pivotal structures on multifusion categories and matched pivotal structures.** Let $\mathcal{C}$ be a fusion category over a field $\mathbf{k}$ of characteristic zero, and $\mathcal{M}$ be a semisimple indecomposable $\mathcal{C}$-module category. Then $\mathcal{D} := \mathcal{C}^{*\mathrm{op}}_{\mathcal{C}\oplus\mathcal{M}}$ is a multifusion category. Namely, we have $\mathcal{D} = \mathcal{D}_{11} \oplus \mathcal{D}_{12} \oplus \mathcal{D}_{21} \oplus \mathcal{D}_{22}$, where $\mathcal{D}_{11} = \mathcal{C}$, $\mathcal{D}_{12} = \mathcal{M}$, $\mathcal{D}_{21} = \mathcal{M}^{\vee} = \mathrm{Fun}_{\mathcal{C}}(\mathcal{M}, \mathcal{C})$ (which is the opposite category of $\mathcal{M}$ viewed as a right $\mathcal{C}$-module via duality), and $\mathcal{D}_{22} = \mathcal{C}^{*\mathrm{op}}_{\mathcal{M}}$.

PROPOSITION 2.19. *(i) Any pivotal structure on $\mathcal{D}$ gives rise to a pivotal structure on $\mathcal{C}$ matched to $\mathcal{M}$ and a module trace on $\mathcal{M}$.*

*(ii) Let $a$ be a pivotal structure on $\mathcal{C}$ matched to $\mathcal{M}$, and fix the corresponding vectors $\mathbf{m} = (m_i)$, $\overline{\mathbf{m}} = (\overline{m}_i)$ using a pivotal normalization. Then the category $\mathcal{D}$ has a pivotal structure which on $\mathcal{D}_{11}$ is defined by $a$, on $\mathcal{D}_{22}$ by $\overline{a}$, and the left traces on $\mathcal{D}_{12}$ and $\mathcal{D}_{21}$ are given by $\mathbf{m}$, $\overline{\mathbf{m}}$, respectively. Moreover, this pivotal structure is spherical if and only if $a$ is a spherical structure and the normalization of $\mathbf{m}$, $\overline{\mathbf{m}}$ is spherical.*

PROOF. (i) The pivotal structure on $\mathcal{C}$ is given by the restriction of the pivotal structure on $\mathcal{D}$ to $\mathcal{D}_{11}$, while the module trace on $\mathcal{M}$ is the left trace on $\mathcal{D}_{12}$.

(ii) In $\mathcal{D}$ we have $M_j \otimes M_i^* = \underline{\mathrm{Hom}}_{\mathcal{C}}(M_i, M_j)$ and $M_i^* \otimes M_j = \underline{\mathrm{Hom}}_{\mathcal{C}^*_{\mathcal{M}}}(M_i, M_j)$, where $M_i^*$ is $M_i$ viewed as an object of $\mathcal{M}^{\vee}$. Therefore, the statement follows from Proposition 2.18. □

**2.7. Existence of matched pivotal structures.** The following question is raised in [**Sch1**].

QUESTION 2.20. ([**Sch1**], end of Section 3) Let $\mathcal{C}$ be a fusion category over $\mathbf{k}$ (and $\mathrm{char}(\mathbf{k}) = 0$). Let $\mathcal{M}$ be a semisimple indecomposable $\mathcal{C}$-module category. Does $\mathcal{C}$ admit a pivotal structure matched to $\mathcal{M}$? In other words, is $S^2$ a conjugation by a trivial grouplike element of the corresponding semisimple weak Hopf algebra $H$?

By Proposition 2.19(i), a positive answer to Question 2.20 follows from the well known conjecture that any multifusion category admits a pivotal structure (see [**EGNO**], p.180).

Note that the answer is "yes" for spherical fusion categories $\mathcal{C}$ when $\mathcal{C}^*_{\mathcal{M}}$ has no invertible objects of order 2; in this case any spherical structure on $\mathcal{C}$ is automatically matched to $\mathcal{M}$, as $\chi(a, \mathcal{M}) = \mathbf{1}$ by Example 2.13. For instance, this happens if $\mathcal{C}$ has an odd Frobenius-Perron dimension $d$ (i.e., such that the norm $N(d)$ of $d$ is odd). Indeed, in this case the dual category $\mathcal{C}^*_{\mathcal{M}}$ (which has the same dimension $d$) cannot have invertible objects of order 2, as the order of an invertible object must divide $d$ ([**EGNO**], Theorem 7.17.6).

On the other hand, if $\mathcal{C}$ is not semisimple, $\mathrm{char}(\mathbf{k}) = 0$ and $\mathcal{M} = \mathrm{Vec}$, then a matched pivotal structure does not exist, since in the corresponding Hopf algebra $S^2 \neq \mathrm{Id}$ (and any trivial grouplike element equals 1 in this case).

## 3. Eigenvalues of $S^2$

**3.1. The case of a matched pivotal structure.** Let $\mathcal{C}$ be a finite tensor category, $\mathcal{M}$ an indecomposable semisimple $\mathcal{C}$-module category, and $a$ a pivotal structure on $\mathcal{C}$ matched to $\mathcal{M}$. Let $H$ be the corresponding weak Hopf algebra, and let us compute the eigenvalues of $S^2$ on $H$.

As explained above, we have $S^2(x) = g_a x g_a^{-1}$, where $g_a = yS(y)^{-1}$ is the pivotal element; in particular $S^2$ is semisimple. Let $m_i \in \mathbf{k}$ be the projections of $y$ to the summands of $A$.

PROPOSITION 3.1. *The characteristic polynomial of* $S^2 : H \to H$ *has the form*

$$\chi(z) = \prod_{i,j,k,l \in I} (z - \lambda_{ijkl})^{n_{ijkl}}, \tag{2}$$

*with*

$$\lambda_{ijkl} := \frac{m_j m_l}{m_i m_k}$$

*and*

$$n_{ijkl} := \sum_{q,r \in J} N_{qi}^j C_{qr} N_{rl}^k,$$

*where* $C = (C_{qr})$ *is the Cartan matrix of* $\mathcal{C}$.

PROOF. Let $F : \mathcal{C} \to \mathrm{End}\mathcal{M}$ be the monoidal functor corresponding to $\mathcal{M}$, and $\overline{F} : \mathcal{C} \to \mathrm{Vec}$ its composition with the forgetful functor. Let $P_r$ be the projective cover of $X_r$. Then the functor $\overline{F}$ is represented by the object $P := \oplus_{r \in J} \overline{F}(X_r^*) \otimes P_r$. Thus

$$H = \mathrm{End}\overline{F} = \mathrm{End}P = \oplus_{q,r \in J} \overline{F}(X_q) \otimes \mathrm{Hom}(P_q, P_r) \otimes \overline{F}(X_r^*).$$

We may view $H$ as a bimodule over $A_s \otimes A_t$, using left and right multiplication by elements of $A_s$ and $A_t$. Let $e_i, i \in A$ be the primitive idempotents of $A_s$, so that $S(e_i)$ are the primitive idempotents of $A_t$. Then for $X \in \mathcal{C}$ we have

$$e_j F(X) e_i = \mathrm{Hom}(M_j, X \otimes M_i).$$

Thus,

$$V_{ijkl} := e_j S(e_k) H e_i S(e_l) =$$
$$\oplus_{q,r \in J} \mathrm{Hom}(M_j, X_q \otimes M_i) \otimes \mathrm{Hom}(P_q, P_r) \otimes \mathrm{Hom}(M_l, X_r^* \otimes M_k).$$

The operator $S^2$ acts on the space $V_{ijkl}$ by the scalar $\frac{m_j m_l}{m_i m_k}$, and

$$\dim V_{ijkl} = \sum_{q,r \in J} N_{qi}^j C_{qr} N_{rl}^k = n_{ijkl}.$$

This implies the statement since $H = \oplus_{i,j,k,l} V_{ijkl}$. □

EXAMPLE 3.2. Let $T_n$ be the Taft Hopf algebra of dimension $n^2$ generated by a grouplike element $g$ such that $g^n = 1$ and an element $x$ such that $x^n = 0$ and $gxg^{-1} = qx$, where $q$ is a primitive $n$-th root of unity ($\mathrm{char}(\mathbf{k}) = 0$). The coproduct is defined by $\Delta(x) = 1 \otimes x + x \otimes g$. Then $T_n$ contains the group algebra $\mathbf{k}[\mathbb{Z}/n]$ generated by $g$. So we have a tensor functor $\mathrm{Res} : \mathcal{C} = \mathrm{Rep}\, T_n \to \mathcal{M} = \mathrm{Rep}\, \mathbb{Z}/n$, in particular giving $\mathcal{M}$ a structure of a semisimple $\mathcal{C}$-module. We have $S(x) = -xg^{-1}$, so $S^2(x) = gxg^{-1}$. This implies that the element $g$ defines a pivotal structure on $\mathcal{C}$.

We claim that this pivotal structure is matched to $\mathcal{M}$. Indeed, this follows immediately from the fact that the tensor functor $\mathrm{Res} : \mathcal{C} \to \mathcal{M}$ admits a splitting $\mathcal{M} \to \mathcal{C}$ coming from the natural homomorphism $T_n \to \mathbf{k}[\mathbb{Z}/n]$. The corresponding numbers $m_i$ equal $q^i$ (for the natural labeling of objects). It is clear that the dimension character occurs in $\mathrm{Gr}(\mathcal{M})$ with multiplicity 1, since $\mathrm{Gr}(\mathcal{C}) = \mathrm{Gr}(\mathcal{M})$.

Thus in this case by Proposition 3.1 the operator $S^2$ has eigenvalues $q^i$ on $H$, with equal multiplicities $n^3$.

COROLLARY 3.3. *If $\mathcal{C}$ is semisimple over a field of characteristic zero then the characteristic polynomial of $S^2$ on $H$ is given by Proposition 3.1 with $m_i$ determined uniquely up to scaling by the formula*

$$N_r\mathbf{m} = \dim(X_r)\mathbf{m}.$$

PROOF. This follows from Proposition 3.1 and Corollary 2.16. □

EXAMPLE 3.4. Let $\mathcal{C}$ be a pivotal fusion category and $\mathcal{M} = \mathcal{C}$. Then $m_i = d_i$, where $d_i = \dim(X_i)$. Thus the characteristic polynomial of $S^2$ on $H$ is

$$\chi(z) = \prod_{i,j,k,l\in J} (z - \lambda_{ijkl})^{n_{ijkl}},$$

where $n_{ijkl} = \dim \mathrm{Hom}(X_j \otimes X_k, X_i \otimes X_l)$ and $\lambda_{ijkl} = \frac{d_j d_k}{d_i d_l}$ is the eigenvalue of $S^2$ on $\mathrm{Hom}(X_j \otimes X_k, X_i \otimes X_l)$.

Note that this polynomial does not depend on the choice of a pivotal structure. Indeed, any two pivotal structures $a, a'$ differ by a tensor automorphism $b$ of the identity functor, so the corresponding dimensions are related by the formula $d_i' = d_i b_i$, where $b_i = b|_{X_i}$. But if $n_{ijkl} \neq 0$ then $b_j b_k = b_i b_l$, as desired.

**3.2. The case of a general fusion category over a field of characteristic zero.** Let us now compute the eigenvalues of $S^2$ on $H$ in the case when $\mathcal{C}$ is any fusion category over a field $\mathbf{k}$ of characteristic zero and $\mathcal{M}$ is any indecomposable semisimple $\mathcal{C}$-module category (without assuming that $\mathcal{C}$ has a pivotal structure). Recall that we have a multifusion category $\mathcal{D} = \mathcal{C}^{*\mathrm{op}}_{\mathcal{C}\oplus\mathcal{M}}$. Let $\widetilde{\mathcal{D}}$ be its pivotalization, defined [**EGNO**], Subsection 7.21. Then $\widetilde{\mathcal{D}}$ is a spherical category, with $\widetilde{\mathcal{D}}_{11} = \widetilde{\mathcal{C}}$ (the pivotalization of $\mathcal{C}$) and $\widetilde{\mathcal{D}}_{12} =: \widetilde{\mathcal{M}}$, which is an indecomposable semisimple $\widetilde{\mathcal{C}}$-module category.

DEFINITION 3.5. We will call $\widetilde{\mathcal{M}}$ the pivotalization of $\mathcal{M}$.

Note that $\widetilde{\mathcal{M}}$ has a canonical forgetful functor $\widetilde{\mathcal{M}} \to \mathcal{M}$ compatible with the forgetful functor $\widetilde{\mathcal{C}} \to \mathcal{C}$. Also, by Proposition 2.9(i), the canonical pivotal structure on $\widetilde{\mathcal{C}}$ is matched to $\widetilde{\mathcal{M}}$.

For convenience assume that $\mathbf{k} = \mathbb{C}$ (this is not essential). Let $X_r^{\pm}$ be the simple objects of $\widetilde{\mathcal{C}}$ of positive, respectively negative dimensions which map to $X_r \in \mathcal{C}$ under the forgetful functor. Similarly, let $M_i^{\pm}$ be the simple objects of $\widetilde{\mathcal{M}}$ of positive, respectively negative dimensions which map to $M_i \in \mathcal{M}$ under the forgetful functor.

Let $\nu_i = |M_i|^2 > 0$ be the Müger squared norms of the simple objects of $\mathcal{M} = \mathcal{D}_{12}$ (recall that they can be defined for any multifusion category, see [**EGNO**], Definition 7.21.2). Then $m_i^{\pm} = \dim M_i^{\pm}$ are given by the formula

$$m_i^{\pm} = \pm\sqrt{\nu_i}.$$

Let $N_{ri}^{j\pm} = \dim \mathrm{Hom}(X_r^+ \otimes M_i^+, M_j^\pm)$. Let

$$n_{ijkl}^+ := \sum_{r\in J} (N_{ri}^{j+} N_{rl}^{k+} + N_{ri}^{j-} N_{rl}^{k-}),$$

$$n_{ijkl}^- := \sum_{r\in J} (N_{ri}^{j-} N_{rl}^{k+} + N_{ri}^{j+} N_{rl}^{k-}).$$

Let

$$\lambda_{ijkl} = \sqrt{\frac{\nu_j \nu_k}{\nu_i \nu_l}}.$$

One of our main results is the following theorem.

THEOREM 3.6. *The characteristic polynomial of $S^2$ on $H$ is*

$$\chi(z) = \prod_{i,j,k,l\in I} (z - \lambda_{ijkl})^{n_{ijkl}^+} (z + \lambda_{ijkl})^{n_{ijkl}^-}.$$

PROOF. Let $\widetilde{H}$ be the weak Hopf algebra attached to $(\widetilde{\mathcal{C}}, \widetilde{\mathcal{M}})$. We have

$$\widetilde{H} = \oplus_{i,j,k,l,s_1,s_2,s_3,s_4} V_{ijkl}^{s_1 s_2 s_3 s_4},$$

where $s_p = \pm$ for $p = 1, 2, 3, 4$ and

$$V_{ijkl}^{s_1 s_2 s_3 s_4} := \oplus_{r\in J, s=\pm} \mathrm{Hom}(M_j^{s_1}, X_r^s \otimes M_i^{s_2}) \otimes \mathrm{Hom}(M_l^{s_3}, (X_r^s)^* \otimes M_k^{s_4}).$$

Moreover, it is easy to see that the space $V_{ijkl}^{s_1 s_2 s_3 s_4}$ depends up to a natural isomorphism only on $s = s_1 s_2 s_3 s_4$, so we will denote it by $V_{ijkl}^s$. We have $\dim V_{ijkl}^s = n_{ijkl}^s$. By Proposition 3.1, $S^2$ acts on $V_{ijkl}^+$ by the scalar $\lambda_{ijkl}$ and on $V_{ijkl}^-$ by $-\lambda_{ijkl}$.

Now let

$$V_{ijkl} = \oplus_{r\in J} \mathrm{Hom}(M_j, X_r \otimes M_i) \otimes \mathrm{Hom}(M_l, X_r^* \otimes M_k).$$

Then $H = \oplus_{i,j,k,l} V_{ijkl}$, and we have a natural isomorphism $V_{ijkl} \cong V_{ijkl}^+ \oplus V_{ijkl}^-$. Moreover, it is easy to show that the squared antipode $S^2$ of $H$ acts by $\lambda_{ijkl}$ on $V_{ijkl}^+$ and by $-\lambda_{ijkl}$ on $V_{ijkl}^-$. This implies the theorem. □

## 4. Eigenvalues of $S^2$ for dynamical quantum groups at roots of 1

**4.1. The $\mathfrak{sl}_2$-case.** If the category $\mathcal{C}$ is not semisimple, then the dimension character may occur more than once in the $\mathrm{Gr}(\mathcal{C}) \otimes \mathbf{k}$-module $\mathrm{Gr}(\mathcal{M}) \otimes \mathbf{k}$, which leads to the non-uniqueness (even up to scaling) of a solution $\mathbf{m}$ of the equations $N_r \mathbf{m} = \dim(X_r)\mathbf{m}$.

As an example consider $\mathcal{C} = \mathrm{Rep}\, \mathfrak{u}_q(\mathfrak{sl}_2)$ over $\mathbb{C}$, where $\mathfrak{u}_q(\mathfrak{sl}_2)$ is the small quantum group at a primitive $\ell$-th root of unity $q = e^{2\pi i s/\ell}$, $-\ell/2 < s < \ell/2$, with $\ell$ odd. Namely, $\mathfrak{u}_q(\mathfrak{sl}_2)$ is the Hopf algebra generated by $E, F, K^{\pm 1}$ with defining relations

$$KE = q^2 EK,\ KF = q^{-2} FK,\ EF - FE = \frac{K - K^{-1}}{q - q^{-1}},$$

$$E^\ell = F^\ell = K^\ell - 1 = 0$$

with coproduct given by

$$\Delta(E) = E \otimes 1 + K \otimes E,\ \Delta(F) = F \otimes K^{-1} + 1 \otimes F,\ \Delta(K) = K \otimes K,$$

and the counit and antipode given by

$$\varepsilon(E) = \varepsilon(F) = 0,\ \varepsilon(K) = 1,\ S(E) = -K^{-1}E,\ S(F) = -FK,\ S(K) = K^{-1}.$$

Thus, $S^2(x) = K^{-1}xK$, so $\mathcal{C}$ has a pivotal structure defined by the element $K^{-1}$. In this case, $\mathrm{Gr}(\mathcal{C})$ has a basis $X_1 = \mathbf{1}, ..., X_\ell$, where $X_i$ is the irreducible $i$-dimensional representation, and the tensor product rule for $X_i$ is given in [**EGNO**], Subsection 5.8:

$$X_2X_i = X_{i+1} + X_{i-1},\ i \le \ell - 1; X_2X_\ell = 2X_1 + 2X_{\ell-1}.$$

Thus $\mathrm{Gr}(\mathcal{C})$ is generated by $X = X_2$, and we have $X_j = P_j(X)$, where $P_j$ is the Chebysheff polynomial of the second kind defined by $P_j(2\cos\theta) = \frac{\sin j\theta}{\sin\theta}$ (note that it has integer coefficients). This implies that $Q(X) = 0$, where

$$Q(X) = XP_\ell(X) - 2P_{\ell-1}(X) - 2.$$

Writing $X = z + z^{-1}$, we get

$$(z + z^{-1})(z^{\ell-1} + ... + z^{-\ell+1}) - 2(z^{\ell-2} + ... + z^{-\ell+2}) - 2 = 0,$$

i.e.,

$$z^\ell + z^{-\ell} - 2 = 0.$$

This polynomial has double roots at all $\ell$-th roots of 1, which means that $Q$ has double roots at $\alpha_j = 2\cos(\pi i j/\ell)$, $j = 1, ..., (\ell-1)/2$, and a simple root at $\alpha_0 = 2$, i.e.,

$$Q(x) = (x - \alpha_0)\prod_{i=1}^{\frac{\ell-1}{2}}(x - \alpha_j)^2,$$

So $\mathrm{Gr}(\mathcal{C}) = \mathbb{Z}[x]/(Q)$ and the categorical dimension character corresponds to evaluating at $x = \alpha_s$ (where $\alpha_s = \alpha_{-s}$ if $s < 0$), while the Frobenius-Perron dimension character corresponds to evaluating at $x = \alpha_0$.

Recall that we have a cyclic group $\mathbb{Z}/\ell$ inside $\mathfrak{u}_q(\mathfrak{sl}_2)$ generated by the element $K$. Thus, $\mathrm{Gr}(\mathcal{C})$ acts naturally on $\mathrm{Gr}(\mathrm{Rep}\,\mathbb{Z}/\ell) = \mathbb{Z}[\chi]/(\chi^\ell - 1)$ by $x \circ f = (\chi + \chi^{-1})f$. It is easy to see that this module is just the associated graded of $\mathrm{Gr}(\mathcal{C})$ with respect to the (2-step) radical filtration. In particular, the categorical dimension character (unlike the Frobenius-Perron dimension character) occurs in $\mathrm{Gr}(\mathcal{M}) \otimes \mathbb{C}$ with multiplicity 2.

Let us compute the possible vectors $\mathbf{m}$. The equations for the coordinates $m_i$ of $\mathbf{m}$ are

$$m_{j+1} + m_{j-1} = (q + q^{-1})m_j,$$

where $j \in \mathbb{Z}/\ell$. Hence

$$m_j = aq^j - bq^{-j},$$

$a, b \in \mathbb{C}$. Moreover, we must have $m_j \neq 0$ for all $j$, hence $\Lambda^\ell \neq 1$, where $\Lambda = a/b \in \mathbb{CP}^1$. Thus, up to scaling we have a 1-parameter family of possibilities parametrized by $\Lambda$.

It turns out that all these possibilities are realized. Namely, let $\Lambda \in \mathbb{C}^*$ be not an $\ell$-th root of 1, and let $\mathcal{J}_\Lambda(\lambda)$ be the dynamical twist defined in [**EN**], Example 5.1.10, and let $H_{\mathcal{J}_\Lambda} = H_\Lambda$ be the corresponding weak Hopf algebra ([**EN**], Subsection 4.2). Namely, $H_\Lambda = \mathrm{End} A \otimes \mathfrak{u}_q(\mathfrak{sl}_2)$ as an algebra, where $A = \mathbb{C}[\mathbb{Z}/\ell]$ is the base, with coproduct twisted by $\mathcal{J}_\Lambda$. It corresponds to a $\mathcal{C}$-module category $\mathcal{M}_\Lambda$ such that $\mathrm{Gr}(\mathcal{M}_\Lambda) = \mathrm{Gr}(\mathcal{M})$ as a $\mathrm{Gr}(\mathcal{C})$-module. By Theorem 5.3.2 of [**EN**], this weak Hopf algebra is self-dual, i.e., $H_\Lambda \cong H_\Lambda^{*\mathrm{cop}}$, so that $\mathcal{C}_\mathcal{M}^{*\mathrm{op}} \cong \mathcal{C}^{\mathrm{op}} \cong \mathcal{C}$ (as $\mathcal{C}$ is braided). Since $\mathcal{C}$ does not have nontrivial invertible objects, we get $\chi(a, \mathcal{M}_\Lambda) = \mathbf{1}$ (where $a$ is the unique ribbon pivotal structure on $\mathcal{C}$), i.e., $a$ is matched to $\mathcal{M}_\Lambda$.

Thus, $S^2$ on $H_\Lambda$ is the conjugation by the trivial grouplike element $g = yS(y)^{-1}$, where $y \in A_s \cong \mathbb{C}[\mathbb{Z}/\ell]$.

Let us compute $g$.

PROPOSITION 4.1. *One has*

$$g = yS(y)^{-1},$$

*where* $y = (y_1, ..., y_\ell) \in A_s$ *and*

$$y_i = \Lambda q^i - q^{-i}.$$

*In particular, the characteristic polynomial of* $S^2$ *on* $H_\Lambda$ *is*

$$\chi(z) = \prod_{i,j,k,l \in \mathbb{Z}/\ell} \left( z - \frac{(\Lambda q^j - q^{-j})(\Lambda q^k - q^{-k})}{(\Lambda q^i - q^{-i})(\Lambda q^l - q^{-l})} \right)^\ell.$$

The proof of Proposition 4.1 is given in the next subsection.

Thus, unlike the semisimple case, the eigenvalues of $S^2$ are nontrivial functions of the continuous parameter $\Lambda$.

In particular, consider the limiting points $\Lambda = 0, \infty$. At these points, there exist limiting weak Hopf algebras $H_0$, $H_\infty$ which correspond to the module category $\mathcal{M} = \mathcal{M}_0 = \mathrm{Rep}(\mathbb{Z}/\ell)$ over $\mathfrak{u}_q(\mathfrak{sl}_2)$ and a similar category $\mathcal{M}_\infty$ for $q$ replaced by $q^{-1}$, obtained by twisting of $\mathcal{M}_0$ by the R-matrix of $\mathfrak{u}_q(\mathfrak{sl}_2)$. In these cases, the characteristic polynomial of $S^2$ on $H_\Lambda$ is

$$\chi(z) = (z^\ell - 1)^{\ell^4}.$$

**4.2. The case of an arbitrary simple Lie algebra.** Proposition 4.1 is a special case of Proposition 4.2 below, which applies to any simple Lie algebra $\mathfrak{g}$ over $\mathbb{C}$. Namely, let $\ell$ be an odd positive integer coprime to the determinant of the Cartan matrix of $\mathfrak{g}$. Let $T$ be a maximal torus of the simply connected group $G$ corresponding to $\mathfrak{g}$, $\mathbb{T} \subset T$ the subgroup of elements of order dividing $\ell$, and $\Lambda \in T$. Let $q$ be a primitive $\ell$-th root of unity. Let $\mathfrak{u}_q(\mathfrak{g})$ be the corresponding small quantum group, [**L**], with the coproduct convention of [**EN**], Subsection 5.1.[2] For a positive root $\alpha$ of $\mathfrak{g}$, let $\Lambda_\alpha = \alpha(\Lambda)$. Assume that $\Lambda_\alpha^\ell \neq 1$ for any $\alpha$. Then we can define the dynamical twist $\mathcal{J}_\Lambda(\lambda)$ ([**EN**], Proposition 5.1.9), which is a $T$-invariant function $\mathcal{J}_\Lambda : \mathbb{T}^\vee \to \mathfrak{u}_q(\mathfrak{g})^{\otimes 2}$ on the character group $\mathbb{T}^\vee$ of $\mathbb{T}$. Let $H_{\mathcal{J}_\Lambda} = H_\Lambda$ be the corresponding weak Hopf algebra ([**EN**], Subsection 4.2). Namely, $H_\Lambda = \mathrm{End} A \otimes \mathfrak{u}_q(\mathfrak{g})$ as an algebra, where $A = \mathbb{C}[\mathbb{T}]$ is the base, with coproduct twisted by $\mathcal{J}_\Lambda$. It corresponds to a $\mathcal{C}$-module category $\mathcal{M}_\Lambda$ such that $\mathrm{Gr}(\mathcal{M}_\Lambda) = \mathrm{Gr}(\mathcal{M})$, as a $\mathrm{Gr}(\mathcal{C})$-module, where $\mathcal{M} = \mathrm{Rep}(\mathbb{T})$. By Theorem 5.3.2 of [**EN**], this weak Hopf algebra is self-dual, i.e., $H_\Lambda \cong H_\Lambda^{*\mathrm{op}}$. Since $\mathcal{C}$ does not have nontrivial invertible objects, we get $\chi(a, \mathcal{M}_\Lambda) = \mathbf{1}$ (where $a$ is the ribbon pivotal structure on $\mathcal{C}$ defined by the element $q^{-2\rho}$, with $\rho$ being the half-sum of positive roots of $\mathfrak{g}$), i.e., $a$ is matched to $\mathcal{M}_\Lambda$. Thus, $S^2$ on $H_\Lambda$ is the conjugation by the trivial grouplike element $g = yS(y)^{-1}$, where $y \in A_s \cong \mathbb{C}[\mathbb{T}]$.

PROPOSITION 4.2. *One has*

$$g = yS(y)^{-1},$$

[2]Note that in the definition of the small quantum group in [**EN**], 5.1, the relations $E_\alpha^\ell = F_\alpha^\ell = 0$ are accidentally omitted for non-simple roots $\alpha$. However, this does not affect the arguments in [**EN**].

*where* $y = (y_\lambda) \in A_s$, $\lambda \in \mathbb{T}^\vee$ *and*

$$y_\lambda = \prod_{\alpha>0} (\Lambda_\alpha q^{(\lambda,\alpha)} - q^{-(\lambda,\alpha)}),$$

*where* $(,)$ *is the inner product on* $\mathbb{T}^\vee$ *with values in* $\mathbb{Z}/\ell$ *defined by the Catran matrix of* $\mathfrak{g}$. *In particular, the characteristic polynomial of* $S^2$ *on* $H_\Lambda$ *is*

$$\chi(z) = \prod_{\lambda,\mu,\nu,\kappa\in\mathbb{T}^\vee} \left( z - \prod_{\alpha>0} \frac{(\Lambda_\alpha q^\lambda - q^{-\lambda})(\Lambda_\alpha q^\kappa - q^{-\kappa})}{(\Lambda_\alpha q^\mu - q^{-\mu})(\Lambda_\alpha q^\nu - q^{-\nu})} \right)^{\ell^{\dim \mathfrak{g} - 2\operatorname{rank}\mathfrak{g}}}.$$

PROOF. Let $E_{\mu,\lambda}$, $\mu, \lambda \in \mathbb{T}^\vee$ be the basis of matrix units in End$A$, and let $P_\mu, \mu \in \mathbb{T}^\vee$ be the basis of primitive idempotents of $\mathbb{C}[\mathbb{T}]$. Let $S_\Lambda$ be the antipode of $H_\Lambda$. By [**EN**], Remark 4.2.5,

$$S_\Lambda(x) = v^{-1}S(x)v,$$

where

$$v = \sum_{\lambda,\mu} E_{\lambda+\mu,\lambda} \otimes \mathcal{P}(\lambda)P_\mu$$

and $\mathcal{P} = S(\mathcal{J}_\Lambda^{(1)})\mathcal{J}_\Lambda^{(2)} : \mathbb{T}^\vee \to \mathfrak{u}_q(\mathfrak{g})$ (using the notation of [**EN**]). This implies that

$$S^2(x) = v^{-1}S(v)S^2(x)S(v)^{-1}v,$$

and

$$v^{-1}S(v) = \sum_{\lambda,\mu} E_{\lambda\lambda} \otimes \mathcal{P}(\lambda)^{-1}S(\mathcal{P})(\lambda+\mu)P_\mu = \sum_\lambda E_{\lambda\lambda} \otimes \mathcal{P}(\lambda)^{-1}S(\mathcal{P})(\lambda+h),$$

where $S(\mathcal{P})(\lambda+h) := \sum_\mu S(\mathcal{P})(\lambda+\mu)P_\mu$. We have $g = q^{-2\rho}v^{-1}S(v)$. Thus, the proposition follows from Proposition 2.13 in [**EV**] specialized to roots of unity, taking into account that the coproduct in the quantum group used in [**EV**] is opposite to that of [**EN**] (hence $q^{2\rho}$ should be replaced by $q^{-2\rho}$). □

## 5. Acknowledgements.

The author thanks D. Nikshych and V. Ostrik for useful discussions and comments on the draft of this paper, and in particular to V. Ostrik for providing the reference [**Sch1**]. He is also very grateful to G. Schaumann for comments on the draft of the paper and for contributing an appendix.

## Appendix A. Module traces and inner-product structures

by Gregor Schaumann

**A.1. Serre functors.** In [**Sch2**] a different compatibility between pivotal tensor categories and module categories is investigated. The goal of this appendix is to relate module traces and these so called inner-product structures.

Throughout this section let $\mathcal{C}$ be a finite tensor category over $\mathbf{k}$. The right dual $x^*$ of $x \in \mathcal{C}$ has the coevaluation $\mathbf{1} \to x \otimes x^*$ and $\mathcal{C}$ has the dual distinguished invertible object $D$ satisfying $x^{****} \cong D \otimes x \otimes D^{-1}$, see [**EGNO**, Sec. 7.19]. Let ${}_\mathcal{C}\mathcal{M}$ be an exact left $\mathcal{C}$ module category. The inner Hom $\underline{\mathsf{Hom}}(-,-)$ of $\mathcal{M}$ is defined via a family of natural isomorphisms

$$\mathrm{Hom}_\mathcal{C}(x, \underline{\mathsf{Hom}}(m,n)) \cong \mathrm{Hom}_\mathcal{M}(x \otimes m, n)$$

for $m, n \in \mathcal{M}$ and $x \in \mathcal{C}$. This naturally defines a left exact functor $\underline{\mathrm{Hom}}(-,-)\colon \mathcal{M}^{\mathrm{op}} \times \mathcal{M} \to \mathcal{C}$. Since the module category $\mathcal{M}$ is an exact $\mathcal{C}$-module category, the functor $\underline{\mathrm{Hom}}(-,-)$ is also right exact [**EO**, Cor. 3.15 & Prop. 3.16].

This implies the existence of the following functors [**FScS**, Def. 4.21]:

DEFINITION A.1. Let $\mathcal{M}$ be a left $\mathcal{C}$-module. A *right relative Serre functor* on $\mathcal{M}$ is an endofunctor $\mathrm{S}^{\mathrm{r}}_{\mathcal{M}}$ of $\mathcal{M}$ together with a family of isomorphisms

$$\underline{\mathrm{Hom}}(m,n)^* \xrightarrow{\cong} \underline{\mathrm{Hom}}(n, \mathrm{S}^{\mathrm{r}}_{\mathcal{M}}(m)), \tag{3}$$

which are natural in $m, n \in \mathcal{M}$.

Analogously, a *left relative Serre functor* $\mathrm{S}^{\mathrm{l}}_{\mathcal{M}}$ comes with a family of natural isomorphisms

$$^*\,\underline{\mathrm{Hom}}(m,n) \xrightarrow{\cong} \underline{\mathrm{Hom}}(\mathrm{S}^{\mathrm{l}}_{\mathcal{M}}(n), m). \tag{4}$$

The relative Serre functors are twisted module functors [**FScS**, Lemma. 4.22], i.e. there are coherent natural isomorphisms

$$\mathrm{S}^{\mathrm{l}}_{\mathcal{M}}(x \otimes m) \cong {}^{**}x \otimes \mathrm{S}^{\mathrm{l}}_{\mathcal{M}}(m) \tag{5}$$

and $\mathrm{S}^{\mathrm{r}}_{\mathcal{M}}(x \otimes m) \cong x^{**} \otimes \mathrm{S}^{\mathrm{r}}_{\mathcal{M}}(m)$. When no confusion is possible, we will denote $\mathrm{S}^{\mathrm{l}}_{\mathcal{M}}, \mathrm{S}^{\mathrm{r}}_{\mathcal{M}}$ by $\mathrm{S}^{\mathrm{l}}, \mathrm{S}^{\mathrm{r}}$.

The Serre functors are mutually inverse: It follows from the definition that there is a chain of natural isomorphisms

$$\underline{\mathrm{Hom}}(\mathrm{S}^{\mathrm{l}}\mathrm{S}^{\mathrm{r}}(n), m) \cong {}^*\,\underline{\mathrm{Hom}}(m, \mathrm{S}^{\mathrm{r}}(n)) \cong \underline{\mathrm{Hom}}(n,m), \tag{6}$$

for all $n, m \in \mathcal{M}$. Thus we conclude from the enriched Yoneda lemma (see e.g. [**Sch2**, Lemma 4.11]) that $\mathrm{S}^{\mathrm{l}}\mathrm{S}^{\mathrm{r}} \cong \mathrm{id}_{\mathcal{M}}$. Similarly, it follows that $\mathrm{S}^{\mathrm{r}}\mathrm{S}^{\mathrm{l}} \cong \mathrm{id}_{\mathcal{M}}$.

If $\mathcal{M}$ is additionally semisimple, it is also an exact module category over Vect and thus there exists moreover the **k**-*Serre functor* $\mathrm{S}_{\mathbf{k}}$ on $\mathcal{M}$ which is defined via

$$\mathrm{Hom}_{\mathcal{M}}(m,n) \cong \mathrm{Hom}_{\mathcal{M}}(\mathrm{S}_{\mathbf{k}}(n), m)^*. \tag{7}$$

Then also $\mathrm{S}_{\mathbf{k}}$ is an invertible twisted module functor with

$$\mathrm{S}_{\mathbf{k}}(a \otimes m) \cong a^{**} \otimes \mathrm{S}_{\mathbf{k}}(m). \tag{8}$$

The basic relation between the two Serre functors is obtained from the action of the dual distinguished invertible object $D$ of $\mathcal{C}$:

PROPOSITION A.2. *There is a canonical isomorphism of module functors*

$$\mathrm{S}_{\mathbf{k}} \cong D \otimes \mathrm{S}^{\mathrm{l}}. \tag{9}$$

PROOF. From [**FScS**, Eq (4.25)], we obtain that there is a module natural isomorphism $\mathrm{S}_{\mathbf{k}} \cong N^l$, where $N^l$ is the left exact Nakayama functor. On the other hand, Theorem 4.25 in loc. cit. shows that $N^l \cong D \otimes \mathrm{S}^{\mathrm{l}}_{\mathcal{M}}$ as module functors. This concludes the proof. □

The Serre functors allow to describe the double adjoints of module functors. Denote by ${}^{\vee}F, F^{\vee}$ the right and left adjoints of a functor $F$, respectively. They are canonically module functors, if $F$ is a module functor.

PROPOSITION A.3. *Let $F : {}_{\mathcal{C}}\mathcal{M} \to {}_{\mathcal{C}}\mathcal{N}$ be a module functor between exact module categories. There is a natural module isomorphism*

$$F^{\vee} \cong \mathrm{S}^{\mathrm{l}}_{\mathcal{M}} \circ {}^{\vee}F \circ \mathrm{S}^{\mathrm{r}}_{\mathcal{N}}, \tag{10}$$

*which is coherent with respect to the composition of module functors.*

PROOF. By [**EO**], $F$ is exact and thus its adjoints exist. The isomorphism is defined by

$$(11)\quad \begin{aligned}\underline{\mathsf{Hom}}(F^{\vee}(n), m) &\cong \underline{\mathsf{Hom}}(n, F(m)) \cong {}^*\,\underline{\mathsf{Hom}}(Fm, \mathrm{S}^{\mathrm{r}}_{\mathcal{N}}(n)) \\ &\cong {}^*\,\underline{\mathsf{Hom}}(m, {}^{\vee}F \circ \mathrm{S}^{\mathrm{r}}_{\mathcal{N}}(n)) \\ &\cong \underline{\mathsf{Hom}}(\mathrm{S}^{\mathrm{l}}_{\mathcal{M}} \circ {}^{\vee}F \circ \mathrm{S}^{\mathrm{r}}_{\mathcal{N}}(n), m)).\end{aligned}$$

using the enriched Yoneda lemma. That it is a module natural isomorphism as well as its coherence follow from standard arguments. □

Thus it follows that $(F^{\vee})^{\vee} \cong \mathrm{S}^{\mathrm{l}}_{\mathcal{N}} \circ F \circ \mathrm{S}^{\mathrm{r}}_{\mathcal{M}}$.

Using moreover that on ${}_{\mathcal{C}}\mathcal{C}$, $\mathrm{S}^{\mathrm{r}}_{\mathcal{C}} = (-)^{**}$, we conclude that

COROLLARY A.4. *Let $m \in \mathcal{M}$ and $F_m : {}_{\mathcal{C}}\mathcal{C} \to {}_{\mathcal{C}}\mathcal{M}$ the corresponding module functor $F_m(a) = a \otimes m$. The double dual of $F_m$ computes as*

$$(12)\quad (F_m^{\vee})^{\vee} \cong F_{\mathrm{S}^{\mathrm{l}}_{\mathcal{M}}(m)}.$$

For a fixed finite tensor category $\mathcal{C}$ there is a bicategory $\mathsf{Mod}^{\mathrm{ex}}(\mathcal{C})$ with objects exact module categories and morphism categories the categories of module functors. Every 1-morphism in $\mathsf{Mod}^{\mathrm{ex}}(\mathcal{C})$ has a left and right dual by [**EO**] and Proposition A.3 allows to describe the double dual 2-functor $(-)^{**}$ on $\mathsf{Mod}^{\mathrm{ex}}(\mathcal{C})$, which is the identity on objects and the double left adjoint on functors and natural transformations, via the Serre functors: There is a 2-functor $\mathsf{S}$ on $\mathsf{Mod}^{\mathrm{ex}}(\mathcal{C})$ that is defined as the identity on objects and it maps a module functor $F : {}_{\mathcal{C}}\mathcal{M} \to {}_{\mathcal{C}}\mathcal{N}$ to $\mathsf{S}(F) = \mathrm{S}^{\mathrm{l}}_{\mathcal{N}} \circ F \circ \mathrm{S}^{\mathrm{r}}_{\mathcal{M}}$. On module natural transformations it is defined analogously.

COROLLARY A.5. *The 2-functor $(-)^{**}$ on $\mathsf{Mod}^{\mathrm{ex}}(\mathcal{C})$ is canonically isomorphic to the 2-functor $\mathsf{S}$.*

Assume now that $\mathcal{C}$ has a pivotal structure. Then the functors $\mathrm{S}^{\mathrm{l}}, \mathrm{S}^{\mathrm{r}}, \mathrm{S}_{\mathbf{k}}$ are canonical invertible module endofunctors of ${}_{\mathcal{C}}\mathcal{M}$ and it is natural to consider compatibilities between the pivotal structure and ${}_{\mathcal{C}}\mathcal{M}$ as studied in [**Sch1**], [**Sch2**] and the current article.

DEFINITION A.6. Let $(\mathcal{C}, a)$ be a pivotal finite tensor category and ${}_{\mathcal{C}}\mathcal{M}$ an exact module category.

(1) An *inner-product module structure* [**Sch2**, Def. 5.2] on ${}_{\mathcal{C}}\mathcal{M}$ is a module natural isomorphism $\mathrm{S}^{\mathrm{l}} \cong \mathrm{id}_{\mathcal{M}}$.
(2) Assume that $\mathcal{M}$ is semisimple. A *module trace* on $\mathcal{M}$ is a module natural isomorphism $\mathrm{S}_{\mathbf{k}} \cong \mathrm{id}_{\mathcal{M}}$.

It is easy to see that this is precisely the definition of module trace of the current article.

**A.2. Relating the two structures.** We would like to compare these structures in the case where they both apply, namely if $\mathcal{M}$ is semisimple. If $\mathcal{C}$ is additionally a pivotal fusion category, it is a consequence of [**Sch2**, Thm 5.23] that an inner-product structure on ${}_{\mathcal{C}}\mathcal{M}$ is the same as a module trace on ${}_{\mathcal{C}}\mathcal{M}$.

We now clarify the difference between these structures futher using the multitensor category $\mathcal{D} = \mathrm{End}_{\mathcal{C}}({}_{\mathcal{C}}\mathcal{C} \oplus {}_{\mathcal{C}}\mathcal{M})$ as in Section 2.6.

In Proposition 2.19 it is shown that for semisimple $\mathcal{C}$ and $\mathcal{M}$ there is a bijection between pivotal structures on $\mathcal{D}$ and pairs of a pivotal structure on $\mathcal{C}$ and a module trace on $\mathcal{M}$. A generalization of this correspondence to exact module categories over finite tensor categories is:

PROPOSITION A.7. *Let ${}_{\mathcal{C}}\mathcal{M}$ be an exact module category over a finite tensor category. There is a bijection between pivotal structures on $\mathcal{D} = \mathrm{End}_{\mathcal{C}}({}_{\mathcal{C}}\mathcal{C} \oplus {}_{\mathcal{C}}\mathcal{M})$ and pairs of a pivotal structure on $\mathcal{C}$ and an inner-product module structure on $\mathcal{M}$.*

PROOF. First note that a pivotal structure on $\mathcal{D}$ is the same as a pivotal structure on the full sub-bicategory $\mathsf{D}_{\mathcal{C},\mathcal{M}}$ of $\mathsf{Mod}^{\mathrm{ex}}(\mathcal{C})$ which has two objects ${}_{\mathcal{C}}\mathcal{C}$ and ${}_{\mathcal{C}}\mathcal{M}$ and all higher morphisms of $\mathsf{Mod}^{\mathrm{ex}}(\mathcal{C})$ between them.

Assume now that $\mathcal{C}$ is pivotal and ${}_{\mathcal{C}}\mathcal{M}$ has the structure of an inner-product module category. It follows directly from [**Sch2**, Thm 5.13] that $\mathsf{D}_{\mathcal{C},\mathcal{M}}$ is a pivotal bicategory.

On the other hand, if $\mathsf{D}_{\mathcal{C},\mathcal{M}}$ is pivotal, we first obtain a pivotal structure on $\mathcal{C}$. Consider next the module functors $F_m : {}_{\mathcal{C}}\mathcal{C} \to {}_{\mathcal{C}}\mathcal{M}$ , $c \mapsto c \otimes m$ for $m \in \mathcal{M}$. Module natural transformations between $F_m$ and $F_n$ for $m, n \in \mathcal{M}$ are the same as morphisms from $m$ to $n$. By Corollary A.4, the double left dual of $F_m$ is given by $F_{\mathrm{S}^{\mathrm{l}}_{\mathcal{M}}(m)}$, so the pivotal structure on $\mathcal{D}$ gives in particular a trivialisation of $\mathrm{S}^{\mathrm{l}}_{\mathcal{M}}$, thus an inner-product structure.

The two constructions are mutually inverse. □

To describe the analoguos correspondence for module traces on exact module categories, we define a tensor functor $\phi : \mathcal{D} \to \mathcal{D}$ from the action of the dual distinguished object $D$ as follows. Denote by $\mathcal{C}^{\mathrm{op}}$ the tensor category with reversed tensor product and by $\mathcal{C}^{\vee}$ the dual category with reversed arrows.

- On $\mathcal{C} \cong \mathrm{End}_{\mathcal{C}}({}_{\mathcal{C}}\mathcal{C})^{\mathrm{op}}$, $\phi$ is defined by

$$\phi(c) = D^{-1} \otimes c \otimes D, \tag{13}$$

- on $\mathrm{End}_{\mathcal{C}}({}_{\mathcal{C}}\mathcal{M})$, $\phi$ is the identity,
- on $\mathrm{Fun}_{\mathcal{C}}({}_{\mathcal{C}}\mathcal{C}, {}_{\mathcal{C}}\mathcal{M})$, $\phi$ is defined by $F_m \mapsto F_{D^{-1} \otimes m}$ for $F_m \in \mathrm{Fun}_{\mathcal{C}}({}_{\mathcal{C}}\mathcal{C}, {}_{\mathcal{C}}\mathcal{M})$ as defined above,
- on $\mathrm{Fun}_{\mathcal{C}}({}_{\mathcal{C}}\mathcal{M}, {}_{\mathcal{C}}\mathcal{C})$ we define $\phi$ by $\phi(G_n) = G_{D^{-1} \otimes n}$ for $n \in \mathcal{M}^{\vee}$ and $G_n = \underline{\mathsf{Hom}}(n, -) \in \mathrm{Fun}_{\mathcal{C}}({}_{\mathcal{C}}\mathcal{M}, {}_{\mathcal{C}}\mathcal{C})$.

Thus $\phi$ is the quadruple dual on $\mathcal{C}$.

To see that $\phi$ is indeed a monoidal functor, consider the composite $G_n \circ F_m = - \otimes \underline{\mathsf{Hom}}(n, m) : {}_{\mathcal{C}}\mathcal{C} \to {}_{\mathcal{C}}\mathcal{C}$. We compute

$$\begin{aligned} \phi(G_n) \circ \phi(F_m)(\mathbf{1}) &= \underline{\mathsf{Hom}}(D^{-1} \otimes n, D^{-1} \otimes m) \\ &\simeq D^{-1} \otimes \underline{\mathsf{Hom}}(n, m) \otimes D = \phi(G_n \circ F_m)(\mathbf{1}). \end{aligned} \tag{14}$$

On the other hand it follows that $\phi(F_m) \circ \phi(G_n) = F_m \circ G_n$. The remaining coherence data of the 2-functor are obvious and the axioms for a 2-functor follow directly. We can finally formulate:

PROPOSITION A.8. *Let $\mathcal{C}$ be a finite tensor category and ${}_{\mathcal{C}}\mathcal{M}$ a semisimple module category. There is a bijection between monoidal natural isomorphisms $(-)^{**} \cong \phi$ on $\mathcal{D} = \mathrm{End}_{\mathcal{C}}({}_{\mathcal{C}}\mathcal{C} \oplus {}_{\mathcal{C}}\mathcal{M})$ and pairs of a pivotal structure on $\mathcal{C}$ and a module trace on $\mathcal{M}$.*

PROOF. Let $\mathcal{C}$ be pivotal and assume ${}_{\mathcal{C}}\mathcal{M}$ has a module trace. It follows from Proposition A.2 that the module trace induces a module natural isomorphism

$$\mathrm{S}^{\mathrm{l}}_{\mathcal{M}} \cong D^{-1} \otimes - \tag{15}$$

of module endofunctors on ${}_{\mathcal{C}}\mathcal{M}$ (the pivotal structure is needed for the module structure of $D^{-1} \otimes -$). Furthermore, the pivotal structure provides a trivialisation of the relative Serre functor on ${}_{\mathcal{C}}\mathcal{C}$ (which is just the double dual), thus the monoidal functor $\mathsf{S}$ on $\mathcal{D}$ from Corollary A.5 can be described in terms of the action of the dual distinguished object $D$. This produces with Corollary A.5 a monoidal isomorphism to the functor $\phi$.

Assume to the contrary, that a monoidal isomorphism from the double dual to $\phi$ on $\mathcal{D}$ is specified. On $\mathcal{C}$ this produces a monoidal isomorphism from the quadruple dual to the double dual, thus a pivotal structure. Since the double dual of $F_m$ is given by $F_{\mathrm{S}^{\mathrm{l}}_{\mathcal{M}}(m)}$, we moreover obtain an isomorphism $\mathrm{S}^{\mathrm{l}}_{\mathcal{M}} \cong D^{-1} \otimes -$, which is a module natural isomorphism, thus it gives a module trace on ${}_{\mathcal{C}}\mathcal{M}$, again by Proposition A.2. □

There are two main situations in which we have a bijection between the two structures: If $\mathcal{C}$ is unimodular, we use as in Section 2.4 the notation $\overline{a}$ for the "dual" pivotal structure defined by $\overline{a} = \iota \circ a^{-1}$ with $\iota$ the canonical isomorphism to the quadruple dual.

First we obtain as a consequence of Propositions A.7 and A.8:

Corollary A.9. *Let $\mathcal{C}$ be unimodular and ${}_{\mathcal{C}}\mathcal{M}$ a semisimple module category. There are bijections*

$$\begin{aligned} \{(\mathcal{C}, a), {}_{\mathcal{C}}\mathcal{M} \quad \textit{inner product}\} &\leftrightarrow \{\textit{pivotal structures on} \quad \mathcal{D}\} \\ &\leftrightarrow \{(\mathcal{C}, \overline{a}), {}_{\mathcal{C}}\mathcal{M} \quad \textit{module trace}\} \end{aligned} \tag{16}$$

Second we consider the case that the dual category is unimodular.

Proposition A.10. *Assume that $\mathcal{C}^*_{\mathcal{M}}$ is unimodular. Then ${}_{\mathcal{C}}\mathcal{M}$ has a module trace over $(\mathcal{C}, a)$ if and only if it has an inner-product structure over $(\mathcal{C}, a)$.*

Proof. There is an obvious generalization of the Radford theorem, see [**EGNO**, Sec. 7.19], to multitensor categories and applied to $\mathcal{D} = \mathrm{End}_{\mathcal{C}}({}_{\mathcal{C}}\mathcal{C} \oplus {}_{\mathcal{C}}\mathcal{M})$ we obtain isomorphisms

$$m^{****} \simeq D_{\mathcal{C}}^{-1} \otimes m \otimes D_{\mathcal{C}^*_{\mathcal{M}}}, \tag{17}$$

for $m \in \mathcal{M}$ considered under the equivalence $\mathrm{Fun}_{\mathcal{C}}({}_{\mathcal{C}}\mathcal{C}, {}_{\mathcal{C}}\mathcal{M}) \cong \mathcal{M}$ and where $D_{\mathcal{C}^*_{\mathcal{M}}}, D_{\mathcal{C}}$ are the dual distinguished invertible objects of $\mathcal{C}^*_{\mathcal{M}}$ and $\mathcal{C}$, respectively. Suppose that $\mathcal{C}^*_{\mathcal{M}}$ is unimodular and ${}_{\mathcal{C}}\mathcal{M}$ is inner-product, then using Proposition A.7, $m \simeq m^{****} \simeq D_{\mathcal{C}}^{-1} \otimes m$. Hence we obtain that $D_{\mathcal{C}} \otimes - \simeq \mathrm{id}_{\mathcal{M}}$ as $\mathcal{C}$-module functors and with Proposition A.2 we conclude that ${}_{\mathcal{C}}\mathcal{M}$ has a module trace.

Conversely, if ${}_{\mathcal{C}}\mathcal{M}$ has a module trace, Proposition A.8 implies $m^{****} \simeq \phi^2(m) = D_{\mathcal{C}}^{-2} \otimes m$ and Equation (17) implies that $D_{\mathcal{C}} \otimes -$ is isomorphic to $\mathrm{id}_{\mathcal{M}}$ as module functor, thus ${}_{\mathcal{C}}\mathcal{M}$ is also inner-product. □

Next we discuss situations where the two structures disagree. First we have as a direct consequence from Proposition A.2:

Corollary A.11. *If a semisimple module category ${}_{\mathcal{C}}\mathcal{M}$ over a pivotal finite tensor category $(\mathcal{C}, a)$ admits a module trace and the module functor $D \otimes - : {}_{\mathcal{C}}\mathcal{M} \to {}_{\mathcal{C}}\mathcal{M}$ is not isomorphic to the identity functor as a linear functor, ${}_{\mathcal{C}}\mathcal{M}$ does not admit an inner-product module structure over any pivotal structure of $\mathcal{C}$.*

This situation can indeed happen:

EXAMPLE A.12. Let $\mathcal{C} = \operatorname{Rep} T_n$ and $\mathcal{M} = \operatorname{Rep} \mathbb{Z}/n$ be as in the Example 3.2, i.e. ${}_{\mathcal{C}}\mathcal{M}$ has a module trace over $\mathcal{C}$. The dual distinguished invertible object $D$ of $\mathcal{C}$ is the one-dimensional representation with action of $g$ by $q$ and $x$ by 0. It acts non-trivially on $Gr(\mathcal{M})$, thus ${}_{\mathcal{C}}\mathcal{M}$ cannot admit an inner-product structure for any pivotal structure of $\mathcal{C}$.

In general for module categories from tensor functors we have:

EXAMPLE A.13. Let $F : \mathcal{C} \to \mathcal{C}'$ be a pivotal (exact) tensor functor between pivotal finite categories and let $\mathcal{C}'$ be semisimple. Then $\mathcal{C}'$ is a module category over $\mathcal{C}$ that is matched to the given pivotal structure. It has an inner-product structure if and only if $F(D_{\mathcal{C}}) = \mathbf{1}$.

We do not know of an example of the converse kind that has an inner-product structure but cannot possess a matched pivotal structure.

However, reversing Proposition A.10 we obtain:

COROLLARY A.14. *Suppose $\mathcal{C}$ is unimodular and $\mathcal{C}^*_{\mathcal{M}}$ is not and suppose $(\mathcal{C}, a)$ is matched to ${}_{\mathcal{C}}\mathcal{M}$. Then $(\mathcal{C}, \overline{a})$ has an inner-product structure over ${}_{\mathcal{C}}\mathcal{M}$, but it cannot have a module trace.*

PROOF. If ${}_{\mathcal{C}}\mathcal{M}$ has also a module trace over $(\mathcal{C}, \overline{a})$, it follows from Equation (17) that the action of $D_{\mathcal{C}^*_{\mathcal{M}}}$ on ${}_{\mathcal{C}}\mathcal{M}$ is isomorphic to $\mathrm{id}_{\mathcal{M}}$ as module functors, thus $\mathcal{C}^*_{\mathcal{M}}$ is unimodular and we have a contradiction. □

EXAMPLE A.15. Example A.12 provides also a case where Corollary A.14 applies: The category $\mathcal{C} \boxtimes \mathcal{C}^*_{\mathcal{M}}$ is not unimodular and acts on $\mathcal{M}$. The dual category is the center $\mathcal{Z}(\mathcal{C}) = \operatorname{Rep}(\mathfrak{u}_q(\mathfrak{sl}_2)) \boxtimes \operatorname{Rep} \mathbb{Z}/n$ which is unimodular and inherits a pivotal structure $a$ from $\mathcal{C}$. Then $(\mathcal{Z}(\mathcal{C}), a)$ is matched to $\mathcal{M}$, while $(\mathcal{Z}(\mathcal{C}), \overline{a})$ has an inner-product structure but is not matched.

## References

[EGNO] Pavel Etingof, Shlomo Gelaki, Dmitri Nikshych, and Victor Ostrik, *Tensor categories*, Mathematical Surveys and Monographs, vol. 205, American Mathematical Society, Providence, RI, 2015. MR3242743

[ENO1] Pavel Etingof, Dmitri Nikshych, and Viktor Ostrik, *On fusion categories*, Ann. of Math. (2) **162** (2005), no. 2, 581–642, DOI 10.4007/annals.2005.162.581. MR2183279

[ENO2] Pavel Etingof, Dmitri Nikshych, and Victor Ostrik, *Fusion categories and homotopy theory*, Quantum Topol. **1** (2010), no. 3, 209–273, DOI 10.4171/QT/6. With an appendix by Ehud Meir. MR2677836

[EN] Pavel Etingof and Dmitri Nikshych, *Dynamical quantum groups at roots of 1*, Duke Math. J. **108** (2001), no. 1, 135–168, DOI 10.1215/S0012-7094-01-10814-4. MR1831822

[EO] Pavel Etingof and Viktor Ostrik, *Finite tensor categories* (English, with English and Russian summaries), Mosc. Math. J. **4** (2004), no. 3, 627–654, 782–783. MR2119143

[EV] Pavel Etingof and Alexander Varchenko, *Traces of intertwiners for quantum groups and difference equations. I*, Duke Math. J. **104** (2000), no. 3, 391–432, DOI 10.1215/S0012-7094-00-10433-4. MR1781477

[FScS] J. Fuchs, G. Schaumann, and C. Schweigert. math.RT/1612.04561 Eilenberg-Watts calculus for finite categories and a bimodule Radford $S^4$ theorem.

[L] George Lusztig, *Introduction to quantum groups*, Modern Birkhäuser Classics, Birkhäuser/Springer, New York, 2010. Reprint of the 1994 edition. MR2759715

[N1] Dmitri Nikshych, *On the structure of weak Hopf algebras*, Adv. Math. **170** (2002), no. 2, 257–286, DOI 10.1006/aima.2002.2081. MR1932332

[N2] Dmitri Nikshych, *Semisimple weak Hopf algebras*, J. Algebra **275** (2004), no. 2, 639–667, DOI 10.1016/j.jalgebra.2003.09.025. MR2052631

[O] Victor Ostrik, *Module categories, weak Hopf algebras and modular invariants*, Transform. Groups **8** (2003), no. 2, 177–206, DOI 10.1007/s00031-003-0515-6. MR1976459

[Sch1] Gregor Schaumann, *Traces on module categories over fusion categories*, J. Algebra **379** (2013), 382–425, DOI 10.1016/j.jalgebra.2013.01.013. MR3019263

[Sch2] Gregor Schaumann, *Pivotal tricategories and a categorification of inner-product modules*, Algebr. Represent. Theory **18** (2015), no. 6, 1407–1479, DOI 10.1007/s10468-015-9547-6. MR3435098

Department of Mathematics, Massachusetts Institute of Technology, Cambridge, MA 02139, USA

*Email address*: etingof@math.mit.edu

Faculty of Mathematics, University of Vienna, Austria

*Email address*: gregor.schaumann@univie.ac.at

Contemporary Mathematics
Volume **728**, 2019
https://doi.org/10.1090/conm/728/14658

# Cohen-Macaulay invariant subalgebras of Hopf dense Galois extensions

Ji-Wei He and Yinhuo Zhang

Abstract. Let $H$ be a semisimple Hopf algebra, and let $R$ be a noetherian left $H$-module algebra. If $R/R^H$ is a right $H^*$-dense Galois extension, then the invariant subalgebra $R^H$ will inherit the AS-Cohen-Macaulay property from $R$ under some mild conditions, and $R$, when viewed as a right $R^H$-module, is a Cohen-Macaulay module. In particular, we show that if $R$ is a noetherian complete semilocal algebra which is AS-regular of global dimension 2 and $H = \mathbf{k}G$ for some finite subgroup $G \subseteq \mathrm{Aut}(R)$, then all the indecomposable Cohen-Macaulay module of $R^H$ is a direct summand of $R_{R^H}$, and hence $R^H$ is Cohen-Macaulay-finite, which generalizes a classical result for commutative rings. The main tool used in the paper is the extension groups of objects in the corresponding quotient categories.

## 0. Introduction

Motivated by the study of quotient singularities of noncommutative projective schemes, Van Oystaeyen and the authors introduced the concept of Hopf dense Galois extension in [**HVZ2**]. The present paper is a further study of Hopf dense Galois extensions. In this paper, we focus on the Cohen-Macaulay properties of the invariant subalgebra of a Hopf dense Galois extension. Let $H$ be a finite dimensional semisimple Hopf algebra, and let $R$ be a left $H$-module algebra. Let $H^*$ be the dual Hopf algebra. Since $H$ is finite dimensional, $R$ is a right $H^*$-comodule algebra. Denote by $R^H(= R^{coH^*})$ the invariant subalgebra. The algebra extension $R/R^H$ is called a *right $H^*$-dense Galois extension* ([**HVZ2**, Definition 1.1]) if the cokernel of the map

$$\beta : R \otimes_{R^H} R \to R \otimes H^*, r \otimes r' \mapsto (r \otimes 1)\rho(r')$$

is finite dimensional, where $\rho : R \to R \otimes H^*$ is the right $H^*$-comodule structure map.

The concept of a Hopf dense Galois extension is a weaker version of that of a Hopf Galois extension. The main feature of a Hopf dense Galois extension is that there is an equivalence between some quotient categories. More precisely, let $R$

2010 *Mathematics Subject Classification.* Primary 16D90, 16E65, Secondary 16B50.
*Key words and phrases.* Hopf dense Galois extension, quotient category, Cohen-Macaulay module.
The first author was supported by NSFC (Nos. 11571239, 11671351) and ZJNSF (No. LY19A010011).
The second author was supported by an FWO-grant.

be a noetheiran algebra which is a left $H$-module algebra over a finite dimensional semisimple Hopf algebra $H$. Let $\operatorname{Mod} R$ be the category of right $R$-modules and let $\operatorname{Tor} R$ be the full subcategory of $\operatorname{Mod} R$ consisting of torsion $R$-modules (see Section 1). Then we obtain a quotient category $\operatorname{QMod} R = \frac{\operatorname{Mod} R}{\operatorname{Tor} R}$. Set $B = R\#H$. Since $R$ is noetherian, $B$ and the invariant subalgebra $R^H$ are also noetherian. Then we also have quotient category $\operatorname{QMod} B$ and $\operatorname{QMod} R^H$. If $R/R^H$ is right $H^*$-dense Galois extension, then there is an equivalence of abelian categories $\operatorname{QMod} R^H \cong \operatorname{QMod} B$ (cf. [**HVZ2**, Theorem 2.4]). Under this equivalence, we show that the invariant subalgebra $R^H$ inherits many homological properties from $R$. Especially, we have the following results (Theorems 3.5 and 4.5). The terminologies may be found in Section 3.

**Theorem.** *Let $H$ be a finite dimensional semisimple Hopf algebra, and let $R$ be a noetherian left $H$-module algebra. Assume that the $H$-module $R$ is admissible and* $\operatorname{Tor} B$ *is stable. If $R/R^H$ is a right $H^*$-dense Galois extension, then the following statements hold.*

(i) *If $R$ is AS-Cohen-Macaulay of dimension $d$, so is $R^H$.*

(ii) *If $R$ is AS-Cohen-Macaulay of dimension $d$, then $R$, viewed as a right $R^H$-module, is Cohen-Macaulay.*

There are plenty of $H$-module algebras satisfying the assumptions in the theorem above. For example, if $R$ is a noetherian complete semilocal algebra and $H$ is a finite group algebra, then the $H$-module algebra $R$ is admissible and $\operatorname{Tor} B$ is stable (cf. Example 2.5).

We remark that if the algebra $R$ in the above theorem is connected graded and Artin-Schelter Gorenstein, the results were proved in [**KKZ**, Lemma 3.2](b) by using the homological integrals and balanced dualizing complexes. The techniques used in [**KKZ**] can not be applied to our case directly. The main tool for proving of the above theorem is the following observation (Theorem 2.11) which gives a way to compute the extension groups of objects in the quotient category $\operatorname{QMod} R^H$.

**Theorem.** *Assume that the $H$-module algebra $R$ is admissible and* $\operatorname{Tor} B$ *is stable. Let $N$ and $M$ be finitely generated right $B$-modules. Let $\mathcal{N}$ and $\mathcal{M}$ be the corresponding objects of $N$ and $M$ in the quotient categories. Then we have*

(i) *for each $i \geq 0$, $\operatorname{Ext}^i_{\operatorname{QMod} R}(\mathcal{N}, \mathcal{M})$ is a right $H$-module,*

(ii) *there are isomorphisms $\operatorname{Ext}^i_{\operatorname{QMod} B}(\mathcal{N}, \mathcal{M}) \cong \operatorname{Ext}^i_{\operatorname{QMod} R}(\mathcal{N}, \mathcal{M})^H$, $\forall\, i \geq 0$.*

If $R$ is further a noetherian complete semilocal algebra which is also AS-regular of global dimension 2, and $G \subseteq \operatorname{Aut}(R)$ is a finite subgroup, then the invariant subalgebra $R^G$ is Cohen-Macaulay-finite. Indeed, we have the following result (Theorem 5.4), which is a noncommutative version of a classical result for commutative rings due to Auslander [**A**].

**Theorem.** *Let $R$ be a noetherian complete semilocal algebra, and let $G \subseteq \operatorname{Aut}(R)$ be a finite subgroup. Assume $\operatorname{char} \mathbf{k} \nmid |G|$ and $R/R^G$ is a right $\mathbf{k}G^*$-dense Galois extension and $R$ is AS-regular of global dimension 2. Then the following statements hold.*

(i) *$R^G$ is AS-Cohen-Macaulay of dimension 2.*

(ii) *A finitely generated right $R^G$-module $M$ is Cohen-Macaulay if and only if $M \in \operatorname{add}(R_{R^G})$, where $\operatorname{add}(R_{R^G})$ is the subcategory of $\operatorname{Mod} R^G$ consisting of all the direct summands of finite direct sums of $R_{R^G}$.*

When the algebra $R$ is graded, the above result was obtained by several authors (see Remark 5.5). However, the methods applied in the graded case can not be applied to nongraded case directly. Our method depends on the relations of extension groups of objects in the quotient categories over $R^G$ and $R\#\mathbf{k}G$ respectively, which is different from the methods used in the graded case.

Throughout, $\mathbf{k}$ is a field. All the algebras and modules are over $\mathbf{k}$. Unadorned $\otimes$ means $\otimes_{\mathbf{k}}$. We refer to [**HVZ2**] for basic properties of Hopf dense Galois extensions.

## 1. Hom-sets of the quotient category

Let $B$ be a (right) noetherian algebra. We denote by $\operatorname{Mod} B$ the category of right $B$-modules. Let $M$ be a right $B$-module. We say that a submodule $K$ of $M$ is *cofinite* if $M/K$ is finite dimensional. For an element $m \in M$, we say that $m$ is a torsion element if $mB$ is finite dimensional. Let $\Gamma_B(M) = \{m \in M | m \text{ is a torsion element}\}$. Then $\Gamma_B(M)$ is a submodule of $M$. Note that $\Gamma_B$ is in fact a functor from $\operatorname{Mod} B$ to itself. The functor $\Gamma_B$ can be represented as $\Gamma_B = \varinjlim \operatorname{Hom}_B(B/K, -)$, where $K$ runs over all the cofinite right ideals of $B$. In the sequel, if there is no risk of confusion, we will omit the subscript in the functor $\Gamma_B$, and simply write it as $\Gamma$.

For a right $B$-module, if $\Gamma(M) = 0$, then we say that $M$ is *torsion free*. If $\Gamma(M) = M$, then $M$ is called a *torsion module*. Let $\operatorname{Tor} B$ be the full subcategory of $\operatorname{Mod} B$ consisting of torsion modules. One sees that $\operatorname{Tor} B$ is a localizing subcategory of $\operatorname{Mod} B$. Hence we have an abelian quotient category $\operatorname{QMod} B := \frac{\operatorname{Mod} B}{\operatorname{Tor} B}$. The natural projection functor is denoted by $\pi : \operatorname{Mod} B \to \operatorname{QMod} B$, which has a right adjoint functor $\omega : \operatorname{QMod} B \to \operatorname{Mod} B$. For $N, M \in \operatorname{Mod} B$, we write $\mathcal{N}$ and $\mathcal{M}$ for $\pi(N)$ and $\pi(M)$ respectively in the quotient category. The morphisms in $\operatorname{QMod} B$ are defined to be the set

$$\operatorname{Hom}_{\operatorname{QMod} B}(\mathcal{N}, \mathcal{M}) = \varinjlim \operatorname{Hom}_B(N', M/\Gamma(M)),$$

where $N'$ runs over all the cofinite submodules of $N$, and the direct system is induced by the inclusion maps of the cofinite submodules.

If $M$ is a torsion free module, then its injective envelope is torsion free as well. The injective envelope of a torsion module is not torsion in general. If $\operatorname{Tor} B$ is closed under taking injective envelope, then we say that $\operatorname{Tor} B$ is *stable*. For example, if $B$ is a noetherian commutative local algebra, then $\operatorname{Tor} B$ is stable.

LEMMA 1.1. *Let $N$ be a finitely generated right $B$-module, and let $E$ be an injective torsion $B$-module. Then $\varinjlim \operatorname{Hom}_B(N', E) = 0$, where $N'$ runs over all the cofinite submodule of $N$.*

PROOF. For every cofinite submodule $N'$ of $N$, applying the functor $\operatorname{Hom}_B(-, E)$ to the exact sequence $0 \to N' \to N \to N/N' \to 0$, we obtain an exact sequence $0 \to \operatorname{Hom}_B(N/N', E) \to \operatorname{Hom}_B(N, E) \to \operatorname{Hom}_B(N', E) \to 0$. Take the direct limit over all the cofinite submodules $N' \subseteq N$, we obtain an exact sequence

$$0 \to \varinjlim \mathrm{Hom}_B(N/N', E) \to \mathrm{Hom}_B(N, E) \to \varinjlim \mathrm{Hom}_B(N', E) \to 0.$$

For each $B$-module morphism $f : N \to E$, since $N$ is finitely generated, $\mathrm{im}\, f$ is finite dimensional. Hence $\ker f$ is cofinite. Since $f$ factors through $N/\ker f$, we see that the morphism $\varinjlim \mathrm{Hom}_B(N/N', E) \to \mathrm{Hom}_B(N, E)$ is an epimorphism, and hence $\varinjlim \mathrm{Hom}_B(N', E) = 0$. □

LEMMA 1.2. *Assume that* $\mathrm{Tor}\, B$ *is stable. For* $N, M \in \mathrm{Mod}\, B$, *if* $N$ *is finitely generated, then* $\mathrm{Ext}^i_{\mathrm{QMod}\, B}(\mathcal{N}, \mathcal{M}) = \varinjlim \mathrm{Ext}^i_B(N', M)$ *for all* $i \geq 0$, *where* $N'$ *runs over all the cofinite submodules of* $N$.

PROOF. Take a minimal injective resolution of $M$ as follows

$$0 \to M \to I^0 \to I^1 \to \cdots \to I^n \to \cdots. \tag{1.2.1}$$

Since $\mathrm{Tor}\, B$ is stable, for each $n \geq 0$, $I^n = F^n \oplus E^n$ where $F^n$ is a torsion free module and $E^n$ is a torsion module. Note that if $F$ is an injective torsion free module, then $\mathcal{F} = \pi(F)$ is an injective object in $\mathrm{QMod}\, B$. Applying the projection functor to the exact sequence (1.2.1), we obtain an injective resolution of $\mathcal{M}$ in $\mathrm{QMod}\, B$:

$$0 \to \mathcal{M} \to \mathcal{F}^0 \to \mathcal{F}^1 \to \cdots \to \mathcal{F}^n \to \cdots. \tag{1.2.2}$$

Applying the functor $\mathrm{Hom}_{\mathrm{QMod}\, B}(\mathcal{N}, -)$ to (1.2.2), we have

$$0 \to \mathrm{Hom}_{\mathrm{QMod}\, B}(\mathcal{N}, \mathcal{F}^0) \to \cdots \to \mathrm{Hom}_{\mathrm{QMod}\, B}(\mathcal{N}, \mathcal{F}^n) \to \cdots,$$

which is equivalent to the following complex

$$0 \to \varinjlim \mathrm{Hom}_B(N', F^0) \to \cdots \to \varinjlim \mathrm{Hom}_B(N', F^n) \to \cdots. \tag{1.2.3}$$

By Lemma 1.1, the complex (1.2.3) is equivalent to

$$0 \to \varinjlim \mathrm{Hom}_B(N', F^0 \oplus E^0) \to \cdots \to \varinjlim \mathrm{Hom}_B(N', F^n \oplus E^n) \to \cdots. \tag{1.2.4}$$

Taking the cohomology of (1.2.4), we obtain the desired result since the direct limit is exact. □

## 2. Hom-sets of the quotient categories of smash products

Throughout this section, let $R$ be a noetherian algebra, $H$ a finite dimensional semisimple Hopf algebra acting on $R$ so that $R$ is a left $H$-module algebra. Let $B = R\#H$ be the smash product. We may view $R$ as a left $B$-module by setting the left $B$-action

$$(a\#h) \cdot r = a(h \cdot r), \ \forall\, a, r \in R, h \in H, \tag{2.0.1}$$

where $h \cdot r$ is the left $H$-action on $R$. Similarly, we may view $R$ as a right $B$-module by setting the right $B$-action

$$r \cdot (a\#h) = (S^{-1}h_{(1)} \cdot r)(h_{(2)} \cdot a), \ \forall\, a, r \in R, h \in H, \tag{2.0.2}$$

where $S$ is the antipode of $H$ and $\Delta(h) = h_{(1)} \otimes h_{(2)}$.

Let $M$ be a right $B$-module. For every element $x \in M$, we see that $xB$ is finite dimensional if and only if $xR$ is finite dimensional since $H$ is finite dimensional.

Hence we see as a right $B$-module $\Gamma_R(M) = \Gamma_B(M)$. From this observation, we obtain the following fact.

LEMMA 2.1. *Let $M$ be a right $B$-module. Then as right $R$-modules, $R^i\Gamma_R(M) \cong R^i\Gamma_B(M)$ for all $i \geq 0$.*

PROOF. Let $0 \to M \to I^0 \to I^1 \to \cdots$ be an injective resolution of $M$ in the category of right $B$-modules. By [**HVZ1**, Proposition 2.6], each $I^i$ is injective as a right $R$-module. Hence $R^i\Gamma_R(M)$ is the $i$th cohomology of the complex $\Gamma_R(I^{\cdot})$. Since $\Gamma_R(I^i) = \Gamma_B(I^i)$ for all $i \geq 0$, as complexes of right $R$-modules, $\Gamma_R(I^{\cdot}) = \Gamma_B(I^{\cdot})$. Thus we have $R^i\Gamma_R(M) = R^i\Gamma_B(M)$. □

Let $N$ be a right $R$-module. We define a right $B$-module $N\#H = N \otimes H$ whose right $B$-action is given by $(n \otimes h)(a \otimes g) = n(h_{(1)}a) \otimes h_{(2)}g$ for $n \in N, g, h \in H$.

LEMMA 2.2. *The following statements hold.*

(i) *If* $\operatorname{Tor} B$ *is stable, then so is* $\operatorname{Tor} R$.

(ii) *Let $A$ and $A'$ be two noetherian algebras. If $F : \operatorname{Mod} A \longrightarrow \operatorname{Mod} A'$ is an equivalence of abelian categories, then the restriction of $F$ to* $\operatorname{Tor} A$ *induces an equivalence* $\operatorname{Tor} A \longrightarrow \operatorname{Tor} A'$.

PROOF. (i) Let $N$ be a torsion $R$-module. Set $M = N\#H$. Then $M$ is $B$-torsion module. Let $I$ be the injective envelope of $M$. Since $\operatorname{Tor} B$ is stable, $I$ is a torsion $B$-module. By [**HVZ1**, Proposition 2.6], $I$ is injective as a right $R$-module. We view $N$ as an $R$-submodule of $M$ through the inclusion map $N \to M : n \mapsto n\#1$. Then $N$ is a submodule of the torsion injective module $I$. Hence the injective envelope of $N$ must be torsion.

(ii) This is classical. □

PROPOSITION 2.3. *Let $H$ be a finite dimensional semisimple and cosemisimple Hopf algebra. Let $R$ be a noetherian left $H$-module algebra. Then* $\operatorname{Tor} R$ *is stable if and only if* $\operatorname{Tor} B$ *is stable.*

PROOF. We only need to prove the "only if" part. Note that $B = R\#H$ is a right $H$-comodule algebra. It is a left $H^*$-module algebra. The invariant subalgebra $B^{H^*} = R$. Since $B$ is a right $H$-Galois extension of $R$, $R$ and $B\#H^*$ are Morita equivalent (cf. [**CFM**, Theorem 1.2]). Since $\operatorname{Tor} R$ is stable, $\operatorname{Tor} B\#H^*$ is stable by Lemma 2.2(ii). By Lemma 2.2(i), $\operatorname{Tor} B$ is stable. □

Let $N$ and $M$ be right $B$-modules. Note that $\operatorname{Hom}_R(N, M)$ has a right $H$-module action defined as follows: for $f \in \operatorname{Hom}_R(N, M), h \in H$,

$$(2.3.1) \qquad (f \leftharpoonup h)(n) = f(nS(h_{(1)}))h_{(2)}, \quad \forall\, n \in N.$$

Under this $H$-action on $\operatorname{Hom}_R(N, M)$, we have the following isomorphism [**CFM**]

$$(2.3.2) \qquad \operatorname{Hom}_B(N, M) \cong \operatorname{Hom}_R(N, M)^H.$$

The right $H$-action on $\operatorname{Hom}_R(N, M)$ may be extended to the extension groups by choosing a projective resolution of $N$ (or injective resolution of $M$) in $\operatorname{Mod} B$. Moreover, we have the following isomorphisms [**HVZ1**]

$$(2.3.3) \qquad \operatorname{Ext}^i_B(N, M) \cong \operatorname{Ext}^i_R(N, M)^H, \ \forall\, i \geq 0.$$

DEFINITION 2.4. We say that the left $H$-module algebra $R$ is (right) *admissible* if for any finitely generated right $B$-module $N$, and any cofinite $R$-submodule $K$ of $N$, there is a cofinite $B$-submodule $K'$ of $N$ such that $K' \subseteq K$.

There are several natural classes of admissible noetherian $H$-module algebras.

EXAMPLE 2.5. Let $R$ be a noetherian semilocal algebra, that is, $R/J$ is a finite dimensional semisimple algebra where $J$ is the Jacobson radical of $R$. Let $G$ be a finite group which acts on $R$ so that $R$ is a left $G$-module algebra. Let $B = R\#\mathbf{k}G$. We know that $J$ is stable under the $G$-action. Let $N$ be a finitely generated right $B$-module, and let $K$ be an $R$-submodule such that $\overline{N} := N/K$ is finite dimensional. Since $\overline{N}$ is finite dimensional, there is an integer $k$ such that $\overline{N}J^k = 0$. Hence $NJ^k \subseteq K$. For $x \in N$, $r \in J^k$ and $g \in G$, we have $(xr)g = (xg)(g^{-1}(r)) \in NJ^k$. Hence $NJ^k$ is a $B$-submodule of $N$. Since $N/(NJ^k)$ is a finitely generated module over $R/J^k$, it is finite dimensional. Hence the $G$-module algebra $R$ is admissible.

If $R$ is also complete with respect to the $J$-adic topology and $\operatorname{char}\mathbf{k} \nmid |G|$, then $\operatorname{Tor} B$ is stable. Indeed, a right $B$-module $M$ is torsion if and only if $M$ is torsion as a right $R$-module. Since $R$ is complete, by [**Ja**, Theorem 1.1], $J$ has the Artin-Rees property which insures that the injective envelope of a torsion $R$-module is still torsion. Since $\mathbf{k}G$ is both semisimple and cosemisimple, $\operatorname{Tor} B$ is stable by Proposition 2.3.

REMARK 2.6. In Definition 2.4, if $R$ is a graded algebra and the $H$-action is homogeneous, then we say that the left graded $H$-module algebra $R$ is admissible if the above conditions hold for finitely generated graded modules.

EXAMPLE 2.7. Let $R = R_0 \oplus R_1 \oplus \cdots$ be a graded noetherian algebra such that $\dim R_i < \infty$ for all $i \geq 0$. Let $H$ be a finite dimensional semisimple Hopf algebra which acts on $R$ homogeneously so that $R$ is an $H$-module algebra. Set $B = R\#H$. Let $N$ be a graded finitely generated right $B$-module, and $K$ a graded $R$-submodule of $N$ such that $N/K$ is finite dimensional. Let $\mathfrak{m} = \oplus_{i\geq 1} R_i$. Since $N/K$ is finite dimensional, there is an integer $k$ such that $(N/K)\mathfrak{m}^k = 0$. Hence $N\mathfrak{m}^k \subseteq K$. Since $\mathfrak{m}$ is stable under the $H$-action, $N\mathfrak{m}^k$ is a $B$-submodule of $N$. As $R$ is noetherian, $R/\mathfrak{m}^k$ is finite dimensional for all $k \geq 0$. Hence $N/(N\mathfrak{m}^k)$ is finite dimensional. It follows that the left $H$-module algebra $R$ is admissible in the graded sense.

LEMMA 2.8. *Let $R$ be an admissible $H$-module algebra. Assume* $\operatorname{Tor} R$ *is stable. Let $N$ be a finitely generated right $B$-module, and $M$ a right $R$-module. Then*

$$\operatorname{Hom}_{\operatorname{QMod} R}(\mathcal{N}, \mathcal{M}) = \varinjlim \operatorname{Hom}_R(K, M),$$

*where $K$ runs over all the cofinite $B$-submodules of $N$.*

PROOF. By Lemma 1.2, $\operatorname{Hom}_{\operatorname{QMod} R}(\mathcal{N}, \mathcal{M}) = \varinjlim \operatorname{Hom}_R(K, M)$ where the limit runs over all the cofinite $R$-submodules $K$ of $N$. Since the $H$-module algebra $R$ is assumed to be admissible, the direct system formed by all the cofinite $R$-submodules of $N$ and the direct system formed by all the cofinite $B$-submodules of $N$ are cofinal. Hence $\varinjlim \operatorname{Hom}_R(K, M) = \varinjlim \operatorname{Hom}_R(K', M)$, where on the left hand the direct limit runs over all the cofinite $R$-submodules $K$ of $N$, and on the right hand the direct limit runs over all the cofinite $B$-submodules $K'$ of $N$. □

PROPOSITION 2.9. *Let $R$ be an admissible $H$-module algebra. Assume* $\operatorname{Tor} R$ *is stable. Let $N$ and $M$ be finitely generated right $B$-modules. Then there is a natural right $H$-module action on* $\operatorname{Hom}_{\operatorname{QMod} R}(\mathcal{N}, \mathcal{M})$. *Moreover, with this $H$-module* *structure we have*

$$\mathrm{Hom}_{\mathrm{QMod}\,B}(\mathcal{N},\mathcal{M}) \cong \mathrm{Hom}_{\mathrm{QMod}\,R}(\mathcal{N},\mathcal{M})^H,$$

*where the isomorphism is functorial in $M$.*

PROOF. By Lemma 2.8, $\mathrm{Hom}_{\mathrm{QMod}\,R}(\mathcal{N},\mathcal{M}) = \varinjlim \mathrm{Hom}_R(K,M)$, where each $K$ in the direct system is a $B$-submodule of $N$. By (2.3.1), $\mathrm{Hom}_R(K,M)$ is a right $H$-module for each cofinite $B$-submodule $K$. It is easy to see that the direct system is compatible with the right $H$-module actions. Hence $\mathrm{Hom}_{\mathrm{QMod}\,R}(\mathcal{N},\mathcal{M})$ is a right $H$-module.

For each $B$-submodule $K$ of $N$, we have $\mathrm{Hom}_B(K,M) \cong \mathrm{Hom}_R(K,M)^H$ by (2.3.2). Since $H$ is semisimple, the functor $(\ )^H$ commutates with taking direct limits. Hence

$$\begin{aligned}
\mathrm{Hom}_{\mathrm{QMod}\,B}(\mathcal{N},\mathcal{M}) &= \varinjlim \mathrm{Hom}_B(K,M) \\
&\cong \varinjlim \mathrm{Hom}_R(K,M)^H \\
&\cong (\varinjlim \mathrm{Hom}_R(K,M))^H \\
&\cong \mathrm{Hom}_{\mathrm{QMod}\,R}(\mathcal{N},\mathcal{M})^H.
\end{aligned}$$

If $M'$ is another $B$-module and there is a right $B$-module morphism $f : M \to M'$, then it is easy to see that the direct systems and the right $H$-module structures on $\mathrm{Hom}_R(K,M)$ and $\mathrm{Hom}_R(K,M')$ are compatible with the morphism $f$. Hence the isomorphism is functorial in $M$. □

Next we show that the proposition above can be extended to extension groups, that is, there are natural $H$-module actions on extension groups $\mathrm{Ext}^i_{\mathrm{QMod}\,R}(\mathcal{N},\mathcal{M})$ if the left $H$-module $R$ is admissible and $\mathrm{Tor}\,B$ is stable.

For the rest of this section, the $H$-module algebra $R$ is admissible and $\mathrm{Tor}\,B$ is stable. Let $M$ be a finitely generated right $B$-module. Take a minimal injective resolution of $M$ as follows:

$$0 \to M \to I^0 \to I^1 \to \cdots \to I^k \to \cdots. \tag{2.9.1}$$

Since $\mathrm{Tor}\,B$ is stable, the injective envelope of a torsion module is still torsion. Then each injective module in the above sequence has a decomposition $I^i = E^i \oplus T^i$ such that $E^i$ is a torsion free $B$-module and $T^i$ is a torsion $B$-module. Write $I^{\cdot}$ for the complex obtained from (2.9.1) by dropping $M$ on the left in the beginning. Similar to the discussions in [**AZ**, Section 7, P.271], we see that there are complexes $E^{\cdot}$, $T^{\cdot}$ and a morphism $f : E^{\cdot}[-1] \to T^{\cdot}$ such that $I^{\cdot} = \mathit{cone}(f)$. Applying the projection functor $\pi : \mathrm{Mod}\,B \to \mathrm{QMod}\,B$ to the resolution (2.9.1) and noticing that $E^i$ is torsion free injective for all $i \geq 0$, we obtain an injective resolution of $\mathcal{M}$ in $\mathrm{QMod}\,B$:

$$0 \to \mathcal{M} \to \mathcal{E}^0 \to \mathcal{E}^1 \to \cdots \to \mathcal{E}^i \to \cdots. \tag{2.9.2}$$

Note that any right $B$-module is also a right $R$-module. Moreover, a right $B$-module $K$ is torsion free if and only if it is torsion free as an $R$-module. We see that $E^i$ is also torsion free when it is viewed as an $R$-module, and $T^i$ is torsion as an $R$-module for all $i \geq 0$. By [**HVZ1**, Proposition 2.6], $E^i$ is also injective as a right $R$-module for each $i \geq 0$. Hence we have the following observation.

LEMMA 2.10. *Assume that* $\operatorname{Tor} B$ *is stable. The sequence (2.9.2) is an injective resolution of* $\mathcal{M}$ *in* $\operatorname{QMod} B$. *If* $\mathcal{M}$ *and* $\mathcal{E}^i$ $(i \geq 0)$ *are viewed as objects in* $\operatorname{QMod} R$ *(by an abuse of the notions), then (2.9.2) is also an injective resolution of* $\mathcal{M}$ *in* $\operatorname{QMod} R$.

THEOREM 2.11. *Assume that* $R$ *is an admissible* $H$*-module algebra and* $\operatorname{Tor} B$ *is stable. Let* $N$ *and* $M$ *be finitely generated right* $B$*-modules. Then for each* $i \geq 0$, $\operatorname{Ext}^i_{\operatorname{QMod} R}(\mathcal{N}, \mathcal{M})$ *is a right* $H$*-module, moreover, we have the following isomorphisms*

$$\operatorname{Ext}^i_{\operatorname{QMod} B}(\mathcal{N}, \mathcal{M}) \cong \operatorname{Ext}^i_{\operatorname{QMod} R}(\mathcal{N}, \mathcal{M})^H, \ \forall\, i \geq 0.$$

PROOF. By Lemma 2.10, the sequence (2.9.2) is an injective resolution of $\mathcal{M}$ when viewed as an object in $\operatorname{QMod} R$. Applying $\operatorname{Hom}_{\operatorname{QMod} R}(\mathcal{N}, -)$ to (2.9.2), we obtain

$$0 \to \operatorname{Hom}_{\operatorname{QMod} R}(\mathcal{N}, \mathcal{E}^0) \to \operatorname{Hom}_{\operatorname{QMod} R}(\mathcal{N}, \mathcal{E}^1) \to \cdots \to \operatorname{Hom}_{\operatorname{QMod} R}(\mathcal{N}, \mathcal{E}^i) \to \cdots .$$

By Proposition 2.9, each component of the complex above is a right $H$-module and the differential is compatible with the right $H$-module structures. Taking the cohomology of the complex above, we obtain that $\operatorname{Ext}^i_{\operatorname{QMod} R}(\mathcal{N}, \mathcal{M})$ is a right $H$-module for each $i \geq 0$.

Now applying $\operatorname{Hom}_{\operatorname{QMod} B}(\mathcal{N}, -)$ to the sequence (2.9.2), we obtain the following complex

$$(2.11.1) \quad 0 \to \operatorname{Hom}_{\operatorname{QMod} B}(\mathcal{N}, \mathcal{E}^0) \to \\ \to \operatorname{Hom}_{\operatorname{QMod} B}(\mathcal{N}, \mathcal{E}^1) \to \cdots \to \operatorname{Hom}_{\operatorname{QMod} B}(\mathcal{N}, \mathcal{E}^i) \to \cdots .$$

By Proposition 2.9, the complex (2.11.1) is isomorphic to the following complex

$$(2.11.2) \quad 0 \to \operatorname{Hom}_{\operatorname{QMod} R}(\mathcal{N}, \mathcal{E}^0)^H \to \\ \to \operatorname{Hom}_{\operatorname{QMod} R}(\mathcal{N}, \mathcal{E}^1)^H \to \cdots \to \operatorname{Hom}_{\operatorname{QMod} R}(\mathcal{N}, \mathcal{E}^i)^H \to \cdots .$$

By Lemma 2.10, (2.9.2) is also an injective resolution of $\mathcal{M}$ in $\operatorname{QMod} R$. Noticing that the functor $(\ )^H$ commutes with taking the cohomology, we obtain that the $i$th cohomology of the complex (2.11.2) is equal to $\operatorname{Ext}^i_{\operatorname{QMod} R}(\mathcal{N}, \mathcal{M})^H$, which is isomorphic to $\operatorname{Ext}^i_{\operatorname{QMod} B}(\mathcal{N}, \mathcal{M})$, the $i$th cohomology of (2.11.1) since (2.9.2) is also an injective resolution of $\mathcal{M}$ in $\operatorname{QMod} B$. Then we obtain the desired isomorphisms. □

## 3. Invariant subalgebras of Hopf dense Galois extensions

Throughout this section, $H$ is a finite dimensional semisimple Hopf algebra and $R$ is a noetherian $H$-module algebra. As before, set $B := R\#H$ and $A = R^H$.

SETUP 3.1. We assume that the $H$-module algebra $R$ satisfies the following conditions:

(i) the $H$-module algebra $R$ is admissible and $\operatorname{Tor} B$ is stable (e.g. $R$ is a complete semilocal algebra, cf. Example 2.5);

(ii) $R/A$ is a right $H^*$-dense Galois extension.

As we have known in the beginning of Section 2, $R$ is a $B$-$A$-bimodule. Then we have a functor

$$- \otimes_B R : \operatorname{Mod} B \longrightarrow \operatorname{Mod} A,$$

which induces a functor between the corresponding quotient categories:

$$- \otimes_{\mathcal{B}} \mathcal{R} : \mathrm{QMod}\, B \longrightarrow \mathrm{QMod}\, A.$$

We recall a result obtained in [**HVZ2**].

LEMMA 3.2. [**HVZ2**, Theorem 2.4] *Assume that Setup* 3.1*(ii) holds. Then the functor*

$$- \otimes_{\mathcal{B}} \mathcal{R} : \mathrm{QMod}\, B \longrightarrow \mathrm{QMod}\, A$$

*is an equivalence of abelian categories.*

Let $S$ be a noetherian algebra. Recall from Section 1 that we have a torsion functor $\Gamma_S : \mathrm{Mod}\, S \to \mathrm{Mod}\, S$. The $i$-th right derived functor of $\Gamma_S$ is denoted by $R^i\Gamma_S$. If there is no risk of confusion, we will drop the subscript $S$. Note that we have $\Gamma = \varinjlim \mathrm{Hom}_S(S/K, -)$, and $R^i\Gamma = \varinjlim \mathrm{Ext}^i_S(S/K, -)$, where $K$ runs over all the cofinite $S$-submodules of $S$, and the direct system is induced by the inclusion maps of cofinite submodules.

DEFINITION 3.3. We say that a noetherian algebra $S$ is (right) *AS-Cohen-Macaulay* of dimension $d$ ($< \infty$) if $R^i\Gamma(S) = 0$ for all $i \neq d$ and $R^d\Gamma(S) \neq 0$.

Assume $S$ is AS-Cohen-Macaulay of dimension $d$. A finitely generated right $S$-module $M$ is said to be a *Cohen-Macaulay* module, if $R^i\Gamma(M) = 0$ for all $i \neq d$.

The definition of an AS-Cohen-Macaulay module is a generalization of the commutative version (cf. [**CH**, **Z**]). The letters "AS" stand for Artin-Schelter, because the property in the definition is also a generalization of that of Artin-Schelter Gorenstein algebras (cf. [**VdB**, Remark 8.5]).

The next lemma is a nongraded version of [**AZ**, Proposition 7.2].

LEMMA 3.4. *Let $S$ be a noetherian algebra such that* $\mathrm{Tor}\, S$ *is stable. For a finitely generated right $S$-module, we have*

(i) $\mathrm{Ext}^i_{\mathrm{QMod}\, S}(\mathcal{S}, \mathcal{M}) \cong R^{i+1}\Gamma(M)$ *for* $i > 0$, *and*

(ii) *an exact sequence* $0 \to \Gamma(M) \to M \xrightarrow{\phi} \mathrm{Hom}_{\mathrm{QMod}\, S}(\mathcal{S}, \mathcal{M}) \to R^1\Gamma(M) \to 0$, *where $\phi$ is the natural map* $M = \mathrm{Hom}_S(S, M) \xrightarrow{\pi} \mathrm{Hom}_{\mathrm{QMod}\, S}(\mathcal{S}, \mathcal{M})$ *induced by the projection functor* $\pi : \mathrm{Mod}\, S \to \mathrm{QMod}\, S$.

PROOF. For any cofinite right $S$-submodule $K$ of $S$, we have an exact sequence $0 \to K \to S \to S/K \to 0$, which implies isomorphisms

$$\mathrm{Ext}^i_S(K, M) \cong \mathrm{Ext}^{i+1}_S(S/K, M), \ \forall\, i \geq 1,$$

and an exact sequence

$$0 \to \mathrm{Hom}_S(S/K, M) \to M \xrightarrow{\phi} \mathrm{Hom}_S(K, M) \to \mathrm{Ext}^1_S(S/K, M) \to 0,$$

where $\phi$ is the natural map induced by the inclusion map $K \hookrightarrow S$. Taking direct limit on all the cofinite submodules of $S$, by Lemma 1.2 and $R^i\Gamma(M) = \varinjlim \mathrm{Ext}^i_S(S/K, -)$ for all $i \geq 0$, we obtain the desired results. $\square$

THEOREM 3.5. *Assume that $R$ and $H$ satisfy the conditions in Setup* 3.1. *If $R$ is AS-Cohen-Macaulay of dimension $d$, then so is $R^H$.*

PROOF. Set $A := R^H$. By Lemma 3.2, the functor $-\otimes_{\mathcal{B}} \mathcal{R} : \mathrm{QMod}\, B \longrightarrow \mathrm{QMod}\, A$ is an equivalence of abelian categories. Under this equivalence, the object $\mathcal{R} \in \mathrm{QMod}\, B$ corresponds to $\mathcal{A} \in \mathrm{QMod}\, A$. Then we have

$$\mathrm{Ext}^i_{\mathrm{QMod}\, A}(\mathcal{A}, \mathcal{A}) \cong \mathrm{Ext}^i_{\mathrm{QMod}\, B}(\mathcal{R}, \mathcal{R}), \ \forall\, i \geq 0. \tag{3.5.1}$$

Combining (3.5.1) with Theorem 2.11, we obtain

$$\mathrm{Ext}^i_{\mathrm{QMod}\, A}(\mathcal{A}, \mathcal{A}) \cong \mathrm{Ext}^i_{\mathrm{QMod}\, R}(\mathcal{R}, \mathcal{R})^H, \ \forall\, i \geq 0. \tag{3.5.2}$$

Assume $d \geq 2$. Then $\mathrm{Ext}^i_{\mathrm{QMod}\, R}(\mathcal{R}, \mathcal{R}) = 0$ for all $i \neq d-1$ and $i \geq 1$ by Lemma 3.4(i), and hence $\mathrm{Ext}^i_{\mathrm{QMod}\, A}(\mathcal{A}, \mathcal{A}) = 0$ for all $i \neq d-1$ and $i \geq 1$. By Lemma 3.4(i) again, $R^{i+1}\Gamma_A(A) = 0$ for $i \neq d-1$ and $i \geq 1$. Since $R^i\Gamma_R(R) = 0$ for $i \neq d$, Lemma 3.4(ii) implies that the natural map $R \to \mathrm{Hom}_{\mathrm{QMod}\, R}(\mathcal{R}, \mathcal{R})$ is an isomorphism. Applying the functor $(\ )^H$ on this isomorphism, we see that the natural map $A \to \mathrm{Hom}_{\mathrm{QMod}\, A}(\mathcal{A}, \mathcal{A})$ is an isomorphism. Hence $R^i\Gamma_A(A) = 0$ for $i = 0, 1$.

Now assume $d = 1$. Similar to the above discussions, we have $R^i\Gamma_A(A) = 0$ for $i \geq 2$. Since $\Gamma_R(R) = 0$, Lemma 3.4(ii) implies that the natural map $R \xrightarrow{\phi} \mathrm{Hom}_{\mathrm{QMod}\, R}(\mathcal{R}, \mathcal{R})$ is injective. Applying the functor $(\ )^H$ to this map, we obtain that the natural map $A \xrightarrow{\phi} \mathrm{Hom}_{\mathrm{QMod}\, A}(\mathcal{A}, \mathcal{A})$ is injective. Hence $\Gamma_A(A) = 0$.

Assume $d = 0$. By (3.5.1), (3.5.2) and Lemma 3.4(i), we have $R^i\Gamma_A(A) = 0$ for $i \geq 2$. Since $R^1\Gamma_R(R) = 0$, Lemma 3.4(ii) implies that the natural map $R \to \mathrm{Hom}_{\mathrm{QMod}\, R}(\mathcal{R}, \mathcal{R})$ is an epimorphism. Applying the functor $(\ )^H$ to this map, we obtain that the natural map $A \xrightarrow{\phi} \mathrm{Hom}_{\mathrm{QMod}\, A}(\mathcal{A}, \mathcal{A})$ is an epimorphism. Hence $R^1\Gamma_A(A) = 0$. □

## 4. Cohen-Macaulay property of $R$ as an $R^H$-module

Keep the same notions as in Section 3. We recall some properties of modules over a Hopf algebra.

Let $X$ be a right $H$-module. The tensor product $X \otimes H$ has two right $H$-module structures. The first one is the diagonal action of $H$, that is, $(x \otimes h) \cdot g = x \cdot g_{(1)} \otimes hg_{(2)}$ for $x \in X$, $g, h \in H$. To avoid possible confusion, the tensor product $X \otimes H$ with the diagonal action of $H$ will be denoted by $X \hat{\otimes} H$. The other $H$-action is the multiplication of elements of $H$ on the right of $X \otimes H$, that is, $(x \otimes h) \cdot g = x \otimes hg$ for $x \in X$ and $g, h \in H$. This right $H$-module structure will be denoted by $X \otimes H_\bullet$. The following lemma is well known.

LEMMA 4.1. *Let $X$ be a right $H$-module. Define maps*

$$\varphi : X \hat{\otimes} H \longrightarrow X \otimes H_\bullet, \ x \otimes h \mapsto xS(h_{(1)}) \otimes h_{(2)},$$

*and*

$$\psi : X \otimes H_\bullet \longrightarrow X \hat{\otimes} H, \ x \otimes h \mapsto xh_{(1)} \otimes h_{(2)}.$$

*Then $\varphi$ and $\psi$ are $H$-module isomorphisms which are inverse to each other.*

Let $X$ be a right $B$-module. Recall from the beginning of Section 2 that $R$ is a right $B$-module by the right $B$-action (2.0.2). Then $\mathrm{Hom}_R(X, R)$ is a right $H$-module. Note that $B$ itself is a right $B$-module. Hence $\mathrm{Hom}_R(X, B)$ is a right $H$-module. The next lemma should be well known. Since it is difficult to find a reference, we include the proof.

LEMMA 4.2. *Let $X$ be a right $B$-module. Then we have an isomorphism of right $H$-modules*

$$\mathrm{Hom}_R(X,R)\hat{\otimes}H \cong \mathrm{Hom}_R(X,R\#H).$$

PROOF. Firstly, note that the smash product may be written as $H\#R$ with the multiplication defined by

$$(h\#r)(g\#r') = hg_{(2)}\#(S^{-1}g_{(1)}r)r'.$$

There is an isomorphism of algebras:

$$\xi : H\#R \to R\#H,\ h\#r \mapsto h_{(1)}r\#h_{(2)}.$$

Define a linear map

$$\zeta : \mathrm{Hom}_R(X,R)\hat{\otimes}H \longrightarrow \mathrm{Hom}(X,R\#H)$$

by $\zeta(f\otimes h)(x) = h_{(1)}f(x)\#h_{(2)}$ for $f \in \mathrm{Hom}_R(X,R)$, $x\in X$ and $h \in H$. We first show that $\zeta(f\otimes h)$ is a right $R$-module morphism for every $f\in \mathrm{Hom}_R(X,R)$ and $h\in H$. For $r\in R$ and $x \in X$, we have

$$\begin{aligned}
\zeta(f\otimes h)(xr) &= h_{(1)}f(xr)\#h_{(2)}\\
&= h_{(1)}(f(x)r)\#h_{(2)}\\
&= (h_{(1)}f(x))(h_{(2)}r)\#h_{(3)}\\
&= (h_{(1)}f(x)\#h_{(2)})r\\
&= \zeta(f\otimes h)(x)r.
\end{aligned}$$

Hence the image of $\zeta$ is indeed in $\mathrm{Hom}_R(X,R\#H)$. Therefore we obtain a map (still use the same notion):

$$\zeta : \mathrm{Hom}_R(X,R)\hat{\otimes}H \longrightarrow \mathrm{Hom}_R(X,R\#H).$$

We next check that $\zeta$ is a right $H$-module morphism. For $f\in \mathrm{Hom}_R(X,R)$, $h,g\in H$ and $x\in X$, we have

$$\begin{aligned}
\zeta(f\otimes h)\leftharpoonup g(x) &= \big(\zeta(f\otimes g)(xS(g_{(1)}))\big)\,g_{(2)}\\
&= \big(h_{(1)}f(xS(g_{(1)}))\#h_{(2)}\big)\,g_{(2)}\\
&= h_{(1)}f(xS(g_{(1)}))\#h_{(2)}g_{(2)}.
\end{aligned}$$

On the other hand, we have

$$\begin{aligned}
\zeta\left(\left(f\otimes h\right)\cdot g\right)(x) &= \zeta(f\leftharpoonup g_{(1)}\otimes hg_{(2)})(x)\\
&= h_{(1)}g_{(2)}f\leftharpoonup g_{(1)}(x)\#h_{(2)}g_{(3)}\\
&= h_{(1)}g_{(3)}S^{-1}(g_{(2)})f(xS(g_{(1)}))\#h_{(2)}g_{(4)}\\
&= h_{(1)}f(xS(g_{(1)}))\#h_{(2)}g_{(2)},
\end{aligned}$$

where the third equality follows from the definitions of the right $B$-module action (2.0.2) on $R$ and the right $H$-action on $\mathrm{Hom}_R(X,R)$. Therefore, $\zeta$ is a right $H$-module morphism.

Since $H\#R$ is a finitely generated free $R$-module, it follows that

$$\theta : \mathrm{Hom}_R(X,R)\otimes H \longrightarrow \mathrm{Hom}_R(X,H\#R),\ f\otimes h\mapsto [x\mapsto h\otimes f(x)]$$

is a linear isomorphism. Thus $\zeta$, equal to $\mathrm{Hom}_R(X,\xi)\circ\theta$, is an isomorphism since $\xi$ is an isomorphism. □

LEMMA 4.3. *Let $X$ be a right $B$-module. For each $i \geq 0$, we have an $H$-module isomorphism*

$$\mathrm{Ext}^i_R(X, R\#H) \cong \mathrm{Ext}^i_R(X, R)\hat{\otimes}H.$$

*Moreover, the isomorphism is functorial in $X$.*

PROOF. Take a projective resolution of the right $B$-module $X$ as follows:

$$\cdots \longrightarrow P^i \longrightarrow \cdots \longrightarrow P^1 \longrightarrow P^0 \longrightarrow X \longrightarrow 0. \tag{4.3.1}$$

By [**HVZ1**, Proposition 2.5], each $P^i$ is also projective as a right $R$-module. Apply the functor $\mathrm{Hom}_R(-, B)$ on (4.3.1) to obtain the following complex

$$0 \longrightarrow \mathrm{Hom}_R(P^0, B) \longrightarrow \mathrm{Hom}_R(P^1, B) \longrightarrow \cdots \longrightarrow \mathrm{Hom}_R(P^i, B) \longrightarrow \cdots. \tag{4.3.2}$$

By Lemma 4.2, the complex (4.3.2) can be written in the following way

$$\begin{aligned} 0 \longrightarrow \mathrm{Hom}_R(P^0, R)\hat{\otimes}H \longrightarrow \\ \longrightarrow \mathrm{Hom}_R(P^1, R)\hat{\otimes}H \longrightarrow \cdots \longrightarrow \mathrm{Hom}_R(P^i, R)\hat{\otimes}H \longrightarrow \cdots. \end{aligned} \tag{4.3.3}$$

Comparing the cohomology of (4.3.2) and (4.3.3), we obtain the desired isomorphisms. It follows from the general homology theory that the isomorphism is functorial in $X$. $\square$

PROPOSITION 4.4. *Let $N$ be a finitely generated right $B$-module. Assume Setup 3.1(i). For each $i \geq 0$, we have*

$$\mathrm{Ext}^i_{\mathrm{QMod}\,B}(\mathcal{N}, \mathcal{B}) \cong \mathrm{Ext}^i_{\mathrm{QMod}\,R}(\mathcal{N}, \mathcal{R}).$$

PROOF. By Lemma 1.2 and the assumption that the $H$-module algebra $R$ is admissible, $\mathrm{Ext}^i_{\mathrm{QMod}\,R}(\mathcal{N}, \mathcal{B}) \cong \varinjlim \mathrm{Ext}^i_R(N', B)$ where $N'$ runs over all the cofinite $B$-submodules of $N$. By Lemma 4.3, $\mathrm{Ext}^i_R(N', B) \cong \mathrm{Ext}^i_R(N', R)\hat{\otimes}H$ for all cofinite $B$-submodules $N'$ of $N$. Taking the direct limits over all the cofinite $B$-submodules of $N$, we obtain

$$\mathrm{Ext}^i_{\mathrm{QMod}\,R}(\mathcal{N}, \mathcal{B}) \cong \mathrm{Ext}^i_{\mathrm{QMod}\,R}(\mathcal{N}, \mathcal{R})\hat{\otimes}H. \tag{4.4.1}$$

Combining Theorem 2.11 and Lemma 4.1, we have

$$\mathrm{Ext}^i_{\mathrm{QMod}\,B}(\mathcal{N}, \mathcal{B}) \cong \left(\mathrm{Ext}^i_{\mathrm{QMod}\,R}(\mathcal{N}, \mathcal{R}) \otimes H_\bullet\right)^H. \tag{4.4.2}$$

The right hand side of (4.4.2) is isomorphic to $\mathrm{Ext}^i_{\mathrm{QMod}\,R}(\mathcal{N}, \mathcal{R})$ as a vector space. $\square$

Now we arrive at our main result of this section.

THEOREM 4.5. *Assume that $R$ and $H$ satisfy the conditions in Setup* 3.1. *If $R$ is AS-Cohen-Macaulay of dimension $d$, then $R$, viewed as a right $R^H$-module, is Cohen-Macaulay in the sense of Definition* 3.3.

PROOF. Set $A := R^H$. Since $R/A$ is a right $H^*$-dense Galois extension, the functor $-\otimes_{\mathcal{B}} \mathcal{R} : \mathrm{QMod}\,B \to \mathrm{QMod}\,A$ is an equivalence of abelian categories (cf. Lemma 3.2). Note that $R$ may be viewed both as a right $A$-module and as a right $B$-module. Under the equivalence $-\otimes_{\mathcal{B}}\mathcal{R}$, $\mathcal{R}$ when viewed as an object in $\mathrm{QMod}\,A$ corresponds to $\mathcal{B} \in \mathrm{QMod}\,B$, and $\mathcal{A} \in \mathrm{QMod}\,A$ corresponds to $\mathcal{R}$ which is viewed as an object in $\mathrm{QMod}\,B$. Hence for $i \geq 0$, we have

$$\mathrm{Ext}^i_{\mathrm{QMod}\,B}(\mathcal{R}, \mathcal{B}) \cong \mathrm{Ext}^i_{\mathrm{QMod}\,A}(\mathcal{A}, \mathcal{R}).$$

By Proposition 4.4, we obtain

$$\mathrm{Ext}^i_{\mathrm{QMod}\,R}(\mathcal{R},\mathcal{R}) \cong \mathrm{Ext}^i_{\mathrm{QMod}\,A}(\mathcal{A},\mathcal{R}). \tag{4.5.1}$$

The rest of the proof is almost the same as that of Theorem 3.5 according to the value of $d$. We omit the similar narratives. □

REMARK 4.6. Under the assumptions in Theorem 4.5, the isomorphism (4.5.1) implies that

$$R^i\Gamma_{R^H}(R) \cong R^i\Gamma_R(R), \text{ for all } i \geq 0. \tag{4.6.1}$$

If the algebra $R$ is a noetherian complete semilocal algebra and some further conditions are satisfied, then (4.6.1) follows from [**WZ1**, Theorem 2.9]. When the algebra $R$ is connected graded, the isomorphism (4.6.1) follows from a more general result [**AZ**, Theorem 8.3(3)].

## 5. Finite group actions on noetherian complete semilocal algebras

We define that a noetherian algebra $\Lambda$ is *semilocal* if $\Lambda/J(\Lambda)$ is a finite dimensional semisimple algebra, where $J(\Lambda)$ is the Jacobson radical of $\Lambda$. We say that $\Lambda$ is *complete* if $\Lambda$ is complete with respect to the $J(\Lambda)$-adic topology, or equivalently, $\Lambda = \varprojlim \Lambda/J(\Lambda)^n$.

Recall from [**WZ2**] that a noetherian semilocal algebra $\Lambda$ is called a right *AS-Gorenstein* algebra if $\Lambda_\Lambda$ has finite injective dimension $d$, $\mathrm{Ext}^i_\Lambda(\Lambda/J(\Lambda),\Lambda) = 0$ for $i \neq d$, and $\mathrm{Ext}^d_\Lambda(\Lambda/J(\Lambda),\Lambda) \cong \Lambda/J(\Lambda)$ as a left $\Lambda$-module. Left AS-Gorenstein algebras are defined similarly. If $\Lambda$ is both left and right AS-Gorenstein, then we simply say that $\Lambda$ is *AS-Gorenstein*. If, furthermore, $\Lambda$ has finite global dimension, then we say that $\Lambda$ is *AS-regular*. We remark that our definition of AS-Gorenstein algebras is a little stronger than that in [**WZ2**].

Let $\Lambda$ be a noetherian semilocal algebra. For any cofinite right ideal $K$ of $\Lambda$, there is an integer $n$ such that $(\Lambda/K)J(\Lambda)^n = 0$. Then $J(\Lambda)^n \subseteq K$. Hence the direct system defined by all the cofinite right ideals of $S$ and the direct system defined by $\{J(\Lambda)^n | n \geq 1\}$ are cofinal. Therefore the torsion functor defined in this paper coincides with that in [**WZ1**, **WZ2**]. That is, we have the following form of the torsion functor.

LEMMA 5.1. *Let $\Lambda$ be a noetherian semilocal algebra. Then*

$$\Gamma_\Lambda \cong \lim_{n\to\infty} \mathrm{Hom}_\Lambda(\Lambda/J(\Lambda)^n, -).$$

In the rest of this section, we assume that $R$ is a noetherian complete semilocal algebra with Jacobson radical $J$. Let $G \subseteq \mathrm{Aut}(R)$ be a finite subgroup. Then $R$ becomes a left $G$-module algebra. As before, set $B = R\#\mathbf{k}G$ and $A = R^G$. Example 2.5 shows that the $G$-module algebra $R$ is admissible and $\mathrm{Tor}\,B$ is stable. Note that $J$ is stable under the $G$-action. Let $\mathfrak{m} = J\#\mathbf{k}G$. Then $\mathfrak{m}$ is a cofinite ideal of $B$, and $\mathfrak{m}^n = J^n\#\mathbf{k}G$ for all $n \geq 1$. In general, $\mathfrak{m}$ is not the Jacobson radical of $B$. Besides, for a general semisimple Hopf algebra $H$, an $H$-action on $R$ does not imply that the $H$ acts on the Jacobson radical $J$ stably. Some discussions on the stability of the $H$-action on the Jacobson radical, and on the Jacobson radical of the smash product may be found in [**LMS**].

Similar to Lemma 5.1, we also have

$$\Gamma_B \cong \lim_{n\to\infty} \mathrm{Hom}_B(B/\mathfrak{m}^n, -). \tag{5.1.1}$$

Recall that the *depth* of a right module $M$ over a noetherian algebra $\Lambda$ is defined as follows

$$\operatorname{depth}_\Lambda(M) = \min\{i | R^i\Gamma_\Lambda(M) \neq 0\}.$$

We have the following Auslander-Buchsbaum formula, which is a consequence of [**WZ1**, Theorem 0.1(1)]. We include the proof for the completeness. We remark that our definition of the depth of a module is equivalent to that in [**WZ1**] when the algebra $\Lambda$ is a notherian complete semilocal algebra (see also, Lemma 5.3 below).

LEMMA 5.2. [**WZ1**, Theorem 0.1(1)] *Let $R$ be a noetherian complete semilocal algebra which is also AS-Gorenstein of injective dimension $d$. For a finitely generated right $B$-module $M$ with finite projective dimension, we have* $\operatorname{projdim}_B(M) + \operatorname{depth}_B(M) = d$.

PROOF. Note that $M$ is also a finitely generated right $R$-module. By Lemma 2.1, $\operatorname{depth}_B(M) = \operatorname{depth}_R(M)$. By [**WZ1**, Theorem 0.1(1)], we have

$$\operatorname{projdim}_R(M) = d - \operatorname{depth}_R(M) = d - \operatorname{depth}_B(M). \tag{5.2.1}$$

Since a projective right $B$-module is projective as a right $R$-module, $\operatorname{projdim}_R(M) \leq \operatorname{projdim}_B(M)$. On the other hand, set $p = \operatorname{projdim}_R(M)$. Let $\cdots \to P^i \xrightarrow{\delta^i} \cdots \xrightarrow{\delta^1} P^0 \to M \to 0$ be a projective resolution of the right $B$-module $M$. Then the $p$th syzygy $\ker \delta^{p-1}$ of the resolution is projective as an $R$-module since $\operatorname{projdim}_R(M) = p$. By [**HVZ1**, Proposition 2.5], $\ker \delta^{p-1}$ is projective as a right $B$-module, and hence $\operatorname{projdim}_B(M) \leq p$. Therefore $\operatorname{projdim}_R(M) = \operatorname{projdim}_B(M)$. By the equalities (5.2.1), $\operatorname{projdim}_B(M) + \operatorname{depth}_B(M) = d$. □

For the proof of the main result of this section, we need the following lemma.

LEMMA 5.3. *Let $A$ be a noetherian algebra. Let $M_A$ be a finitely generated module. If $R^i\Gamma_A(M) = 0$ for $i < d$, then $\operatorname{Ext}^i_A(T, M) = 0$ for all finite dimensional right $A$-module $T$.*

PROOF. Take a minimal injective resolution of $M$ as follows

$$0 \to M \to I^0 \xrightarrow{\delta^0} I^1 \xrightarrow{\delta^1} \cdots \xrightarrow{\delta^{i-1}} I^i \xrightarrow{\delta^i} \cdots.$$

We claim that each $I^i$ is torsion free for all $i < d$. Since $\Gamma_A(M) = 0$, $M$ is torsion free. Hence $I^0$ is torsion free. Now suppose that there is some $i < d$ such that $I^i$ is not torsion free. Assume that $k < d$ is the smallest integer such that $\Gamma_A(I^k) \neq 0$. Since the injective resolution is minimal, $\ker \delta^k$ is essential in $I^k$. Hence $\ker \delta^k \cap \Gamma_A(I^k) \neq 0$. Then the kernel of the restriction $\delta^k : \Gamma_A(I^k) \to \Gamma_A(I^{k+1})$ is nontrivial. Therefore $R^k\Gamma_A(M) \neq 0$, a contradiction. Hence for each $i < d$, $I^i$ is torsion free. Therefore, for any finite dimensional module $T$, $\operatorname{Hom}_A(T, I^i) = 0$ for all $i < d$, which implies that $\operatorname{Ext}^i_A(T, M) = 0$. □

It is not hard to see that if a noetherian complete semilocal algebra is AS-Gorenstein with injective dimension $d$, then it is AS-Cohen-Macaulay of dimension $d$ in the sense of Definition 3.3 (cf. [**WZ1**]).

THEOREM 5.4. *Let $R$ be a noetherian complete semilocal algebra, and let $G \subseteq \operatorname{Aut}(R)$ be a finite subgroup. Assume that $R/R^G$ is a right $\mathbf{k}G^*$-dense Galois extension and that $R$ is AS-regular of global dimension 2. Set $A = R^G$. Then the following statements hold.*

(i) *$A$ is AS-Cohen-Macaulay of dimension 2.*

(ii) *A finitely generated right $A$-module $M$ is Cohen-Macaulay if and only if $M \in \operatorname{add}(R_A)$, where $\operatorname{add}(R_A)$ is the subcategory of* $\operatorname{Mod} A$ *consisting of all the direct summands of finite direct sums of $R_A$.*

PROOF. The statement (i) is a special case of Theorem 3.5.

(ii) By Theorem 4.5, $R_A$ is a Cohen-Macaulay module. Hence any module in $\operatorname{add} R_A$ is Cohen-Macaulay.

Conversely, note that the functor $-\otimes_{\mathcal{B}} \mathcal{R} : \operatorname{QMod} B \to \operatorname{QMod} A$ in Lemma 3.2 is induced by the functor $-\otimes_B R : \operatorname{Mod} B \to \operatorname{Mod} A$ (cf. [**HVZ2**, Sections 2 and 3]), that is, we have the following commutative diagram

$$
\begin{array}{ccc}
\operatorname{Mod} B & \xrightarrow{-\otimes_B R} & \operatorname{Mod} A \\
\downarrow{\scriptstyle\pi} & & \downarrow{\scriptstyle\pi} \\
\operatorname{QMod} B & \xrightarrow{-\otimes_{\mathcal{B}} \mathcal{R}} & \operatorname{QMod} A.
\end{array}
\tag{5.4.1}
$$

Let $F = -\otimes_B R$ and $\mathcal{F} = -\otimes_{\mathcal{B}} \mathcal{R}$. The functor $F$ has a right adjoint functor $F' := \operatorname{Hom}_A(R, -)$ (cf. [**HVZ2**, Section 2]), where $\operatorname{Hom}_A(-,-)$ is the Hom-functor in the category of left $A$-modules. Moreover, $FF' \cong \operatorname{id}_{\operatorname{Mod} A}$. Hence $F'$ is fully faithful and $\operatorname{Hom}_B(F'X, F'Y) \xrightarrow{F} \operatorname{Hom}_A(X, Y)$ is an isomorphism for any $X, Y \in \operatorname{Mod} A$.

Let $N_A$ be a nonzero finitely generated Cohen-Macaulay module. Then we obtain that the map $\operatorname{Hom}_B(F'(R), F'(N)) \xrightarrow{F} \operatorname{Hom}_A(R, N)$ is an isomorphism. Since $R/A$ is a right $\mathbf{k}G^*$-dense Galois extension and $R$ is AS-regular of global dimension 2, the natural map $R\#\mathbf{k}G \to \operatorname{Hom}_A(R,R), r\#h \mapsto [r' \mapsto r(h\cdot r')]$ is an isomorphism of algebras by [**HVZ2**, Theorem 3.10]. Hence as a right $B$-module, $F'(R) = \operatorname{Hom}_A(R,R)$ is isomorphic to $B$. Set $M = F'(N) = \operatorname{Hom}_A(R,N) \in \operatorname{Mod} B$.

From the above commutative diagram (5.4.1), we obtain the following commutative diagram

$$
\begin{array}{ccc}
\operatorname{Hom}_B(B, M) & \xrightarrow{F} & \operatorname{Hom}_A(R, N) \\
\downarrow{\scriptstyle\pi} & & \downarrow{\scriptstyle\pi} \\
\operatorname{Hom}_{\operatorname{QMod} B}(\mathcal{B}, \mathcal{M}) & \xrightarrow{\mathcal{F}} & \operatorname{Hom}_{\operatorname{QMod} A}(\mathcal{R}, \mathcal{N}).
\end{array}
\tag{5.4.2}
$$

The top map in the diagram (5.4.2) is an isomorphism. By Lemma 3.2, the bottom map is also an isomorphism. Consider the exact sequence

$$0 \to K \to R \to R/K \to 0,$$

where $K$ is a cofinite submodule of $R$. Applying $\operatorname{Hom}_A(-, N)$ to the above sequence, we obtain the following exact sequence

$$0 \to \operatorname{Hom}_A(R/K, N) \to \operatorname{Hom}_A(R, N) \to \operatorname{Hom}_A(K, N) \to \operatorname{Ext}^1_A(R/K, N) \tag{5.4.3}$$

Since $N$ is Cohen-Macaulay and $R/K$ is finite dimensional, $\operatorname{Hom}_A(R/K, N) = 0$ and $\operatorname{Ext}^1_A(R/K, N) = 0$ by Lemma 5.3. Hence (5.4.3) implies that the natural map

$$\operatorname{Hom}_A(R, N) \to \operatorname{Hom}_A(K, N)$$

is an isomorphism for all cofinite submodules $K$ of $R$.

Taking the direct limits on both sides of the isomorphism above over all the cofinite submodules $K$ of $R$, we obtain that the projection map

$$\mathrm{Hom}_A(R,N) \xrightarrow{\pi} \mathrm{Hom}_{\mathrm{QMod}\,A}(\mathcal{R},\mathcal{N})$$

is an isomorphism.

In summary, we obtain that the top map, the bottom map and the right vertical map in the diagram (5.4.2) are all isomorphisms. Hence the left vertical map in the diagram (5.4.2) is also an isomorphism.

By Lemma 5.1, the exact sequence $0 \to \mathfrak{m}^n \to B \to B/\mathfrak{m} \to 0$ implies the exact sequence

$$0 \to \Gamma_B(M) \to \mathrm{Hom}_B(B,M) \xrightarrow{\pi} \mathrm{Hom}_{\mathrm{QMod}\,B}(\mathcal{B},\mathcal{M}) \to R^1\Gamma_B(M) \to 0.$$

Since the projection map $\pi$ is an isomorphism, $\Gamma_B(M) = R^1\Gamma_B(M) = 0$. Since $N$ is finitely generated, by [**HVZ2**, Theorem 3.8(i)], $\mathcal{M}$ is in $\mathrm{qmod}\,B$. Hence there is a finitely generated $B$-module $M'$ and a $B$-module morphism $f : M \to M'$ such that both $\ker f$ and $\mathrm{coker} f$ are torsion modules. Since $\Gamma_B(M) = 0$, $f$ is injective. Hence $M$ itself is finitely generated. Since $R$ is of global dimension 2, the global dimension of $B$ is 2 as well (cf. [**L**]). By Lemma 5.2, $M$ is a projective $B$-module. Hence $M$ is a direct summand of a free module $B^{\oplus n}$. Therefore $N = F(M)$ is a direct summand of $F(B)^{\oplus n} = (R_A)^{\oplus n}$, that is, $N \in \mathrm{add}(R_A)$. □

REMARK 5.5. The above theorem may be viewed as a nongraded version of [**Jo**, Section 3], [**Ue1**, Corollary 1.3] (see also [**Ue2**, Proposition 5.2]) and [**CKWZ**, Theorem 4.4]. Our method is also valid for noetherian graded algebras (cf. Example 2.7), however our method is quite different from those in [**Jo**, **Ue1**, **CKWZ**].

**Acknowledgments.** We thank the referee for the useful comments and suggestions.

## References

[A] Maurice Auslander, *Rational singularities and almost split sequences*, Trans. Amer. Math. Soc. **293** (1986), no. 2, 511–531, DOI 10.2307/2000019. MR816307

[AZ] M. Artin and J. J. Zhang, *Noncommutative projective schemes*, Adv. Math. **109** (1994), no. 2, 228–287, DOI 10.1006/aima.1994.1087. MR1304753

[CH] Olgur Celikbas and Henrik Holm, *Equivalences from tilting theory and commutative algebra from the adjoint functor point of view*, New York J. Math. **23** (2017), 1697–1721. MR3741857

[CFM] M. Cohen, D. Fischman, and S. Montgomery, *Hopf Galois extensions, smash products, and Morita equivalence*, J. Algebra **133** (1990), no. 2, 351–372, DOI 10.1016/0021-8693(90)90274-R. MR1067411

[CKWZ] K. Chan, E. Kirkman, C. Walton, and J. J. Zhang, *McKay correspondence for semisimple Hopf actions on regular graded algebras, I*, J. Algebra **508** (2018), 512–538, DOI 10.1016/j.jalgebra.2018.05.008. MR3810305

[Ja] Arun Vinayak Jategaonkar, *Morita duality and Noetherian rings*, J. Algebra **69** (1981), no. 2, 358–371, DOI 10.1016/0021-8693(81)90210-6. MR617084

[Jo] Peter Jørgensen, *Finite Cohen-Macaulay type and smooth non-commutative schemes*, Canad. J. Math. **60** (2008), no. 2, 379–390, DOI 10.4153/CJM-2008-018-0. MR2398754

[HVZ1] Ji-Wei He, Fred Van Oystaeyen, and Yinhuo Zhang, *Hopf algebra actions on differential graded algebras and applications*, Bull. Belg. Math. Soc. Simon Stevin **18** (2011), no. 1, 99–111. MR2809906

[HVZ2] Ji-Wei He, Fred Van Oystaeyen, and Yinhuo Zhang, *Hopf dense Galois extensions with applications*, J. Algebra **476** (2017), 134–160, DOI 10.1016/j.jalgebra.2016.12.014. MR3608147

[KKZ] E. Kirkman, J. Kuzmanovich, and J. J. Zhang, *Gorenstein subrings of invariants under Hopf algebra actions*, J. Algebra **322** (2009), no. 10, 3640–3669, DOI 10.1016/j.jalgebra.2009.08.018. MR2568355

[L] Gongxiang Liu, *A note on the global dimension of smash products*, Comm. Algebra **33** (2005), no. 8, 2625–2627, DOI 10.1081/AGB-200064346. MR2159492

[LMS] V. Linchenko, S. Montgomery, and L. W. Small, *Stable Jacobson radicals and semiprime smash products*, Bull. London Math. Soc. **37** (2005), no. 6, 860–872, DOI 10.1112/S0024609305004662. MR2186719

[VdB] Michel van den Bergh, *Existence theorems for dualizing complexes over non-commutative graded and filtered rings*, J. Algebra **195** (1997), no. 2, 662–679, DOI 10.1006/jabr.1997.7052. MR1469646

[Ue1] Kenta Ueyama, *Noncommutative graded algebras of finite Cohen-Macaulay representation type*, Proc. Amer. Math. Soc. **143** (2015), no. 9, 3703–3715, DOI 10.1090/proc/12527. MR3359563

[Ue2] Kenta Ueyama, *Graded maximal Cohen-Macaulay modules over noncommutative graded Gorenstein isolated singularities*, J. Algebra **383** (2013), 85–103, DOI 10.1016/j.jalgebra.2013.02.022. MR3037969

[WZ1] Q.-S. Wu and J. J. Zhang, *Dualizing complexes over noncommutative local rings*, J. Algebra **239** (2001), no. 2, 513–548, DOI 10.1006/jabr.2000.8689. MR1832904

[WZ2] Q.-S. Wu and J. J. Zhang, *Homological identities for noncommutative rings*, J. Algebra **242** (2001), no. 2, 516–535, DOI 10.1006/jabr.2001.8817. MR1848957

[Z] Majid Rahro Zargar, *On the relative Cohen-Macaulay modules*, J. Algebra Appl. **14** (2015), no. 3, 1550042, 7, DOI 10.1142/S0219498815500425. MR3275579

Department of Mathematics, Hangzhou Normal University, Hangzhou Zhejiang 310036, China

*Email address*: jwhe@hznu.edu.cn

Department of Mathematics and Statistics, University of Hasselt, Universitaire Campus, 3590 Diepenbeek, Belgium

*Email address*: yinhuo.zhang@uhasselt.be

Contemporary Mathematics
Volume **728**, 2019
https://doi.org/10.1090/conm/728/14659

# Bialgebra coverings and transfer of structure

Aaron Lauve and Mitja Mastnak

Abstract. We introduce the bicategory of bialgebras with coverings (which can be thought of as coalgebra-indexed families of morphisms), and provide a motivating application to the transfer of formulas for primitives and antipode. Additionally, we study properties of this bicategory and various subbicategories, and describe some universal constructions. Finally, we generalize Nichols' result on bialgebra quotients of Hopf algebra, which gives conditions on when the resulting bialgebra quotient is a Hopf algebra.

## Introduction

We define and study the bicategory of bialgebras with coverings. Its construction was, in large part, inspired by an idea due to Grunefelder and Paré [**4**] of indexing families of algebra morphisms by coalgebras. In this note we focus on the algebraic aspects of the theory. Combinatorial applications will be explored further in [**8**].

The study of this bicategory was originally motivated by applications to combinatorial properties of $\mathbb{N}$-graded connected bialgebras, but we hope that this work will also lead to further developments to the general theory of Hopf algebras. A case in point is a natural generalization of Nichols' result [**12**], which states that a bialgebra quotient of a Hopf algebra that is either finite dimensional or cocommutative is automatically a Hopf algebra (see Theorems 4.1 and 4.5). It should also serve as a nice new example of a bicategory (and also a double category; see the remark of Section 2.1).

This work began as an attempt to understand the primitives and antipode of the Hopf algebra $\mathbf{\Pi}$ of symmetric functions in noncommuting variables [**9**]. The formulas found there for expressing primitives and the antipode in terms of a distinguished generating set of $\mathbf{\Pi}$ point to the Hopf algebra OMP of ordered multiset partitions (see Example 14). The Hopf algebra OMP is the free algebra generated by all finite subsets of natural numbers and has comultiplication given by partitioning a set in all possible ways. Even though OMP is, in some sense, much bigger then $\mathbf{\Pi}$, it is much easier to get a handle on the formulas for computing its primitives and the antipode in terms of its generators (finite subsets of $\mathbb{N}$). There is a nice

2010 *Mathematics Subject Classification.* 18D05, 16T05, 05E05, 16T30.

*Key words and phrases.* measuring, combinatorial Hopf algebras, set partitions, ordered multiset partitions, antipode, primitives, convolution algebra.

The first author is partially supported by NSA grant H98230-12-1-0286.

The second author is partially supported by NSERC Discovery Grant 371994-2014.

family of bialgebra homomorphisms from OMP to $\mathbf{\Pi}$ that jointly covers it; and this family can then be used to transport the structure formulas from OMP to $\mathbf{\Pi}$.

This can be formalized as follows: if $B$ and $A$ are bialgebras and $C$ a coalgebra, then we say that $f\colon B \otimes C \to A$ is a partial covering if it is a measuring (*i.e.*, it corresponds to an algebra map from $B$ to the convolution algebra $\mathrm{Hom}(C, A)$) as well as a coalgebra map. We say that $f$ is a covering if it is also surjective. The transfer of structure discussed above can then be formalized as follows (cf. Theorem 2.4):

*Let $f\colon B \otimes C \to A$ be a covering. Let $\iota\colon A \to B \otimes C$ be any linear section of $f$, that is, $f \circ \iota = \mathrm{id}$. Then the following hold.*

(1) *If $p \in B$ is primitive then for any $c \in C$, $f(p, c)$ is primitive in $A$.*
(2) *If $A$ and $B$ are Hopf algebras, then their antipodes are related by the formula $S_A = f \circ (S_B \otimes \mathrm{id}) \circ \iota$.*

Given two partial coverings $f\colon B \otimes C_f \to A$ and $g\colon B \otimes C_g \to A$, we may also consider morphisms between them. These are given by coalgebra maps $\mathbf{t}\colon C_f \to C_g$ such that $f = g \circ (\mathrm{id} \otimes \mathbf{t})$ and lead naturally to the bicategory of bialgebras with partial coverings. This bicategory context allows us to define a concept of a covering-equivalence of bialgebras and Hopf algebras. We hope that this will serve as a nice organizing principle in classification efforts for various classes of Hopf algebras (in particular, but not limited to, combinatorial Hopf algebras).

This paper is split into four sections:

In Section 1 we introduce several motivating examples and the bicategory of algebras with measurings (*i.e.*, we briefly forget about the coalgebra structure on $B$ and $A$ and we merely demand that $f\colon B \otimes C \to A$ is a measuring). Again, this follows the seminal work of Grunenfelder and Paré [**4**]; though here we realize a bicategory structure by making $C$ a variable (in [**4**] the choice of $C$ was fixed).

In Section 2 we define the bicategories of bialgebras with coverings and partial coverings and give additional motivating examples. We adapt Sweedler's construction of the universal measuring coalgebra to construct a universal partial covering coalgebra. Additionally, we define and discuss the notion of equivalent coverings. We conclude the section by proving the theorem on transfer of structure discussed above.

In Sections 3 and 4, we discuss properties of subcategories of interest. In the former, we focus on cocommutative and graded coverings. We also identify weak initial objects for coverings of graded connected cocommutative bialgebras. In Section 4, we focus on coverings by Hopf algebras. We prove the following: if a Hopf algebra $B$ covers a bialgebra $A$ and if either $A$ is finite dimensional or $B$ is pointed, then $A$ is a Hopf algebra. This is the above mentioned generalization of Nichols' result.

*Notation.* Throughout, we suppress the unit map $u\colon \Bbbk \to A$ for algebras, identifying $1_{\Bbbk}$ with $1_A$ (and, e.g., writing $u \circ \varepsilon(a)$ as $\varepsilon(a)$ for bialgebra elements $a$).

## 1. Measurings

If $A$ and $B$ are algebras and $C$ a coalgebra, then we say that a map $f\colon B \otimes C \to A$ is a *measuring* (or measures $B$ to $A$) if it corresponds to a unital algebra map from $B$ to the convolution algebra $\mathrm{Hom}_k(C, A)$. More explicitly, $f$ is a measuring if, for all $b, b' \in B$ and $c \in C$, it satisfies

$$(1.1) \qquad f(bb' \otimes c) = \sum_{(c)} f(b \otimes c_1) f(b' \otimes c_2),$$

$$(1.2) \qquad f(1 \otimes c) = \varepsilon(c).$$

Measurings were first studied by Sweedler [**17**].

**1.1. Motivating examples.** It was Grunenfelder and Paré [**4**] who first observed that measurings can be viewed as $C$-indexed families of morphisms from $B$ to $A$. From this perspective, it is natural to abuse notation and write $f(b, c)$ for $f(b \otimes c)$ and we do so freely in what follows. We are most interested in the cases where the family $\{f(\cdot, c)\}_{c \in C}$ jointly spans $A$, and the most interesting of these come when no single $f(\cdot, c)$ is surjective. We now illustrate with several examples (and one non example).

EXAMPLE 1. Let $A, B$ be algebras over $\Bbbk$, and $f\colon B \to A$ an algebra map. Give $\Bbbk$ the coalgebra structure induced by letting 1 be group-like. Then $\overline{f}\colon B \otimes \Bbbk \to A$ given by $\overline{f}(b, 1) = f(a)$ is a measuring. Putting $f = \mathrm{id}$, deduce that every algebra has at least one surjective measuring, albiet not a particularly interesting one.

EXAMPLE 2. Identify the group algebra of $\mathbb{Z}$ with $\Bbbk[z, z^{-1}]$ and let $A$ be a Hopf algebra. Now let $C = \mathsf{coalg}(A)$. We define a measuring $f\colon \Bbbk[z, z^{-1}] \otimes C \to A$ by putting $f(z, \cdot) = \mathrm{id}$, $f(z^{-1}, \cdot) = S$, and $f(z^n, \cdot) = \mathrm{id}^{*n}$, where $\mathrm{id}^{*n}$ stands for the $n$-th convolution power of $\mathrm{id} \in \mathrm{Hom}_{\Bbbk}(A, A)$. In more detail, we have that

$$f(z^n, a) = \begin{cases} a_1 \cdots a_n & n > 0 \\ \mu\varepsilon(a) & n = 0 \\ S(a_1) \cdots S(a_{-n}) & n < 0 \end{cases}.$$

Already we catch a glimpse of the possibilities beyond ordinary morphisms: if the algebra $A$ above is noncommutative, then we find it is jointly spanned by mappings from a commutative one. We may ask under what conditions is the $f$ above a coalgebra map. This happens if and only if $A$ is cocommutative. (Assuming $A$ is cocommutative, checking the coalgebra map condition is straightforward; the reverse implication follows from Lemma 3.3, noting that $\Bbbk[z, z^{-1}]$ is Hopf.)

Above and below, we employ a useful method of building measurings. If a linear map $f(\cdot, c)$ is defined for generators of $B$, then we may *decree* $f$ to satisfy (1.1) and (1.2). Checking that this respects the relations among those generators will frequently be left to the reader.

EXAMPLE 3. Let $\Bbbk[p_1, p_2, \ldots]$ and $\Bbbk\langle H_1, H_2, \ldots\rangle$ denote, respectively, the free commutative and noncommutative algebras on countable generators, one in each degree. These are presentations of what are known in algebraic combinatorics as the rings of *symmetric functions* (Sym) and *noncommutative symmetric functions* (NSym), respectively. Recall that for any positive integer $n$, we say $\lambda$ is a partition of $n$ (written $\lambda \vdash n$), if $\lambda$ is an tuple of positive integers $(\lambda_1 \geq \ldots \geq \lambda_r)$ summing to $n$ (written $|\lambda| = \lambda_1 + \cdots + \lambda_r = n$). We take $(0)$ to be the unique partition of 0. Then Sym has basis $\{p_\lambda := p_{\lambda_1} p_{\lambda_2} \cdots p_{\lambda_r} \mid \lambda \vdash n, n \in \mathbb{N}\}$, identifying $p_{(0)}$ with 1. Let each $p_k$ be primitive so that Sym is Hopf (being the universal envelope of a countably infinite dimensional abelian Lie algebra) and put $C = \mathsf{coalg}(\mathrm{Sym})$. The map $f\colon \mathrm{NSym} \otimes C \to \mathrm{Sym}$ given on generators by $f(H_k, p_\lambda) = \delta_{k,|\lambda|} p_\lambda$ induces a surjective measuring.

In the preceding example, one can view the map $f$ as *degree preserving:* taking $\mathsf{deg}(b \otimes c) = \mathsf{deg}\, c$, we have $\mathsf{deg}\, f(b \otimes c) = \mathsf{deg}(b \otimes c)$. Such maps will be important in applications of our framework to combinatorics, but we do not insist on this property at present. See Sections 3.2 and 3.3.

EXAMPLE 4. Recall that for any positive integer $n$, we say $\gamma$ is a composition of $n$ (written $\gamma \vDash n$), if $\gamma$ is an ordered tuple of positive integers $(\gamma_1, \ldots, \gamma_r)$ summing to $n$ (written $|\gamma| = \gamma_1 + \cdots + \gamma_r = n$). We take $(0)$ to be the unique composition of 0. The space $\Bbbk\{M_\gamma \mid \gamma \vDash n, n \in \mathbb{N}\}$ becomes an algebra under the shuffle product: $M_\alpha \cdot M_\beta = \sum_{\gamma \in \alpha \sqcup\!\sqcup \beta} M_\gamma$. See [**14**], [**6**], [**16**, Ch. 7] for details. (Here we identify $M_{(0)}$ with 1.) This is a presentation of what is known as the ring QSym of *quasisymmetric functions*. It is a graded connected Hopf algebra with (noncocommutative) deconcatenation coproduct: if $\gamma = (\gamma_1, \gamma_2, \ldots, \gamma_r)$, then $\Delta(M_\gamma) = \sum_{0 \le i \le r} M_{(\gamma_1, \ldots, \gamma_i)} \otimes M_{(\gamma_{i+1}, \ldots, \gamma_r)}$. (Here we take $\gamma_0 = \gamma_{r+1} = 0$.) The map $f \colon k[z] \otimes \mathrm{QSym} \to \mathrm{QSym}$ given by $f(z^n, M_\gamma) = \delta_{n,r}\, M_{\gamma_1} \cdots M_{\gamma_r}$ is a measuring.

The above resembles Example 2, but there are important differences: (1) here the mapping is a length-graded convolution power (again, useful for applications to combinatorics); and (2) the measuring algebra $B$ is now only a bialgebra, not a Hopf algebra—this distinction will become important in Proposition 3.4. The above also represents our first example of a family of maps that is not jointly surjective, as QSym is not generated by the monomial quasisymmetric functions $M_k$. (The proper subalgebra they generate is well-known to be isomorphic to Sym.)

**1.2. The bicategory of algebras with measurings.** Here we define a bicategory $\mathbf{Malg}_\Bbbk$ of algebras with measurings, and collect some of its elementary properties.

*Objects:* algebras over $\Bbbk$.

1-*Cells:* morphisms $f \colon B \to A$ are identified with the set $\mathsf{Meas}(B, A)$ of measurings $f \colon B \otimes C_f \to A$.

*Identities:* $\mathrm{id} \colon A \to A$ is given by $C_{\mathrm{id}} = \Bbbk$.

*Composition:* for $f \colon B \to A$ and $g \colon D \to B$ we define their composite $h = fg \colon D \to A$ by $C_h = C_g \otimes C_f$ and $h = f \circ (g \otimes \mathrm{id})$.

2-*Cells:* if $f, g \colon B \to A$ are 1-cells, then a 2-cell $\mathbf{t} \colon f \to g$ is a coalgebra map $\mathbf{t} \colon C_f \to C_g$ making the following diagram commute.

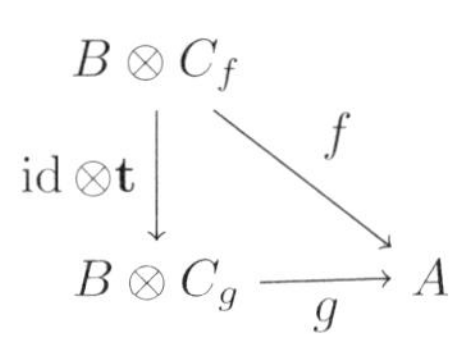


*Identities:* if $f \colon B \to A$ is a 1-cell, then $\mathrm{id} \colon f \to f$ is the identity map on $C_f$.

*Horizontal composition:* for $f, f' \in \mathsf{Meas}(B, A)$ and $g, g' \in \mathsf{Meas}(D, B)$, and 2-cells $\mathbf{s} \colon f \to f'$ and $\mathbf{t} \colon g \to g'$, we define $\mathbf{s} \circ \mathbf{t} \colon fg \to f'g'$ so as to make the diagram at left below commute.

*Vertical composition:* for $f, g, h \in \mathsf{Meas}(B, A)$ and 2-cells $\mathbf{t}\colon f \to g$, $\mathbf{s}\colon g \to h$, we define $\mathbf{st}\colon f \to h$ so as to make the diagram at right below commute.

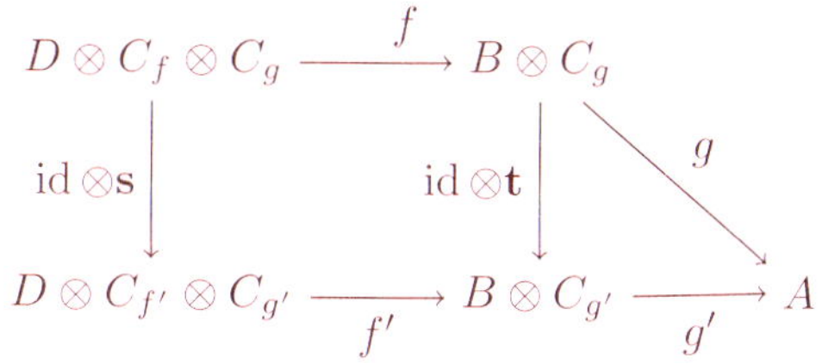

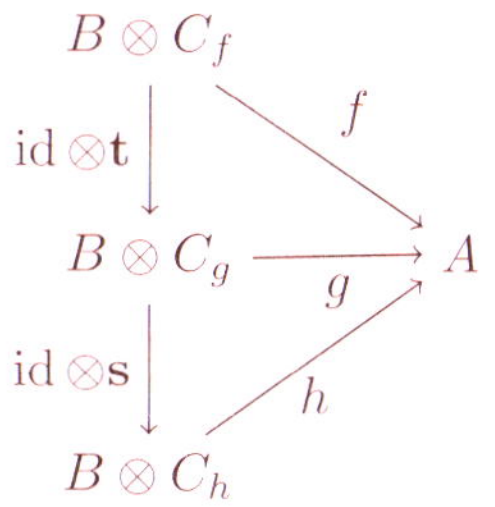

The 1-cell category $\mathsf{Meas}(B, A)$ inherits many of the properties of the category of coalgebras. In particular, pushouts and direct sums (categorical coproducts) exist. In fact, this holds more generally for any slice category of a category with colimits, cf. [**11**, Ch. V]. As the theme of this work is computation, we show how to construct these objects explicitly for related categories in Section 2.1.

## 2. Partial Coverings and Transfer of Structure

If $A, B$ are bialgebras and $C$ a coalgebra, then we say that a map $f\colon B \otimes C \to A$ is a *partial covering* if it is a measuring as well as a coalgebra map. We say that $f$ is a *covering* if it is also surjective. In what follows, when we wish to deemphasize the indexing coalgebra, we write $B \nrightarrow A$ to indicate that $B$ partially covers $A$.

### 2.1. The bicategories of bialgebras with coverings and partial coverings.

We define bicategories **PCbialg** and **Cbialg** of bialgebras with partial coverings and bialgebras with coverings analogous to the bicategory of algebras with measurings.

REMARK. The bicategories **Malg**, **PCbialg**, and **Cbialg** can also be viewed as double categories with horizontal morphisms being (algebra or bialgebra) homomorphisms and vertical homomorphisms being measurings or (partial) coverings. We do not explore this viewpoint any further here.

*Direct Sums:* Let $B$, $A$ be bialgebras. If $f\colon B \otimes C_f \to B$, $g\colon B \otimes C_g \to A$ are (partial) coverings, define the measuring $(h, C_h)$ by putting $C_h := C_f \oplus C_g$ and $h := f \circ \iota_f + g \circ \iota_g$. (Here, $\iota_f, \iota_g$ are the canonical injections of $C_f, C_g$ into $C_f \oplus C_g$.) One checks that $h$ is a partial covering. If either $f$ or $g$ is a covering, then so is $h$.

*Pushouts:* Let $B$, $A$ be bialgebras, and consider coverings $f, g, h \in \mathsf{PCov}(B, A)$ and morphisms $\mathbf{s}\colon f \to g$, $\mathbf{t}\colon f \to h$ between them. The pushout (fibered coproduct) is the partial covering $(k, C_k)$ completing the diagram

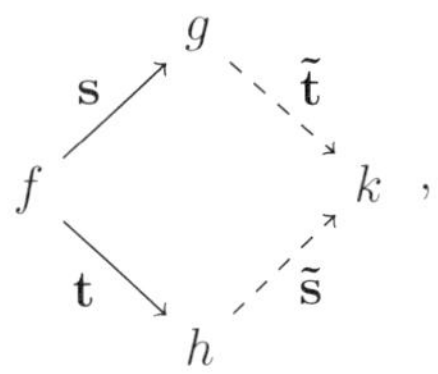


where $C_k = C_g \oplus_{C_f} C_h := (C_g \oplus C_h)/I$, where $I = \{(\mathbf{s}(x), -\mathbf{t}(x)) \mid x \in C_f\}$ is the pushout in the category of coalgebras. The partial covering $k\colon B \otimes C_k \to A$ is given by $k(b \otimes ((x,y)+I)) = g(b \otimes x) + h(b \otimes y)$. If $f, g, h$ are coverings, then so is $k$. If $\mathbf{s}$ is surjective, then so is $\tilde{\mathbf{s}}$. Similarly for $\mathbf{t}, \tilde{\mathbf{t}}$.

EXAMPLE 5 (1, continued). Suppose $\Xi$ is a jointly surjective family of bialgebra maps from $B$ to $A$ and $C = k\Xi$ is the free pointed coalgebra on $\Xi$. Define a bilinear mapping $f\colon B \otimes C \to A$ by $f(b, \xi) = \xi(b)$, for $b \in B$, $\xi \in \Xi$. It is straightforward to check that $f$ is a covering. (In particular, bialgebra maps are coverings.)

EXAMPLE 6 (2, continued). Let $A$ be a group algebra $\Bbbk G$. It is straightforward to check that the measuring $f$ of Example 2 is a coalgebra map.

EXAMPLE 7 (3, continued). The algebra NSym $= \Bbbk\langle H_1, H_2, \ldots\rangle$ becomes a Hopf algebra with coproduct on generators given by $\Delta H_k = \sum_{i+j=k} H_i \otimes H_k$. (In fact, as shown in [**1**], it is the graded dual of the Hopf algebra QSym.) We check that the measuring $f$ of Example 3 is a coalgebra map:

$$
\begin{aligned}
\left(f^{\otimes 2}\Delta\right)(H_k \otimes p_\lambda) &= f^{\otimes 2} \sum_{\substack{i+j=k\\ \mu\sqcup\tau=\lambda}} (H_i \otimes p_\mu) \otimes (H_j \otimes p_\tau)\\
&= \sum_{\substack{i+j=k\\ \mu\sqcup\tau=\lambda}} \delta_{i,|\mu|}\, p_\mu \otimes \delta_{j,|\tau|}\, p_\tau\\
&= \delta_{k,|\lambda|} \sum_{\mu\sqcup\tau=\lambda} p_\mu \otimes p_\tau \;=\; (\Delta f)(H_k \otimes p_\lambda).
\end{aligned}
$$

EXAMPLE 8. We give another covering of Sym. Give $\Bbbk[x]$ the structure of Hopf algebra by letting $x$ be primitive. Let $C = \mathsf{coalg}(\mathrm{Sym})$, using the power sum basis $\{p_\lambda\}$ as above. For a partition $\lambda$ of length $r$, put $f(x^n, p_\lambda) = \delta_{n,r}\, n! \cdot p_\lambda$. It is easy to check that $f\colon \Bbbk[x] \otimes C \to \mathrm{Sym}$ is a measuring. To show that $f$ is a partial covering, we verify that it is a coalgebra map.

$$
\begin{aligned}
\left(f^{\otimes 2}\Delta\right)(x^n \otimes p_\lambda) &= f^{\otimes 2} \sum_{\substack{i+j=n\\ I\sqcup J=[r]}} \binom{n}{i} \left(x^i \otimes p_{\lambda_I}\right) \otimes \left(x^j \otimes p_{\lambda_J}\right)\\
&= \sum_{\substack{i+j=n\\ I\sqcup J=[r]}} \binom{n}{i} \delta_{i,|I|}\, \delta_{j,|J|}\, i!j! \cdot p_{\lambda_I} \otimes p_{\lambda_J}\\
&= \delta_{n,r}\, n! \sum_{I\sqcup J=[r]} p_{\lambda_I} \otimes p_{\lambda_J} \;=\; (\Delta f)(x^n \otimes p_\lambda).
\end{aligned}
$$

**2.2. Universal partial coverings.** If $A$ and $B$ are algebras then Sweedler [**17**] constructs a universal measuring coalgebra $\mathcal{M}(B, A)$ and a universal measuring $\Omega\colon B \otimes \mathcal{M}(B, A) \to A$ satisfying the following universal property:

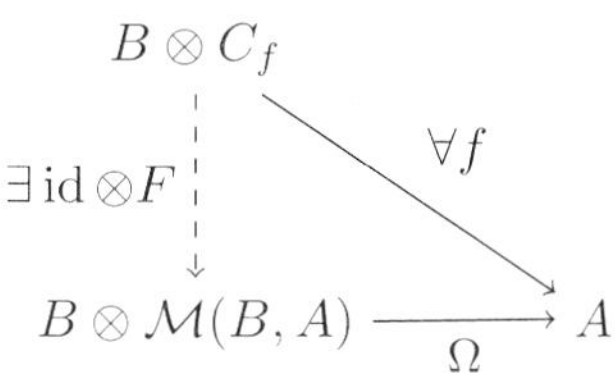


That is, if $f \colon B \otimes C \to A$ is a measuring, then there exists a unique coalgebra map $F \colon C \to \mathcal{M}(B, A)$ such that $f = \Omega \circ (\mathrm{id} \otimes F)$.

This universal object may be viewed as a generalization of Sweedler's finite dual $B^\circ$ (taking $A = \Bbbk$). In the recent literature, one finds other generalizations [**7, 13**], the former of which shares some overlap with our work (see the remark following Proposition 2.2). In Theorem 2.1, we adapt Sweedler's proof to construct universal partial coverings. We remark that the universal object $\mathcal{C}(B, A)$ created there is naturally a subcoalgebra of Sweedler's $\mathcal{M}(B, A)$.

THEOREM 2.1. *Let $A$ and $B$ be bialgebras. Then there exists a universal covering coalgebra $\mathcal{C}(B, A)$ and a universal partial covering $\Omega \colon B \otimes \mathcal{C}(B, A) \to A$ with the following universal property: if $f \colon B \otimes C \to A$ is any partial covering, then there exits a unique coalgebra map $F \colon C \to \mathcal{C}(B, A)$ such that $F = \Omega \circ (\mathrm{id} \otimes F)$.*

PROOF. Let $C$ denote the cofree coalgebra on the space $L(A, B)$ of linear maps from $B$ to $A$ and let $p \colon C \to L(B, A)$ be the canonical projection. We define a linear map $G \colon B \otimes C \to A$ by $G = \mathrm{ev} \circ (\mathrm{id} \otimes p)$, that is, $G(b \otimes c) = p(c)(b)$. Now we define $\mathcal{C}(B, A)$ to be the largest subcoalgebra of $C$ such that $G|_{B \otimes \mathcal{C}(B,A)}$ is a partial covering and we define $\Omega = G|_{B \otimes \mathcal{C}(B,A)}$. The existence of such a coalgebra follows from the following observation: if $(X_i)_{i \in I}$ is a family of subcoalgebras of $C$ such that $G|_{B \otimes X_i}$ are partial coverings, then for $X = \sum_i X_i$, the map $G|_{B \otimes X}$ is also a partial covering (as being a partial covering is described in terms of homogeneous linear equations). Now suppose that $f \colon B \otimes C_f \to A$ is a partial covering. Let $q \colon C_f \to L(B, A)$ be the corresponding map, that is, $q(c)(b) = f(b, c)$. By the cofreeness of $p \colon C \to L(B, A)$, there is a unique coalgebra map $F \colon C_f \to C$ such that $q = p \circ F$. Now note that $G|_{B \otimes F(C_f)}$ is a partial covering and hence $F$ can be viewed as a map from $C_f$ to $\mathcal{C}(B, A)$. Uniqueness of $F$ follows from the construction (coalgebra maps $F$ from $C_f$ to $\mathcal{C}(B, A)$ satisfying $f = \Omega \circ (\mathrm{id} \otimes F)$ correspond injectively to coalgebra maps from $C_f$ to $C$ satisfying $q = p \circ F$). □

REMARK. In [**2, 3**], Grunenfelder and Mastnak considered a generalization of measurings to "bimeasurings" in the case where $B$ is a bialgebra and the algebra $A$ is commutative. Among other things they show that the universal measuring coalgebra $\mathcal{M}(B, A)$ carries a natural structure of a bialgebra. Their argument readily carries over to the present setting of partial coverings. We sketch the proof here.

Suppose $B$ is cocommutative and $A$ is commutative. Define a partial covering $\omega_2 \colon B \otimes \mathcal{C}(B, A) \otimes \mathcal{C}(B, A) \to A$ by $\omega_2 = \mathrm{m}_A(\omega \otimes \omega)(\mathrm{id} \otimes \tau \otimes \mathrm{id})(\Delta_B \otimes \mathrm{id} \otimes \mathrm{id})$ (here $\tau \colon B \otimes \mathcal{C}(B, A) \to \mathcal{C}(B, A) \otimes A$ denotes the usual twist), that is, $\omega_2(b, x \otimes y) = \Omega(b_1, x)\Omega(b_2, y)$. Check that $\omega_2$ is a partial covering (cocommutativity of $B$ is needed to ensure that $\Delta_B$ is a coalgebra map; commutativity of $A$ is needed to ensure that $m_A$ is an algebra map; conclude $\omega_2$ is a measuring). Now define $\mathrm{m} \colon \mathcal{C}(B, A) \otimes \mathcal{C}(B, A) \to \mathcal{C}(B, A)$ to be the unique coalgebra map such that

$\omega_2 = \Omega(\mathrm{id} \otimes \mathrm{m})$. The associativity of m follows from the fact that $\mathrm{m}(\mathrm{id} \otimes \mathrm{m})$ and $\mathrm{m}(\mathrm{m} \otimes \mathrm{id})$ are unique coalgebra maps such that $\omega_3 = \Omega(\mathrm{id} \otimes \mathrm{m}(\mathrm{id} \otimes \mathrm{m}))$, where $\omega_3 \colon B \otimes \mathcal{C}(B, A) \otimes \mathcal{C}(B, A) \otimes \mathcal{C}(B, A) \to A$ is given by $\omega_3(b, x \otimes y \otimes z) = \Omega(b_1, x)\Omega(b_2, y)\Omega(b_3, z)$. The unit $u \colon \Bbbk \to \mathcal{C}(B, A)$ is the unique coalgebra map such that $\omega_0 = \Omega(\mathrm{id} \otimes u)$, where $\omega_0 = \varepsilon \colon B \otimes \Bbbk \to A$. This bialgebra structure on $\mathcal{C}(B, A)$ is universal in the following sense.

PROPOSITION 2.2. *Let $A$ and $B$ be bialgebras with $A$ commutative and $B$ cocommutative. Then the universal covering coalgebra $\mathcal{C}(B, A)$ is a bialgebra satisfying the following universal property: if $C$ is a bialgebra and $f \colon B \otimes C \to A$ is a partial bicovering (that is, it may be viewed as either a partial covering of $A$ by $B$ or a partial covering of $A$ by $C$), then there is a unique bialgebra map $F \colon C \to \mathcal{C}(B, A)$ such that $f = \Omega \circ (\mathrm{id} \otimes F)$.* □

REMARK. The article [**7**] may be relevant here. In the case that $A, B$ are cocommutative, it seems our $\mathcal{C}(B, A)$ reduces to the authors' *universal measuring comonoid in the category of coalgebras $P(B, A)$*. Theorem 10.11 *loc. cit.* provides that if $A$ is furthermore commutative and $B$ is Hopf, then $\mathcal{C}(B, A)$ is also Hopf. We note, in contrast, that $P(B, A)$ is always cocommutative, while that is not the case for $\mathcal{C}(B, A)$; cf. Example 12.

PROPOSITION 2.3. *Let $A, B$ be bialgebras. Then there exits a covering $F \colon B \twoheadrightarrow A$ if and only if the universal partial covering $\Omega \colon B \otimes \mathcal{C}(B, A) \to A$ is surjective.* □

**2.3. Covering equivalence.** We define two 1-cells in $\mathsf{Cov}(B, A)$ to be *covering equivalent* if they factor surjectively through another 1-cell, *i.e.*, $f \sim g$ if there exists a covering $h \in \mathsf{Cov}(B, A)$ and surjective morphisms of coverings $\mathbf{t} \colon f \to h$ and $\mathbf{s} \colon g \to h$. If instead $f, g \in \mathsf{PCov}(B, A)$, we further stipulate that the ranges of $f$, $g$, and $h$ coincide.

This is indeed an equivalence relation. Symmetry is obvious. Transitivity is seen by using the pushout construction: if $f_1, f_2$ factor surjectively through $g$ and $f_2, f_3$ factor surjectively through $h$, then $f_1, f_3$ factor surjectively through the pushout $g \oplus_{f_2} h$.

REMARK. In Section 2.2 we show that given bialgebras $B, A$, there exists a universal partial-covering $\Omega \colon B \otimes \mathcal{C}(B, A) \to A$ that satisfies the property that any other partial covering factors uniquely through $\Omega$. It is easy to see that two partial coverings $f, g \in \mathsf{PCov}(B, A)$ are equivalent if and only if the images of their coalgebras $C_f, C_g$ in $\mathcal{C}(B, A)$ coincide. This observation also gives an alternative prove that the equivalence of coverings above is an equivalence relation.

EXAMPLE 9. Here we identify two partial coverings that are equivalent but not isomorphic. Let $\varphi \colon B \to A$ be a bialgebra homomorphism. Let coalgebras $C, D$ be given by $C = \Bbbk x \oplus \Bbbk y$, $D = \Bbbk z$, where $x, y, z$ are points. Define partial coverings (they are coverings if $\varphi$ is surjective) $f \colon B \otimes D \to A$, $g \colon B \otimes C \to A$ by $f(b \otimes x) = f(b \otimes y) = g(b \otimes z) = \varphi(b)$. Then $f$ factors surjectively through $g$ via the coalgebra map $\mathbf{t} \colon C \to D$ given by $\mathbf{t}(x) = \mathbf{t}(y) = z$.

Let us call a partial covering $f$ *non-degenerate* if the linear map $f(\cdot, x) \colon B \to A$ is nonzero for every nonzero $x \in C_f$. Observe that if $f$ is non-degenerate and factors surjectively through $h$, then $f$ and $h$ are isomorphic. Hence two non-degenerate coverings $f, g$ are equivalent if and only if they are isomorphic.

QUESTION 1. Is everything partial covering equivalent to a non-degenerate one?

**2.4. Transfer of structure.** We discuss some useful applications of coverings.

THEOREM 2.4. *Let* $f\colon B\otimes C\to A$ *be a partial covering, with image* $A'$. *Let* $\iota\colon A'\to B\otimes C$ *be any linear section of* $f$, *that is,* $f\circ\iota=\mathrm{id}$. *Then the following hold.*

(1) *If* $p\in B$ *is primitive then for any* $c\in C$, $f(p,c)$ *is primitive in* $A'$.
(2) *If* $A'$ *and* $B$ *are Hopf algebras, then their antipodes are related by the formula* $S_{A'}=f\circ(S_B\otimes\mathrm{id})\circ\iota$.

PROOF. The proof of (1) is a simple calculation:

$$\begin{aligned}\Delta f(p\otimes c)&=(f\otimes f)\Delta(p\otimes c)\\&=\sum_{(c)}f(p\otimes c_1)\otimes f(1\otimes c_2)+\sum_{(c)}f(1\otimes c_1)\otimes f(p\otimes c_2)\\&=\sum_{(c)}f(p\otimes c_1)\otimes\varepsilon(c_2)+\sum_{(c)}\varepsilon(c_1)\otimes f(p\otimes c_2)\\&=f(p\otimes c)\otimes 1+1\otimes f(p\otimes c).\end{aligned}$$

In the proof of (2), we assume without loss of generality that $A'=A$.

We first show that $S_Af=f(S_B\otimes\mathrm{id})$. We do this by showing that each is the convolution inverse of $f$ in $\mathrm{Hom}_k(B\otimes C,A)$.

Suppose $f(x)=a$. Then

$$\begin{aligned}(S_Af*f)(x)&=\sum_{(x)}S_A(f(x_1))\,f(x_2)=\sum_{(f(x))}S_A(f(x)_1)\,f(x)_2=\sum_{(a)}S_A(a_1)\,a_2\\&=\varepsilon(a)=\varepsilon(x)\,,\end{aligned}$$

since $f$ is a coalgebra map. Similarly, for $b\otimes c\in B\otimes C$, we have

$$\begin{aligned}(f(S_B\otimes\mathrm{id})*f)(b\otimes c)&=\sum_{(b),(c)}f(S_B(b_1)\otimes c_1)f(b_2\otimes c_2)=\sum_{(b)}f(S_B(b_1)b_2\otimes c)\,,\\&=f(\varepsilon(b)\otimes c)=\varepsilon(b)f(1\otimes c)=\varepsilon(b)\varepsilon(c)\\&=\varepsilon(b\otimes c)\,,\end{aligned}$$

where the last equalities on the first and second line use the fact that $f$ measures $B$ to $A$. The proof that $S_Af$ and $f(S_B\otimes\mathrm{id})$ are both right convolution inverses of $f$ is similar.

We conclude that $f$ is convolution invertible and that $S_Af=f(S_B\otimes\mathrm{id})$. This implies that $f(S_B\otimes\mathrm{id})\iota=S_Af\iota=S_A$, as needed. In particular, $f(S_B\otimes\mathrm{id})\iota$ is well-defined and independent of the choice of $\iota$. □

Let $f\colon B\otimes C\to A$ be a covering. Let $P(B)$ and $P(A)$ denote the subspaces of primitive elements in $B$ and $A$. One of the important aspects of coverings is that $f(P(B)\otimes C)\subseteq P(A)$ (just proven above) and hence also $f(\mathcal{U}P(B)\otimes C)\subseteq\mathcal{U}P(A)$ (here $\mathcal{U}P(B)$ stands for the Hopf subalgebra of $B$ generated by $P(B)$; in characteristic 0 it is isomorphic to the universal envelope of the Lie algebra $P(B)$). In a sense this means that $f$ preserves the Lie-theoretic aspects of $B$.

REMARK. In contrast to the preceding paragraph, coverings $f\colon B\twoheadrightarrow A$ need not respect the group-theoretic aspects of $B$. That is, if $g$ is a grouplike element in $B$, then $f(g,\cdot)$ may not even land in the span of the grouplikes of $A$. See Example 2 for one such example.

We now argue that coverings should be viewed as moving between equations in convolution algebras. Fix a measuring $f\colon B\otimes C\to A$. Then any algebra map $\chi\in \mathrm{Alg}(A,\Bbbk)$ gives rise to an element ${}^f\chi\in \mathrm{Alg}(B,C^*)$. Indeed, putting $\langle {}^f\chi(b),c\rangle := (\chi f)(b\otimes c)$, we have

$$\begin{aligned}\langle {}^f\chi(bb'),c\rangle &= (\chi f)(bb'\otimes c) = \chi \sum f(b,c_1)\,f(b',c_2)\\ &= \sum \chi f(b,c_1)\,\chi f(b',c_2) = \langle {}^f\chi(b),c_1\rangle\langle {}^f\chi(b'),c_2\rangle\\ &= \langle {}^f\chi(b)\cdot {}^f\chi(b'),c\rangle.\end{aligned}$$

Now, if $A$ and $B$ are bialgebras then $\mathrm{Alg}(A,\Bbbk)$ and $\mathrm{Alg}(B,C^*)$ are monoids, under the convolution product. So properties of $\chi$ are informed by properties of ${}^f\chi$, and vice versa. We illustrate with some explicit examples.

EXAMPLE 10. Let $B$ be a filtered Hopf algebra. There is a well-known formula for the antipode in the convolution algebra $\mathrm{Hom}(B,B)$ due to Takeuchi. The fact that primitives may be seen as solutions to a similar formula seems to be less appreciated: if $B$ is connected and cocommutative, then the *first Eulerian idempotent* $\mathsf{e}_1 := \log(\mathrm{id}-\varepsilon)\in \mathrm{Hom}(B,B)$ is a projection onto $P(B)$. (See [**10**, Ch. 4] for a proof.)

EXAMPLE 11. If ${}^f\chi$ has order $n$, then the order of $\chi$ divides $n$. Indeed, if $b$ has exponent $n$, then $f(b,c)$ has exponent dividing $n$, for all $c\in C$.

## 3. Cocommutative and Graded Coverings

We investigate subcategories of **PCbialg**.

**3.1. Cocommutative coverings.** We now remark that partial coverings $f\colon B\otimes C\to A$ with $C$ cocommutative are less restrictive than those with $C$ non-cocommutative. Indeed, by the measuring and coalgebra-map properties, we have

$$\begin{aligned}\left(f^{\otimes 2}\Delta\right)(bb'\otimes c) &= f^{\otimes 2}\left[(b_1b_1'\otimes c_1)\otimes(b_2b_2'\otimes c_2)\right]\\ &= f(b_1,c_1)\,f(b_1',c_2)\otimes f(b_2,c_3)\,f(b_2',c_4),\end{aligned}\tag{3.1}$$

while

$$(\Delta f)(bb'\otimes c) = f(b_1,c_1)\,f(b_1',c_3)\otimes f(b_2,c_2)\,f(b_2',c_4).\tag{3.2}$$

Note the twist $(c_2,c_3)\rightsquigarrow(c_3,c_2)$ above, so that $C$ behaves under $f$ somewhat as if it were cocommutative. If $B$ is Hopf, then it behaves precisely so, as we now show.

LEMMA 3.3. *If $f\colon B\otimes C\to A$ is a partial covering with $B$ Hopf, then for all $b,b'\in B$ and $c\in C$ we have that*

$$f(b',c_2)\otimes f(b,c_1) = f(b',c_1)\otimes f(b,c_2).$$

PROOF. In the equality between (3.1) and (3.2), the antipode offers a means to cancel the edges. Using this equality between the fourth and fifth line below, we compute:

$$\begin{aligned} f(b', c_1) \otimes f(b, c_2) &= \varepsilon(b_1 \otimes c_1) f(b'_1, c_2) \otimes f(b_2, c_3) \varepsilon(b'_2 \otimes c_4) \\ &= f(\varepsilon(b_1), c_1) f(b'_1, c_2) \otimes f(b_2, c_3) f(\varepsilon(b'_2), c_4) \\ &= f(S(b_1)b_2, c_1) f(b'_1, c_2) \otimes f(b_3, c_3) f(b'_2 S(b'_3), c_4) \\ &= f(S(b_1), c_1) f(b_2, c_2) f(b'_1, c_3) \otimes f(b_1, c_4) f(b'_2, c_5) f(S(b'_3), c_6) \\ &= f(S(b_1), c_1) f(b_2, c_2) f(b'_1, c_4) \otimes f(b_3, c_3) f(b'_2, c_5) f(S(b'_3), c_6) \\ &= f(S(b_1)b_2, c_1) f(b'_1, c_3) \otimes f(b_3, c_2) f(b'_2 S(b'_3), c_4) \\ &= \varepsilon(b_1)\varepsilon(c_1) f(b'_1, c_3) \otimes f(b_2, c_2) \varepsilon(b'_2)\varepsilon(c_4) \\ &= f(b', c_2) \otimes f(b, c_1). \end{aligned}$$

□

PROPOSITION 3.4. *If $B \twoheadrightarrow A$ is a partial covering with $B$ Hopf, then the image of any cocommutative element of $B$ is cocommutative in $A$.*

PROOF. Applying Lemma 3.3, we may write $\Delta f(b, c)$ as

$$\begin{aligned} f(b, c)_1 \otimes f(b, c)_2 &= f(b_1, c_1) \otimes f(b_2, c_2) = f(b_1, c_2) \otimes f(b_2, c_1) \\ &= f(b_2, c_2) \otimes f(b_1, c_1) = f(b, c)_2 \otimes f(b, c)_1, \end{aligned}$$

which completes the proof. □

Below we provide an example showing that the Hopf condition in Proposition 3.4 (and in Lemma 3.3) is not vacuous. That is, there are interesting coverings by cocommutative $B$'s and non-cocommutative $C$'s. It therefore follows (from Example 12) that the universal covering coalgebra need not always be cocommutative. Whether this statement holds when $B$ is Hopf is an open question.

QUESTION 2. Given any partial covering $f\colon Bp \twoheadrightarrow A$, with $B$ a Hopf algebra, does it factor through a partial covering $f'\colon B \otimes C' \to A$ with $C'$ cocommutative? Equivalently, is the universal covering coalgebra $\mathcal{C}(B, A)$ cocommutative whenever $B$ is Hopf?

Along the same lines, we may also ask the following.

QUESTION 3. Given any partial covering $f\colon B \twoheadrightarrow A$, does there exist a cocommutative $C'$ and partial covering $f'\colon B \otimes C' \to A$ with the same range as $f$? What if we additionally assume that $B$ is a Hopf algebra?

EXAMPLE 12. Let $\{a, b\}$ be the semigroup with $xy = y$ for all $x, y \in \{a, b\}$, and let $M = \{e, a, b\}$ be the corresponding (nonabelian and non-cancelable) monoid. Put $A = (\Bbbk M)^*$ and $C = \mathsf{coalg}(A)$. Writing $\{x_i\}_{i=e,a,b}$ for the dual basis in $A$ and putting $\rho := x_e + x_a + x_b$, one checks that

$$x_i \cdot x_j = \delta_{ij} x_j \ \ (\rho = 1_A), \quad \Delta(x_e) = x_e \otimes x_e, \quad \text{and} \quad \Delta(x_i) = 1 \otimes x_i + x_i \otimes x_e \ (i \neq e).$$

We construct a covering $f\colon B \otimes C \twoheadrightarrow A$.

Give $B := \Bbbk[z]$ the bialgebra structure with $z$ a point, where by *point* we mean a grouplike element. (We adopt this name to avoid the connotation of invertibility.) Define $f$ by sending $(z^n, c)$ to the Sweedler power $\mathrm{id}^{*n}(c)$ for $n \geq 0$ and extending bilinearly. Then $f(z, c) = c$, meaning $f$ is onto. Note that $f$ is a measuring by fiat. We check that $f$ is a coalgebra map. In the present context, (3.1) and (3.2) requires the equality

$$(c_1c_3c_5\cdots)\otimes(c_2c_4c_6\cdots)=(c_1c_2c_3\cdots)\otimes(c_{n+1}c_{n+2}c_{n+3}\cdots). \tag{3.5}$$

This clearly holds for $c=x_e$, so we turn to $x_i$ for $i\neq e$. The general formula for iterated coproducts is

$$\Delta^n(x_i)=1^{\otimes n}\otimes x_i+1^{\otimes n-1}\otimes x_i\otimes x_e+1^{\otimes n-2}\otimes x_i\otimes x_e^{\otimes 2}+\cdots x_i\otimes x_e^{\otimes n}.$$

Hence the only products of $n$ factors (with or without interleaving the odd and even terms) that survive in (3.5) are $1\otimes x_i$ and $x_i\otimes x_e$, and these each occur exactly once on either side.

**3.2. Graded coverings.** If $A,B$ are graded (bi)algebras, then a measuring $f\colon B\otimes C\to A$ is called *graded* if $f(b\otimes c)_n=f(b_n\otimes c)$. We say that $f$ is *locally finite* if for each fixed $c\in C$, $f(b,c)=0$ for all $b$ of sufficiently large degree. Finally, suppose $C$ is also graded. Then we call $f$ *bigraded* if $f(b\otimes c)_n=f(b_n\otimes c_n)$.

PROPOSITION 3.6. *Any graded partial covering $f\colon B\otimes C\to A$ of graded bialgebras may be replaced by a locally finite one.*

PROOF. Let $\mathcal{N}$ denote the coalgebra $\Bbbk\{H_0,H_1,H_2,\ldots\}$ defined by $\Delta(H_k)=\sum_{i+j=k}H_i\otimes H_j$. Put $\bar{C}=C\otimes\mathcal{N}$, and $\bar{f}(b,c\otimes H_n)=f(b,c)_n$, the $n$th homogeneous component of $f(b,c)$. Show that this is an measuring and a coalgebra map. □

EXAMPLE 13 (1, continued). Suppose $A$ is a graded bialgebra. Applying the construction $(f,\Bbbk)\mapsto(\bar{f},\Bbbk\otimes\mathcal{N})$ in the proof above, we see that $\bar{f}$ not only becomes locally finite, but it becomes degree-preserving as well. Thus every graded bialgebra has a degree-preserving measuring that is a covering.

If $U,V,W$ are graded vector spaces, then we define the *diagonal tensor product* by $U\otimes_d V:=U\otimes V/(\bigoplus_{n\neq m}U_n\otimes V_m)$. The space $U\otimes_d V$ is graded by $(u\otimes_d v)_n=u_n\otimes v_n$. Note that bigraded maps from $U\otimes V\to W$ are in bijective correspondence with graded (that is, homogeneous of degree 0) maps from $U\otimes_d V\to W$.

LEMMA 3.7. *If $C,D$ are graded coalgebras, then $C\otimes_d D$ is a graded coalgebra with coproduct given by $\Delta(u\otimes_d v)=(u_1\otimes_d v_1)\otimes(u_2\otimes_d v_2)$ and counit given by $\varepsilon(u\otimes_d v)=\varepsilon(u)\varepsilon(v)$.*

PROOF. Observe that $\bigoplus_{n\neq m}C_n\otimes D_m$ is a coideal in $C\otimes D$. □

The bicategory of graded connected bialgebras with bigraded coverings **Combinat** is now defined in analogy to the category of bialgebras with coverings: we replace tensors by diagonal tensors and coverings by bigraded locally-finite coverings.

REMARK. All bicategories described above also have monoidal structures given by the tensor products of underlying vector spaces.

**3.3. Universal objects in Combinat.** Any cocommutative bialgebra $A$ in **Combinat** has a canonical covering by NSym, defined as follows. For all $a\in A$, put $\mathsf{can}(H_n,a)=a_n$ (the homogeneous component of $a$ of degree $n$), then extend multiplicatively and bilinearly to make $\mathsf{can}$ a measuring. Checking that $\mathsf{can}$ is a coalgebra map is straightforward. Note that this covering is locally finite by construction.

PROPOSITION 3.8. *Given a covering* $f\colon \mathrm{NSym}\otimes C\to A$, *there is a coalgebra map* $\bar f\colon C\to \mathsf{coalg}(A)$ *factoring* $f$ *through* can *if and only if* $C$ *is a locally finite covering.*

PROOF. Put $\bar f(c)=\sum_{n\geq 0} f(H_n,c)$. □

EXAMPLE 14. Let OMP (for *ordered multiset partitions*[1]) denote the free algebra on finite nonempty subsets $K\subseteq\mathbb{N}$. It inherits the structure of Hopf algebra if the coproduct on generators $K$ is defined by $\Delta(K)=\sum_{I\sqcup J=K} I\otimes J$. Identify $\emptyset$ with the unit in OMP. We show that OMP covers NSym, and hence is another weakly initial object in the subcategory of cocommutative objects in **Combinat**.

We take $C$ to be the coalgebra $\mathcal{N}=\Bbbk\{H_0,H_1,H_2,\ldots\}$ from the proof of Proposition 3.6. The map $f\colon \mathrm{OMP}\otimes C\to\mathrm{NSym}$ given on generators by $f(K,H_k)=\delta_{|K|,k}\,k!\,H_k$ induces a measuring, where $|K|$ is the cardinality of the set $K$ and $H_0$ is identified with the unit of NSym.

We check that $f$ is a coalgebra map:

$$\begin{aligned}(f^{\otimes 2}\Delta)\,(K\otimes H_k) &= f^{\otimes 2}\sum_{\substack{I\sqcup J=K\\ i+j=k}}(I\otimes H_i)\otimes(J\otimes H_j)\\ &=\sum_{\substack{I\sqcup J=K\\ i+j=k}}\delta_{|I|,i}\,i!\,H_i\otimes\delta_{|J|,j}\,j!\,H_j\\ &=\delta_{|K|,k}\sum_{i+j=k}\binom{k}{i}\,i!j!\,H_i\otimes H_j \;=(\Delta f)\,(K\otimes H_k).\end{aligned}$$

We have now found two weakly initial objects for the cocommutative part of **Combinat**.

QUESTION 4. Can we drop "weakly" in this bicategory: Is there a sensible equivalence relation on **Combinat** making NSym and OMP equivalent?

QUESTION 5. Are there non-cocommutative or non-graded analogs of NSym or OMP?

## 4. A Nichols Theorem for Hopf coverings

Let $B$ be a Hopf algebra. A result of Nichols [**12**] gives some conditions under which a bialgebra quotient $B/I$ is in fact Hopf. For example, it is Hopf if $B/I$ is commutative or if $B$ is cocommutative or pointed. We now present two analogous results for coverings. (In light of Example 5, these are in fact generalizations of his result.)

THEOREM 4.1. *If a Hopf algebra* $B$ *covers a finite dimensional bialgebra* $A$, *then* $A$ *is also a Hopf algebra.*

PROOF. We consider $A$ as a (faithfully flat, left) $A$-comodule algebra, and study the Galois map $\beta\colon A\otimes A\to A\otimes A$ given by

$$\beta(a,a')=a_1\otimes a_2a'.$$

[1] The reader is cautioned to interpret *ordered multiset partition* as an abbreviation for "ordered partition of a multiset into sets." While the blocks of a multiset partition may have repeated entries, in general, recent combinatorics literature, *e.g.*, [**5**], excludes these cases from consideration.

A result of Schauenburg [**15**] provides that if $\beta$ is bijective, then $A$ is Hopf. We prove that $\beta$ is surjective and therefore bijective.

Let $f\colon B\otimes C\to A$ be the covering and let $\gamma\colon B\otimes C\otimes A\to B\otimes C\otimes A$ be the linear map given by

$$\gamma(b,c,a)=b_1\otimes c_1\otimes f(b_2,c_2)a.$$

Note that $\gamma'\colon B\otimes C\otimes A\to B\otimes C\otimes A$ given by

$$\gamma'(b,c,a)=b_1\otimes c_1\otimes f(S(b_2),c_2)a$$

is a right composition inverse of $\gamma$, so $\gamma$ is surjective. Indeed

$$\begin{aligned}(\gamma\circ\gamma')(b\otimes c\otimes a) &= \gamma(b_1\otimes c_1\otimes f(S(b_2),c_2)a)\\ &= b_1\otimes c_1\otimes f(b_2,c_2)f(S(b_3),c_3)a=b_1\otimes c_1\otimes f(b_2S(b_3),c_2)a\\ &= b_1\otimes c_1\otimes f(\varepsilon(b_2),c_2)a=b_1\otimes c_1\otimes\varepsilon(b_2)\varepsilon(c_2)a=b\otimes c\otimes a.\end{aligned}$$

Now note that $\beta(f\otimes \mathrm{id}_A)=(f\otimes\mathrm{id}_A)\gamma$:

$$\begin{aligned}\beta(f\otimes\mathrm{id}_A)(b,c,a) &= \beta(f(b,c)\otimes a)\\ &= f(b,c)_1\otimes f(b,c)_2a=f(b_1,c_1)\otimes f(b_2,c_2)a\\ &= (f\otimes\mathrm{id}_A)(b_1\otimes c_1\otimes f(b_2,c_2)a)=(f\otimes\mathrm{id}_A)\gamma(b,c,a).\end{aligned}$$

So since $(f\otimes\mathrm{id}_A)\gamma$ is surjective, so must be $\beta$. □

We now consider the case when $A$ is pointed.

LEMMA 4.2. *If $f\colon B\twoheadrightarrow A$ is a partial cover over an algebraically closed field, and $B$ is a pointed Hopf algebra, then its image is pointed in $A$.*

PROOF. Any simple subcoalgebra of $A$ is the image of $B'\otimes C'$ for some simple subcoalgebras $B',C'$. Hence $B'=\Bbbk\{z\}$ for some grouplike $z\in B$. Using Proposition 3.4, we see that $f(B'\otimes C')$ is cocommutative. Hence $f(B\otimes C)$ is pointed. □

LEMMA 4.3. *Suppose $f\colon B\otimes C\to A$ is a partial covering and $z$ is an invertible point in $B$. Then the map $\varphi=f(z,\cdot)\colon C\to A$ is convolution coalgebra map whose convolution inverse $\overline{\varphi}=f(z^{-1},\cdot)$ is also a coalgebra map.*

PROOF. A simple check

$$\begin{aligned}(\overline{\varphi}*\varphi)(c)&=\overline{\varphi}(c_1)\varphi(c_2)\\ &=f(z^{-1},c_1)f(z,c_2)=f(z^{-1}z,c)=f(1,c)=\varepsilon(c).\end{aligned}$$

Similarly for $\varphi*\overline{\varphi}=\varepsilon$. □

LEMMA 4.4. *Assume that $\Bbbk$ is algebraically closed, $C$ a simple coalgebra and $A$ a pointed bialgebra. Assume that $\varphi\colon C\to A$ is a convolution invertible coalgebra map (in $\mathrm{Hom}(C,A)$) such that its convolution inverse $\overline{\varphi}$ is a coalgebra map. Then every point in the range of $\phi$ is invertible.*

PROOF. First note that for all $c \in C$ we have $\varphi(c_2)\overline{\varphi}(c_1) = \varepsilon(c) = \overline{\varphi}(c_2)\varphi(c_1)$. The later equality is proven as follows:

$$\begin{aligned}
\varepsilon(c) &= \overline{\varphi}(c_1)\varepsilon(c_2)\varphi(c_3) = \overline{\varphi}(c_1)(\mathrm{m}\,\Delta(\varepsilon(c_2)))\varphi(c_3) \\
&= \overline{\varphi}(c_1)(\mathrm{m}\,\Delta\varphi(c_2)\overline{\varphi}(c_3))\varphi(c_4) \\
&= \overline{\varphi}(c_1)(\varphi(c_2)\overline{\varphi}(c_4)\varphi(c_3)\overline{\varphi}(c_5))\varphi(c_6) \\
&= (\overline{\varphi}(c_1)\varphi(c_2))\overline{\varphi}(c_4)\varphi(c_3)(\overline{\varphi}(c_5))\varphi(c_6)) \\
&= \varepsilon(c_1)\overline{\varphi}(c_3)\varphi(c_2)\varepsilon(c_4) = \overline{\varphi}(c_2)\varphi(c_1);
\end{aligned}$$

the proof of the former works similarly.

Since $C$ is simple and $\Bbbk$ is algebraically closed, we have that $C$ is isomorphic to a matrix coalgebra $\Bbbk\{e^{ij}\}_{1\leq i,j\leq n} = \mathcal{M}_n(\Bbbk)^*$. Let $A' = \varphi(C)^*$. We identify $A'$ with a subalgebra of $C^* = \mathcal{M}_n(\Bbbk)$ (via $\varphi^* \colon A' \to C^* = \mathcal{M}_n(\Bbbk)$). Let $A' = S \oplus N$ be the Wederburn-Malcev decomposition of $A'$ into its semisimple part $S$ and the radical $N$. Since $A$ and therefore $\varphi(C)$ is pointed (and hence $A'$ is a basic algebra) we have that $S$ is isomorphic to a direct product of copies of the ground field. We additionally assume (using a simultaneous similarity if necessary), that $S$ is a subset of the set of diagonal matrices in $\mathcal{M}_n(\Bbbk)$, and $N$ is a subset of the set of all strictly-upper triangular matrices in $\mathcal{M}_n(\Bbbk)$

We will now show that every $\varphi(e^{ii})$ is invertible, with inverse $\overline{\varphi}(e^{ii})$, from whence the result follows (the above discussion shows that every point in $\varphi(C)$ is in the span of $\{\varphi(e^{ii}) : i = 1\ldots,n\}$. Since the set of all distinct points is linearly independent, we have that every point if $\varphi(C)$ is equal to some $\varphi(e^{ii})$). Since $A'$ is upper triangular we also note that for all $1 \leq i < j \leq n$ we have that $\varphi(e^{ij}) = 0$. We will prove by induction on $i$ that for every $i = 1,\ldots,n$, we have $\varphi(e^{ii})\overline{\varphi}(e^{ii}) = 1 = \overline{\varphi}(e^{ii})\varphi(e^{ii})$ and for all $j > i$, $\overline{\varphi}(e^{ji}) = 0$.

**Base case.** Writing m for multiplication in $A$, we first compute

$$1 = \varepsilon(e^{11}) = \mathrm{m}(\overline{\varphi}\otimes\varphi)\Delta(e^{11}) = \sum_{k=1}^{n}\overline{\varphi}(e^{1k})\varphi(e^{k1}) = \overline{\varphi}(e^{11})\varphi(e^{11}),$$

and

$$1 = \varepsilon(e^{11}) = \mathrm{m}(\varphi\otimes\overline{\varphi})\Delta^{cop}(e^{11}) = \sum_{k=1}^{n}\varphi(e^{k1})\overline{\varphi}(e^{1k}) = \varphi(e^{11})\overline{\varphi}(e^{11}).$$

For $j > 1$, we further have

$$0 = \varepsilon(e^{j1}) = \mathrm{m}(\overline{\varphi}\otimes\varphi)\Delta(e^{j1}) = \sum_{i=1}^{n}\overline{\varphi}(e^{ji})\varphi(e^{i1}) = \overline{\varphi}(e^{j1})\varphi(e^{11}).$$

Since $\varphi(e^{11})$ is invertible we conclude that $\overline{\varphi}(e^{j1}) = 0$.

**Induction step.** Assume that for some $m$, $2 \leq m \leq n$ we have that every $i < m$ satisfies $\varphi(e^{ii})\overline{\varphi}(e^{ii}) = 1 = \overline{\varphi}(e^{ii})\varphi(e^{ii})$ and for all $j > i$, $\overline{\varphi}(e^{ji}) = 0$. Then

$$\begin{aligned}
1 = \varepsilon(e^{mm}) = \mathrm{m}(\overline{\varphi}\otimes\varphi)\Delta(e^{mm}) &= \sum_{k=1}^{n}\overline{\varphi}(e^{mk})\varphi(e^{km}) \\
&= \sum_{k<m}\overline{\varphi}(e^{mk})\varphi(e^{km}) + \overline{\varphi}(e^{mm})\varphi(e^{mm}) + \sum_{k>m}\overline{\varphi}(e^{mk})\varphi(e^{km}) \\
&= 0 + \overline{\varphi}(e^{mm})\varphi(e^{mm}) + 0
\end{aligned}$$

and

$$
\begin{aligned}
1 = \varepsilon(e^{mm}) &= \mathrm{m}(\varphi \otimes \overline{\varphi})\Delta^{cop}(e^{mm}) = \sum_{k=1}^{n} \varphi(e^{km})\overline{\varphi}(e^{mk}) \\
&= \sum_{k<m} \varphi(e^{km})\overline{\varphi}(e^{mk}) + \overline{\varphi}(e^{mm})\varphi(e^{mm}) + \sum_{k>m} \varphi(e^{km})\overline{\varphi}(e^{mk}) \\
&= 0 + \varphi(e^{mm})\overline{\varphi}(e^{mm}) + 0;
\end{aligned}
$$

and for $j > m$, we have

$$
\begin{aligned}
0 = \varepsilon(e^{jm}) &= \mathrm{m}(\overline{\varphi} \otimes \varphi)\Delta(e^{jm}) \\
&= \sum_{i=1}^{n} \overline{\varphi}(e^{ji})\varphi(e^{im}) = \overline{\varphi}(e^{j1})\varphi(e^{11}) \\
&= \sum_{i<m} \overline{\varphi}(e^{ji})\varphi(e^{im}) + \overline{\varphi}(e^{jm})\varphi(e^{mm}) + \sum_{i>m} \overline{\varphi}(e^{ji})\varphi(e^{im}) \\
&= 0 + \overline{\varphi}(e^{jm})\varphi(e^{mm}) + 0.
\end{aligned}
$$

Therefore, since $\varphi(e^{mm})$ is invertible, we get $\overline{\varphi}(e^{jm}) = 0$. □

THEOREM 4.5. *Let $f\colon B \twoheadrightarrow\!\!\!\!\!\!\!+\ A$ be a bialgebra covering. If $B$ is a Hopf algebra with cocommutative coradical, then $A$ is a Hopf algebra as well.*

PROOF. With no loss of generality we can and do assume that $\Bbbk$ is algebraically closed. We use Takeuchi's condition: an antipode exists on $A$ if it may be defined on the coradical of $A$. Looking at that coradical, we use Proposition 3.4 to conclude that $A$ is pointed. Note that every point in $A$ is in the image, under the covering, of some $\Bbbk\{z\} \otimes C \cong C$, where $z$ is a point in $B$ and $C$ is a simple sub-coalgebra of $C_f$. Hence by Lemmas 4.3 and 4.4 we have that every point in $A$ is invertible. □

## Acknowledgements

We would like to thank the referee for a number of useful comments, in particular for pointing us to the references [**7**], [**13**] and for explaining a more categorical way of approaching some of the constructions discussed in the paper: see the discussion at the end of Section 1, the paragraph preceding Theorem 2.1, the remark following Proposition 2.2, and the comment preceding Question 2.

## References

[1] Israel M. Gelfand, Daniel Krob, Alain Lascoux, Bernard Leclerc, Vladimir S. Retakh, and Jean-Yves Thibon, *Noncommutative symmetric functions*, Adv. Math. **112** (1995), no. 2, 218–348, DOI 10.1006/aima.1995.1032. MR1327096

[2] L. Grunenfelder and M. Mastnak, *On bimeasurings*, J. Pure Appl. Algebra **204** (2006), no. 2, 258–269, DOI 10.1016/j.jpaa.2005.04.007. MR2184810

[3] L. Grunenfelder and M. Mastnak, *On bimeasurings. II*, J. Pure Appl. Algebra **209** (2007), no. 3, 823–832, DOI 10.1016/j.jpaa.2006.08.013. MR2298860

[4] Luzius Grünenfelder and Robert Paré, *Families parametrized by coalgebras*, J. Algebra **107** (1987), no. 2, 316–375, DOI 10.1016/0021-8693(87)90093-7. MR885801

[5] J. Haglund, J. B. Remmel, and A. T. Wilson, *The delta conjecture*, Trans. Amer. Math. Soc. **370** (2018), no. 6, 4029–4057, DOI 10.1090/tran/7096. MR3811519

[6] Michael E. Hoffman, *Quasi-shuffle products*, J. Algebraic Combin. **11** (2000), no. 1, 49–68, DOI 10.1023/A:1008791603281. MR1747062

[7] Martin Hyland, Ignacio López Franco, and Christina Vasilakopoulou, *Hopf measuring comonoids and enrichment*, Proc. Lond. Math. Soc. (3) **115** (2017), no. 5, 1118–1148, DOI 10.1112/plms.12064. MR3733560

[8] A. Lauve and M. Mastnak. Transfer of structure in combinatorial Hopf algebras. In preparation.

[9] Aaron Lauve and Mitja Mastnak, *The primitives and antipode in the Hopf algebra of symmetric functions in noncommuting variables*, Adv. in Appl. Math. **47** (2011), no. 3, 536–544, DOI 10.1016/j.aam.2011.01.002. MR2822200

[10] Jean-Louis Loday, *Cyclic homology*, Grundlehren der Mathematischen Wissenschaften [Fundamental Principles of Mathematical Sciences], vol. 301, Springer-Verlag, Berlin, 1992. Appendix E by María O. Ronco. MR1217970

[11] Saunders Mac Lane, *Categories for the working mathematician*, 2nd ed., Graduate Texts in Mathematics, vol. 5, Springer-Verlag, New York, 1998. MR1712872

[12] Warren D. Nichols, *Quotients of Hopf algebras*, Comm. Algebra **6** (1978), no. 17, 1789–1800, DOI 10.1080/00927877808822321. MR508081

[13] Hans-E. Porst and Ross Street, *Generalizations of the Sweedler dual*, Appl. Categ. Structures **24** (2016), no. 5, 619–647, DOI 10.1007/s10485-016-9450-2. MR3546505

[14] Rimhak Ree, *Lie elements and an algebra associated with shuffles*, Ann. of Math. (2) **68** (1958), 210–220, DOI 10.2307/1970243. MR0100011

[15] Peter Schauenburg, *A bialgebra that admits a Hopf-Galois extension is a Hopf algebra*, Proc. Amer. Math. Soc. **125** (1997), no. 1, 83–85, DOI 10.1090/S0002-9939-97-03682-4. MR1363183

[16] Richard P. Stanley, *Enumerative combinatorics. Vol. 2*, Cambridge Studies in Advanced Mathematics, vol. 62, Cambridge University Press, Cambridge, 1999. With a foreword by Gian-Carlo Rota and appendix 1 by Sergey Fomin. MR1676282

[17] Moss E. Sweedler, *Hopf algebras*, Mathematics Lecture Note Series, W. A. Benjamin, Inc., New York, 1969. MR0252485

Department of Mathematics & Statistics, Loyola University Chicago, Chicago, IL 60660 (USA)

*Email address*: `lauve@math.luc.edu`

Department of Mathematics & Computing Science, Saint Mary's University, Halifax, Nova Scotia B3H 3C3 (CANADA)

*Email address*: `mmastnak@cs.smu.ca`

Contemporary Mathematics
Volume **728**, 2019
https://doi.org/10.1090/conm/728/14660

# Classifying braidings on fusion categories

Dmitri Nikshych

ABSTRACT. We show that braidings on a fusion category $\mathcal{C}$ correspond to certain fusion subcategories of the center of $\mathcal{C}$ transversal to the canonical Lagrangian algebra. This allows to classify braidings on non-degenerate braided fusion categories and on those dual to the categories of group-graded vector spaces.

## 1. Introduction

Throughout this article we work over an algebraically closed field $k$ of characteristic 0.

In general, a fusion category $\mathcal{C}$ may have several different braidings or no braidings at all. For example, if $\mathcal{C} = \mathsf{Vec}_G$, the category of finite-dimensional $k$-vectors spaces graded by a finite abelian group $G$, then braidings on $\mathcal{C}$ are parameterized by bilinear forms on $G$. If $G$ is non-Abelian then of course $\mathsf{Vec}_G$ does not admit any braidings.

The goal of this note is to give a convenient parameterization of braidings on an arbitrary fusion category $\mathcal{C}$. We introduce the notion of transversality between algebras and subcategories of a braided fusion category. Then we show that the set of braidings on $\mathcal{C}$ is in bijection with the set of fusion subcategories $\mathcal{B}$ of the center $\mathcal{Z}(\mathcal{C})$ such that $\mathrm{FPdim}(\mathcal{B}) = \mathrm{FPdim}(\mathcal{C})$ and $\mathcal{B}$ is transversal to the canonical Lagrangian algebra of $\mathcal{Z}(\mathcal{C})$. In several interesting situations it is possible to give an explicit parameterization of such subcategories. We do this in two cases: (1) for fusion categories $\mathcal{C}$ admitting a non-degenerate braiding and (2) for group-theoretical categories. In the latter case the parameterization is given in terms of the subgroup lattice of a group and can be conveniently used in concrete computations.

The paper is organized as follows. Section 2 contains some background information and a categorical analogue of Goursat's lemma (Theorem 2.2) for subcategories of tensor products of fusion categories. In Section 3 we introduce transversal pairs of algebras and subcategories and characterize braidings in these terms. In Section 4 we classify braidings on a fusion category $\mathcal{B}$ that already admits a non-degenerate braiding (Theorem 4.1) and consider several examples. We show that with respect to any other braiding the symmetric center of $\mathcal{B}$ remains pointed. In Section 5 we classify braidings on group-theoretical fusion categories (dual to the category $\mathsf{Vec}_G$). As an application we parameterize braidings on the Drinfeld center of $\mathsf{Vec}_G$.

This material is based upon work supported by the National Science Foundation under grant DMS-1801198 and by the NSA grant H98230-16-1-0008.

## 2. Preliminaries

**2.1. Fusion categories.** We refer the reader to [**EGNO**] for a general theory of tensor categories and to [**DGNO**] for braided fusion categories.

A *fusion category* over $k$ is a $k$-linear semisimple rigid tensor category with finitely many isomorphism classes of simple objects, finite-dimensional $\mathsf{Hom}$-spaces, and a simple unit object $\mathbf{1}$. By a *fusion subcategory* of a fusion category $\mathcal{C}$ we always mean a full tensor subcategory. An example of subcategory is the maximal pointed subcategory $\mathcal{C}_{pt} \subset \mathcal{C}$ generated by invertible objects of $\mathcal{C}$. We say that $\mathcal{C}$ is *pointed* if $\mathcal{C} = \mathcal{C}_{pt}$.

We denote $\mathsf{Vec}$ the fusion category of finite-dimensional $k$-vector spaces.

For a fusion category $\mathcal{C}$ let $\mathcal{O}(\mathcal{C})$ denote the set of isomorphism classes of simple objects.

Let $G$ be a finite group. A grading of $\mathcal{C}$ by $G$ is a map $\deg : \mathcal{O}(\mathcal{C}) \to G$ with the following property: for any simple objects $X, Y, Z \in \mathcal{C}$ such that $X \otimes Y$ contains $Z$ one has $\deg Z = \deg X \cdot \deg Y$. We will identify a grading with the corresponding decomposition

$$\mathcal{C} = \bigoplus_{g \in G} \mathcal{C}_g, \tag{1}$$

where $\mathcal{C}_g$ is the full additive subcategory of $\mathcal{C}$ generated by simple objects of degree $g \in G$. The subcategory $\mathcal{C}_1$ is called the *trivial component* of the grading. The grading is called *faithful* if $\deg : \mathcal{O}(\mathcal{C}) \to G$ is surjective.

For any fusion category $\mathcal{C}$ there is a *universal grading* $\mathcal{O}(\mathcal{C}) \to U(\mathcal{C})$ [**GN**], where $U(\mathcal{C})$ is the *universal grading group* of $\mathcal{C}$. Any grading of $\mathcal{C}$ comes from a quotient of $U(\mathcal{C})$. The trivial component of the universal grading is the adjoint fusion subcategory $\mathcal{C}_{ad} \subset \mathcal{C}$ generated by objects $X \otimes X^*$, $X \in \mathcal{O}(\mathcal{C})$.

**2.2. Fiber products of fusion categories.** Let $\mathcal{C}$, $\mathcal{D}$ be fusion categories graded by the same group $G$. The *fiber product* of $\mathcal{C}$ and $\mathcal{D}$ is the fusion category

$$\mathcal{C} \boxtimes_G \mathcal{D} := \bigoplus_{g \in G} \mathcal{C}_g \boxtimes \mathcal{D}_g. \tag{2}$$

Here $\boxtimes$ denotes Deligne's tensor product of abelian categories. Clearly, $\mathcal{C} \boxtimes_G \mathcal{D}$ is a fusion subcategory of $\mathcal{C} \boxtimes \mathcal{D}$ graded by $G$. The trivial component of this grading is $\mathcal{C}_1 \boxtimes \mathcal{D}_1$. When the gradings of $\mathcal{C}$ and $\mathcal{D}$ are faithful one has

$$\mathrm{FPdim}(\mathcal{C} \boxtimes_G \mathcal{D}) = \frac{\mathrm{FPdim}(\mathcal{C})\mathrm{FPdim}(\mathcal{D})}{|G|}. \tag{3}$$

**2.3. Goursat's Lemma for subcategories of the tensor product.** Let $\mathcal{C}$, $\mathcal{D}$ be fusion categories.

DEFINITION 2.1. A *subcategory datum* for $\mathcal{C} \boxtimes \mathcal{D}$ consists of a pair $\mathcal{E} \subset \mathcal{C}$ and $\mathcal{F} \subset \mathcal{D}$ of fusion subcategories, a group $G$, and fixed faithful gradings of $\mathcal{E}$ and $\mathcal{F}$ by $G$.

We will identify subcategory data $(\mathcal{E}, \mathcal{F}, G)$ and $(\mathcal{E}, \mathcal{F}, G')$ if there is an isomorphism $\alpha : G \xrightarrow{\sim} G'$ such that $\mathcal{E}_g = \mathcal{E}_{\alpha(g)}$ and $\mathcal{F}_g = \mathcal{F}_{\alpha(g)}$. When no confusion is likely we will denote a subcategory datum simply by $(\mathcal{E}, \mathcal{F}, G)$ omitting the grading maps.

Given a subcategory datum $(\mathcal{E}, \mathcal{F}, G)$ we can form a fusion subcategory

$$\mathcal{S}(\mathcal{E}, \mathcal{F}, G) := \mathcal{E} \boxtimes_G \mathcal{F} \subset \mathcal{C} \boxtimes \mathcal{D}. \tag{4}$$

It turns out that $\mathcal{S}(\mathcal{E}, \mathcal{F}, G)$ is a typical example of a fusion subcategory of $\mathcal{E} \boxtimes \mathcal{F}$. The following theorem is a categorical analogue of the well known Goursat's Lemma in group theory.

THEOREM 2.2. *Let $\mathcal{C}$, $\mathcal{D}$ be fusion categories. The assignment*

$$(\mathcal{E}, \mathcal{F}, G) \mapsto \mathcal{S}(\mathcal{E}, \mathcal{F}, G) \tag{5}$$

*is a bijection between the set of subcategory data for $\mathcal{C} \boxtimes \mathcal{D}$ and the set of fusion subcategories of $\mathcal{C} \boxtimes \mathcal{D}$.*

PROOF. We need to show that every fusion subcategory $\mathcal{S} \subset \mathcal{C} \boxtimes \mathcal{D}$ is equal to some $\mathcal{S}(\mathcal{E}, \mathcal{F}, G)$ for a unique choice of $(\mathcal{E}, \mathcal{F}, G)$.

Let $\mathcal{E} \subset \mathcal{C}$ be a fusion subcategory generated by all $X \in \mathcal{O}(\mathcal{C})$ such that $X \boxtimes Y \in \mathcal{S}$ for some non-zero $Y \in \mathcal{D}$. Similarly, let $\mathcal{F} \subset \mathcal{D}$ be a fusion subcategory generated by all $Y \in \mathcal{O}(\mathcal{D})$ such that $X \boxtimes Y \in \mathcal{S}$ for some non-zero $X \in \mathcal{C}$.

Let

$$\tilde{\mathcal{E}} := \mathcal{S} \cap (\mathcal{C} \boxtimes \mathsf{Vec}) \subset \mathcal{E} \text{ and } \tilde{\mathcal{F}} := \mathcal{S} \cap (\mathsf{Vec} \boxtimes \mathcal{D}) \subset \mathcal{F}. \tag{6}$$

If $X \in \mathcal{O}(\mathcal{C})$ and $Y \in \mathcal{O}(\mathcal{D})$ are such that $X \boxtimes Y \in \mathcal{S}$ then $(X^* \otimes X) \boxtimes \mathbf{1}$ and $\mathbf{1} \boxtimes (Y^* \otimes Y)$ are objects of $\mathcal{S}$. This means that $\mathcal{E}_{ad} \subset \tilde{\mathcal{E}}$ and $\mathcal{F}_{ad} \subset \tilde{\mathcal{F}}$. Let $H_{\mathcal{E}} \subset U(\mathcal{E})$ and $H_{\mathcal{F}} \subset U(\mathcal{F})$ be the subgroups of the universal groups corresponding to $\tilde{\mathcal{E}}$ and $\tilde{\mathcal{F}}$. We claim that these subgroups are normal. Indeed, let $X \in \mathcal{O}(\tilde{\mathcal{E}})$ and $V \in \mathcal{O}(\mathcal{E})$. Then $X \boxtimes \mathbf{1} \in \mathcal{S}$ and $V \boxtimes U \in \mathcal{S}$ for some $U \in \mathcal{O}(\mathcal{D})$. So $(V^* \boxtimes U^*) \otimes (X \boxtimes \mathbf{1}) \otimes (V \boxtimes U) = (V^* \otimes X \otimes V) \boxtimes (U^* \otimes U) \in \mathcal{S}$ and $V^* \otimes X \otimes V \in \tilde{\mathcal{E}}$. This implies $gxg^{-1} \in H_{\mathcal{E}}$ for all $x \in H_{\mathcal{E}}$ and $g \in U(\mathcal{E})$. Thus, $H_{\mathcal{E}} \subset U(\mathcal{E})$ is normal. Similarly, $H_{\mathcal{F}} \subset U(\mathcal{F})$ is normal.

Hence, subcategories $\mathcal{E}$ and $\mathcal{F}$ have faithful gradings $\deg_{\mathcal{E}} : \mathcal{O}(\mathcal{E}) \to U(\mathcal{E})/H_{\mathcal{E}} =: G_{\mathcal{E}}$ and $\deg_{\mathcal{F}} : \mathcal{O}(\mathcal{F}) \to U(\mathcal{F})/H_{\mathcal{F}} =: G_{\mathcal{F}}$ with trivial components $\tilde{\mathcal{E}}$ and $\tilde{\mathcal{F}}$, respectively.

Let $X \in \mathcal{O}(\mathcal{C})$ and $Y_1, Y_2 \in \mathcal{O}(\mathcal{D})$ be such that that $X \boxtimes Y_1, X \boxtimes Y_2 \in \mathcal{S}$. Then $\mathbf{1} \boxtimes (Y_1^* \otimes Y_2)$ is a subobject of $(X \boxtimes Y_1)^* \otimes (X \boxtimes Y_2)$ and so belongs to $\mathcal{S}$. Therefore, $Y_1^* \otimes Y_2 \in \tilde{\mathcal{F}}$, so $\deg_{\mathcal{F}}(Y_1) = \deg_{\mathcal{F}}(Y_2)$. Similarly, if $X_1, X_2 \in \mathcal{O}(\mathcal{C})$ and $Y \in \mathcal{O}(\mathcal{D})$ are such that that $X_1 \boxtimes Y, X_2 \boxtimes Y \in \mathcal{S}$ then $\deg_{\mathcal{E}}(X_1) = \deg_{\mathcal{E}}(X_2)$.

Therefore, there is a well-defined isomorphism $f : G_{\mathcal{E}} \to G_{\mathcal{F}}$ such that $f(\deg_{\mathcal{E}}(X)) = \deg_{\mathcal{F}}(Y)$ for all $X \in \mathcal{O}(\mathcal{C})$ and $Y \in \mathcal{O}(\mathcal{D})$ such that $X \boxtimes Y \in \mathcal{O}(\mathcal{S})$. This means that $\mathcal{S}$ is a fiber product of $\mathcal{E}$ and $\mathcal{F}$.

It is clear that subcategories $\mathcal{E}$, $\mathcal{F}$ and their gradings are invariants of $\mathcal{S}$. □

REMARK 2.3. Let $(\mathcal{E}_1, \mathcal{F}_1, G_1)$ and $(\mathcal{E}_2, \mathcal{F}_2, G_2)$ be subcategory data for $\mathcal{C} \boxtimes \mathcal{D}$. Then

$$\mathcal{S}(\mathcal{E}_1, \mathcal{F}_1, G_1) \cap \mathcal{S}(\mathcal{E}_2, \mathcal{F}_2, G_2) \cong (\mathcal{E}_1 \cap \mathcal{E}_2) \boxtimes_{G_1 \times G_2} (\mathcal{F}_1 \cap \mathcal{F}_2), \tag{7}$$

where the gradings of $\mathcal{E}_1 \cap \mathcal{E}_2$ and $\mathcal{F}_1 \cap \mathcal{F}_2$ by $G_1 \times G_2$ are such that the $(g_1, g_2)$ components are $(\mathcal{E}_1)_{g_1} \cap (\mathcal{E}_2)_{g_2}$ and $(\mathcal{F}_1)_{g_1} \cap (\mathcal{F}_2)_{g_2}$, respectively (note that these gradings are not faithful in general).

**2.4. Braided fusion categories and their gradings.** Let $\mathcal{B}$ be a braided fusion category with a braiding $c_{X,Y} : X \otimes Y \to Y \otimes X$. Two objects $X, Y$ of $\mathcal{B}$ *centralize* each other if $c_{Y,X} \circ c_{X,Y} = \mathrm{id}_{X\otimes Y}$ and *projectively centralize* each other if $c_{Y,X} \circ c_{X,Y} = \lambda \mathrm{id}_{X\otimes Y}$ for some scalar $\lambda \in k$. For a fusion subcategory $\mathcal{D} \subset \mathcal{B}$ its *centralizer* is

$$\mathcal{D}' = \{Y \in \mathcal{B} \mid Y \text{ centralizes each } X \in \mathcal{D}\}.$$

The symmetric center of $\mathcal{B}$ is $\mathcal{Z}_{sym}(\mathcal{B}) := \mathcal{B} \cap \mathcal{B}'$. We say that $\mathcal{B}$ is *non-degenerate* if $\mathcal{Z}_{sym}(\mathcal{B}) = \mathsf{Vec}$.

For a non-degenerate $\mathcal{B}$ there is a canonical non-degenerate bimultiplicative pairing

$$\langle\, , \rangle : \mathcal{O}(\mathcal{B}_{pt}) \times U(\mathcal{B}) \to k^\times \tag{8}$$

defined by $c_{Y,X}c_{X,Y} = \langle X, g\rangle \,\mathrm{id}_{X\otimes Y}$ for all $X \in \mathcal{O}(\mathcal{B}_{pt})$ and $Y \in \mathcal{B}_g$, $g \in U(\mathcal{B})$. See [**DGNO**, 3.3.4] for details.

PROPOSITION 2.4. *Let $\mathcal{B}$ be a non-degenerate braided fusion category and let $\mathcal{D} \subset \mathcal{B}$ be a fusion subcategory with a faithful grading*

$$\mathcal{D} = \bigoplus_{g\in G} \mathcal{D}_g,$$

*where $G$ is an Abelian group. The centralizer of the trivial component $\mathcal{D}_1$ of $\mathcal{D}$ admits a faithful grading*

$$\mathcal{D}_1' = \bigoplus_{\phi\in \widehat{G}} (\mathcal{D}_1')_\phi,$$

*where $\widehat{G}$ is the group of characters of $G$ and*

$$(\mathcal{D}_1')_\phi = \{X \in \mathcal{B} \mid c_{Y,X} \circ c_{X,Y} = \phi(g) id_{X\otimes Y},\ for\ all\ Y \in \mathcal{D}_g,\ g \in G\}.$$

*The trivial component of this grading is $\mathcal{D}'$.*

PROOF. It follows from [**DGNO**, 3.3] that a simple object $X$ belongs to $\mathcal{D}_1'$ if and only if projectively centralizes every simple $Y \in \mathcal{D}$, i.e., $c_{Y,X} \circ c_{X,Y} = \lambda_Y \mathrm{id}_{X\otimes Y}$ for some $\lambda_Y \in k^\times$. Furthermore, if $Y_1$, $Y_2$ are simple objects lying in $\mathcal{D}_g$ then $\lambda_{Y_1} = \lambda_{Y_2}$. Let us denote the latter scalar by $\phi_X(g)$. It follows from the braiding axioms that the assignment

$$\mathcal{O}(\mathcal{D}_1') \to \widehat{G} : X \mapsto \phi_X$$

is a grading of $\mathcal{D}_1'$ by $\widehat{G}$.

The fact that the trivial component is $\mathcal{D}'$ and the faithfulness of grading follow from the non-degeneracy of $\mathcal{B}$. □

**2.5. Lagrangian algebras in the center.** For any fusion category $\mathcal{C}$ let $\mathcal{Z}(\mathcal{C})$ denote its Drinfeld center.

Let $\mathcal{B}$ be a braided fusion category. A *Lagrangian* algebra in $\mathcal{B}$ is a commutative separable algebra $A$ in $\mathcal{B}$ such that $\mathsf{Hom}_{\mathcal{B}}(A, \mathbf{1}) \cong k$ and $\mathrm{FPdim}(A)^2 = \mathrm{FPdim}(\mathcal{B})$.

Let $I : \mathcal{C} \to \mathcal{Z}(\mathcal{C})$ denote the adjoint of the forgetful functor $F : \mathcal{Z}(\mathcal{C}) \to \mathcal{C}$. Then $I(\mathbf{1})$ is a canonical Lagrangian algebra in $\mathcal{Z}(\mathcal{C})$.

It was explained in [**DMNO**] that any braided equivalence $a : \mathcal{Z}(\mathcal{C}) \xrightarrow{\sim} \mathcal{B}$ gives rise to a Lagrangian algebra $A = a(I(\mathbf{1}))$ in $\mathcal{B}$. Conversely, given a Lagrangian algebra $A \in \mathcal{B}$ there is a braided tensor equivalence $\mathcal{Z}(\mathcal{B}_A) \xrightarrow{\sim} \mathcal{B}$, where $\mathcal{B}_A$ denotes the fusion category of $A$-modules in $\mathcal{B}$.

## 3. Subcategories transversal to a Lagrangian algebra

DEFINITION 3.1. Let $\mathcal{C}$ be a fusion category, let $\mathcal{B} \subset \mathcal{C}$ be a fusion subcategory, and let $A$ be an algebra in $\mathcal{C}$. We will assume that $\mathsf{Hom}_{\mathcal{C}}(A, \mathbf{1}) \cong k$, i.e., that $A$ is a *connected* algebra. We say that $\mathcal{B}$ is *transversal* to $A$ if

$$\mathsf{Hom}_{\mathcal{C}}(X, A) = \mathsf{Hom}_{\mathcal{C}}(X, \mathbf{1}) \tag{9}$$

for all $X \in \mathcal{B}$.

In other words, $\mathcal{B}$ is transversal to $A$ if and only if $\mathsf{Hom}_{\mathcal{C}}(X, A) = 0$ for all non-identity $X \in \mathcal{O}(\mathcal{B})$.

THEOREM 3.2. *Let $\mathcal{C}$ be a fusion category and let $A := I(\mathbf{1})$ be the canonical Lagrangian algebra in $\mathcal{Z}(\mathcal{C})$. Braidings on $\mathcal{C}$ are in bijection with fusion subcategories $\mathcal{B} \subset \mathcal{Z}(\mathcal{C})$ transversal to $A$ and such that $FPdim(\mathcal{B}) = FPdim(\mathcal{C})$.*

PROOF. It is well known that braidings on a fusion category $\mathcal{C}$ are in bijection with sections of the forgetful functor $F : \mathcal{Z}(\mathcal{C}) \to \mathcal{C}$, i.e., with embeddings $\iota : \mathcal{C} \to \mathcal{Z}(\mathcal{C})$ such that $F \circ \iota = \mathrm{id}_{\mathcal{C}}$. The latter correspond to fusion subcategories $\mathcal{B} \subset \mathcal{Z}(\mathcal{C})$ such that the restriction $F|_{\mathcal{B}} : \mathcal{B} \to \mathcal{C}$ is an equivalence. This is equivalent to $\mathrm{FPdim}(\mathcal{B}) = \mathrm{FPdim}(\mathcal{C})$ and $F|_{\mathcal{B}} : \mathcal{B} \to \mathcal{C}$ being injective, i.e., fully faithful.

Note that $F$ is identified with the functor of taking free $A$-modules:

$$\mathcal{Z}(\mathcal{C}) \to \mathcal{Z}(\mathcal{C})_A \cong \mathcal{C} : Z \mapsto A \otimes Z.$$

Observe that

$$\mathsf{Hom}_{\mathcal{C}}(F(Z), \mathbf{1}) \cong \mathsf{Hom}_{\mathcal{Z}(\mathcal{C})_A}(A \otimes Z, A) \cong \mathsf{Hom}_{\mathcal{Z}(\mathcal{C})}(Z, A),$$

for all $Z \in \mathcal{Z}(\mathcal{C})$. The injectivity of $F|_{\mathcal{B}} : \mathcal{B} \to \mathcal{C}$ is equivalent to $\mathsf{Hom}_{\mathcal{C}}(F(Z), \mathbf{1}) = \mathsf{Hom}_{\mathcal{B}}(Z, \mathbf{1})$ for all $Z \in \mathcal{B}$ and, hence, to $A$ and $\mathcal{B}$ being transversal. □

## 4. Braidings on non-degenerate fusion categories

**4.1. Classification of braidings.** Let $\mathcal{B}$ be a fusion category with a non-degenerate braiding $c = \{c_{X,Y}\}$.

Any grading of a fusion category $\mathcal{C}$ by a group $G$ determines a homomorphism

$$h_{\mathcal{C}} : \mathcal{O}(\mathcal{C}_{pt}) \to G.$$

THEOREM 4.1. *The braidings on $\mathcal{B}$ are in bijection with subcategory data $(\mathcal{E}, \mathcal{F}, G)$ such that $\mathcal{E} \vee \mathcal{F} = \mathcal{B}$, $\mathcal{E} \cap \mathcal{F}$ is pointed, and $h_{\mathcal{E}} + h_{\mathcal{F}} : \mathcal{O}(\mathcal{F} \cap \mathcal{E}) \to G$ is an isomorphism. Here $\mathcal{E} \vee \mathcal{F}$ denotes the fusion subcategory of $\mathcal{B}$ generated by $\mathcal{E}$ and $\mathcal{F}$.*

PROOF. We will use the characterization of braidings from Theorem 3.2.

Since $\mathcal{B}$ is non-degenerate, we have $\mathcal{Z}(\mathcal{B}) \cong \mathcal{B} \boxtimes \mathcal{B}^{\mathrm{rev}}$, where $\mathcal{B}^{\mathrm{rev}}$ denotes $\mathcal{B}$ equipped with the reverse braiding $c^{\mathrm{rev}}_{X,Y} := c^{-1}_{Y,X}$. The forgetful functor $F : \mathcal{Z}(\mathcal{B}) \to \mathcal{B}$ is identified with the tensor multiplication $\mathcal{B} \boxtimes \mathcal{B}^{\mathrm{rev}} \to \mathcal{B}$ and the canonical Lagrangian algebra in $\mathcal{Z}(\mathcal{B})$ is

$$A = \oplus_{X \in \mathcal{O}(\mathcal{B})} X^* \boxtimes X.$$

The notion of a subcategory datum for a tensor product of fusion categories was introduced in Definition 2.1. Suppose that $\mathcal{S}(\mathcal{E}, \mathcal{F}, G)$ is transversal to $A$ and is such that $\mathrm{FPdim}(\mathcal{S}(\mathcal{E},\,\mathcal{F},\,G)) = \mathrm{FPdim}(\mathcal{B})$. Since the restriction of $F$ on $\mathcal{S}(\mathcal{E},\,\mathcal{F},\,G)$ is injective we must have

$$\mathrm{FPdim}(F(\mathcal{S}(\mathcal{E},\,\mathcal{F},\,G))) = \mathrm{FPdim}(\mathcal{B}).$$

On the other hand, $\mathrm{FPdim}(F(\mathcal{S}(\mathcal{E},\,\mathcal{F},\,G))) \leq \mathrm{FPdim}(\mathcal{E}\vee\mathcal{F})$, so $\mathcal{E}\vee\mathcal{F}=\mathcal{B}$.

Using [**DGNO**, Lemma 3.38] we get

(10)
$$\mathrm{FPdim}(\mathcal{S}(\mathcal{E},\,\mathcal{F},\,G)) = \frac{\mathrm{FPdim}(\mathcal{E})\mathrm{FPdim}(\mathcal{F})}{|G|} = \frac{\mathrm{FPdim}(\mathcal{E}\vee\mathcal{F})\mathrm{FPdim}(\mathcal{E}\cap\mathcal{F})}{|G|}.$$

It follows from (10) that $\mathrm{FPdim}(\mathcal{E}\cap\mathcal{F}) = |G|$. If $X$ is a non-zero simple object in $\mathcal{E}_g\cap\mathcal{F}_h$ then $X\otimes X^* \in \mathcal{E}_1\cap\mathcal{F}_1$. It follows that $X\otimes X^* = \mathbf{1}$ (since other possibilities contradict the transversality of $\mathcal{S}(\mathcal{E},\,\mathcal{F},\,G)$ and $A$). Thus, $X$ is invertible and $\mathcal{E}\cap\mathcal{F}$ is pointed. For any non-identity $g\in G$ we must have $\mathcal{E}_g\cap\mathcal{F}_{g^{-1}}=0$. This is equivalent to the injectivity of $h_\mathcal{E}+h_\mathcal{F}$. Indeed, otherwise there is a nonzero $X\in\mathcal{E}_g$ such that $X^*\in\mathcal{F}_g$ and $X\boxtimes X^*\in\mathcal{S}(\mathcal{E},\,\mathcal{F},\,G)$, contradicting the transversality assumption.

Since $|\mathcal{O}(\mathcal{E}\cap\mathcal{F})| = |G|$, $h_\mathcal{E}+h_\mathcal{F}$ is an isomorphism.

Conversely, suppose that a datum $(\mathcal{E},\,\mathcal{F},\,G)$ satisfies conditions in the statement of the theorem. By (10), $\mathrm{FPdim}(\mathcal{S}(\mathcal{E},\,\mathcal{F},\,G)) = \mathrm{FPdim}(\mathcal{B})$. We have $\mathcal{E}_g\cap\mathcal{F}_{g^{-1}}=0$ for all $g\in G$, $g\neq e$. Thus, $\mathcal{S}(\mathcal{E},\,\mathcal{F},\,G)$ contains no simple objects of the form $X^*\boxtimes X$ for $X\neq 1$, i.e., $\mathcal{S}(\mathcal{E},\,\mathcal{F},\,G)$ is transversal to $A$. □

REMARK 4.2. Under the conditions of Theorem 4.1, we have $\mathcal{B}\cong\mathcal{E}\boxtimes_G\mathcal{F}$ (as a fusion category) and the corresponding braiding $\tilde{c}$ is given by

$$\tilde{c}_{X_1\boxtimes Y_1,X_2\boxtimes Y_2} = c_{X_1,X_2}\boxtimes c^{-1}_{Y_2,Y_1}$$

for all $X_1\boxtimes Y_1$, $X_2\boxtimes Y_2$ in $\mathcal{B}$.

COROLLARY 4.3. *Let $(\mathcal{E},\,\mathcal{F},\,G)$ be a subcategory datum for $\mathcal{B}\boxtimes\mathcal{B}^{rev}$. Then*

(11)
$$\mathcal{S}(\mathcal{E},\,\mathcal{F},\,G)' = \bigoplus_{\phi\in\widehat{G}} (\mathcal{E}_1')_\phi\boxtimes(\mathcal{F}_1)'_{\phi^{-1}},$$

*where the $\widehat{G}$-gradings on $\mathcal{E}_1'$ and $\mathcal{F}_1'$ are defined as in Proposition* 2.4.

PROOF. For all objects $V,W$ let us denote $\beta_{V,W} := c_{W,V}\circ c_{V,W}$.

Let $X\boxtimes Y$ be an object of $\mathcal{B}\boxtimes\mathcal{B}$ and let $X_g\boxtimes Y_g$ be an object of $\mathcal{S}(\mathcal{E},\,\mathcal{F},\,G)_g$, $g\in G$. Then

$$\beta_{X\boxtimes Y, X_g\boxtimes Y_g} = \beta_{X,X_g}\boxtimes\beta_{Y,Y_g}$$

and so $X\boxtimes Y$ centralizes $X_g\boxtimes Y_g$ if and only if $\beta_{X,X_g}$ and $\beta_{Y,Y_g}$ are mutually inverse scalars. This means that $X$ projectively centralizes $\mathcal{E}$ and centralizes $\mathcal{E}_1$ (respectively, $Y$ projectively centralizes $\mathcal{F}$ and centralizes $\mathcal{F}_1$). Thus,

$$\mathcal{S}(\mathcal{E},\,\mathcal{F},\,G)' = \mathcal{E}_1'\boxtimes_{\widehat{G}}\mathcal{F}_1' = \bigoplus_{\phi\in\widehat{G}}(\mathcal{E}_1')_\phi\boxtimes(\mathcal{F}_1)'_{\phi^{-1}},$$

as required. □

Let $\mathcal{B}(\mathcal{F},\,\mathcal{E},\,G)$ denote the braided fusion category (with underlying fusion category $\mathcal{B}$) corresponding to the datum $(\mathcal{E},\,\mathcal{F},\,G)$ from Theorem 4.1.

COROLLARY 4.4. *We have $\mathcal{B}(\mathcal{E},\,\mathcal{F},\,G)^{rev}\cong\mathcal{B}(\mathcal{E}_1',\,\mathcal{F}_1',\,\widehat{G})$, where the fiber product of $\mathcal{E}_1'$ and $\mathcal{F}_1'$ is as in* (11).

COROLLARY 4.5. *The symmetric center of* $\mathcal{B}(\mathcal{F}, \mathcal{E}, G)$ *has a (not necessarily faithful) grading*

$$\mathcal{Z}_{sym}(\mathcal{B}(\mathcal{F}, \mathcal{E}, G)) = \bigoplus_{(g,\phi)\in G\times\widehat{G}} \mathcal{Z}_{sym}(\mathcal{B}(\mathcal{F}, \mathcal{E}, G))_{(g,\phi)},$$

*where* $\mathcal{Z}_{sym}(\mathcal{B}(\mathcal{F}, \mathcal{E}, G))_{(g,\phi)} \cong (\mathcal{E}_g \cap (\mathcal{E}_1)'_\phi) \boxtimes (\mathcal{F}_g \cap (\mathcal{F}_1)'_{\phi^{-1}})$. *In particular,* $\mathcal{Z}_{sym}(\mathcal{B}(\mathcal{F}, \mathcal{E}, G))$ *is pointed.*

PROOF. The formula for homogeneous components follows from Corollary 4.3. The trivial component of the grading of $\mathcal{Z}_{sym}(\mathcal{B}(\mathcal{F}, \mathcal{E}, G))$ is contained in $\mathcal{E}_1 \boxtimes \mathcal{F}_1$ and so it is equivalent to $\mathsf{Vec}$. Hence, $\mathcal{Z}_{sym}(\mathcal{B}(\mathcal{F}, \mathcal{E}, G))$ is pointed. □

REMARK 4.6. Corollary 4.5 means that if $\mathcal{B}$ has a non-degenerate braiding then other braidings on $\mathcal{B}$ cannot be "too symmetric" as the symmetric center remains pointed. Conversely, if $\mathcal{B}$ has a braiding such that $\mathcal{Z}_{sym}(\mathcal{B})$ is not pointed, then no non-degenerate braidings on $\mathcal{B}$ can exist. In particular, $\mathsf{Rep}(G)$ for a non-abelian $G$ does not admit any non-degenerate braidings (equivalently, there are no modular category structures on $\mathsf{Rep}(G)$).

PROPOSITION 4.7. *Let* $\mathcal{B}$ *be a fusion category that admits a non-degenerate braiding. Then all non-degenerate braidings on* $\mathcal{B}$ *correspond to data* $(\mathcal{E}, \mathcal{F}, G)$ *such that*

$$\mathcal{B} \cap \big((\mathcal{E}_g \cap \mathcal{F}_\phi) \boxtimes (\mathcal{F}_g \cap \mathcal{E}_{\phi^{-1}})\big) = \begin{cases} \mathsf{Vec} & \text{if } g = 1, \phi = 1 \\ 0 & \text{otherwise,} \end{cases} \tag{12}$$

*where we use identification* $\mathcal{B} = \mathcal{E} \boxtimes_G \mathcal{F} \subset \mathcal{E} \boxtimes \mathcal{F}$.

PROOF. Follows Corollary 4.5. □

**4.2. Braidings on unpointed categories.** Let $\mathcal{B}$ be a fusion category with non-degenerate braiding. Suppose that $\mathcal{B}_{pt} = \mathsf{Vec}$, i.e., $\mathcal{B}$ is *unpoitned.* It was shown in [**Mu**] that in this case there is factorization of $\mathcal{B}$ into a direct product of prime subcategories:

$$\mathcal{B} = \mathcal{B}_1 \boxtimes \cdots \boxtimes \mathcal{B}_n, \tag{13}$$

which is unique up to a permutation of factors.

COROLLARY 4.8. *Let* $\mathcal{B}$ *be a fusion category such that* $\mathcal{B}_{pt} = \mathsf{Vec}$. *Suppose that* $\mathcal{B}$ *admits a non-degenerate braiding. Let* (13) *be the prime factorization of* $\mathcal{B}$. *Then all braidings on* $\mathcal{B}$ *are non-degenerate and there are precisely* $2^n$ *such braidings. The corresponding braided fusion categories are*

$$\mathcal{B} = \mathcal{B}_1^{\pm} \boxtimes \cdots \boxtimes \mathcal{B}_n^{\pm}, \tag{14}$$

*where* $\mathcal{B}_i^+ = \mathcal{B}_i$ *and* $\mathcal{B}_i^- = \mathcal{B}_i^{rev}$ *for* $i = 1, \dots, n$.

PROOF. Since, $\mathcal{B}$ is unpointed, according to Remark 4.2 we have $\mathcal{B} \cong \mathcal{E} \boxtimes \mathcal{F}$ as a fusion category. We claim that $\mathcal{E}$ and $\mathcal{F}$ centralize each other with respect to the original braiding of $\mathcal{B}$. Indeed, for all $X \in \mathcal{O}(\mathcal{E})$ and $X \in \mathcal{O}(\mathcal{F})$ the object $X \boxtimes Y$ is simple and, therefore,

$$c_{Y,X} \circ c_{X,Y} = \lambda_{X,Y} \mathrm{id}_{X\boxtimes Y}, \qquad \lambda_{X,Y} \in k^\times.$$

It follows that the map

$$\mathcal{O}(\mathcal{E} \boxtimes \mathcal{F}) \to k^\times : X \boxtimes Y \mapsto \lambda_{X,Y}$$

is a grading of $\mathcal{E}\boxtimes\mathcal{F}$ . But $U(\mathcal{E}\boxtimes\mathcal{F}) \cong \mathcal{O}(\widehat{(\mathcal{E}\boxtimes\mathcal{F})_{pt}})$ is trivial, and so $\lambda_{X,Y} = 1$ for all $X$, $Y$, which proves the claim. It follows that $\mathcal{E}$ and $\mathcal{F}$ must be non-degenerate subcategories of $\mathcal{B}$. By [**DMNO**, Section 2.2] there is a subset $J \subset \{1, \dots, n\}$ such that $\mathcal{E} = \bigoplus_{i\in J} \mathcal{B}_i$ and $\mathcal{F} = \bigoplus_{i\notin J} \mathcal{B}_i$. This implies the statement. □

**4.3. Gauging.** Let $\mathcal{B}$ be a non-degenerate braided fusion category with a braiding $c_{X,Y} : X \otimes Y \to Y \otimes X$. A *gauging* of $\mathcal{B}$ is the following procedure of changing the braiding by a bilinear form $b : U(\mathcal{B}) \times U(\mathcal{B}) \to k^\times$. A new braiding $\tilde{c}_{X,Y} : X \otimes Y \to Y \otimes X$ is defined by

$$\tilde{c}_{X,Y} = b(\deg(X), \deg(Y))\, c_{X,Y},$$

for all $X, Y \in \mathcal{O}(\mathcal{B})$, where deg denotes the degree of a simple object with respect to the universal grading. By definition, gaugings of a given braiding form a torsor over the group of bilinear forms on $U(\mathcal{B})$.

The corresponding embedding $\mathcal{B} \to \mathcal{Z}(\mathcal{B}) = \mathcal{B}\boxtimes\mathcal{B}^{\text{rev}}$ is given by $X \mapsto (X\otimes V_X)\boxtimes V_X^*$ for all $X \in \mathcal{O}(\mathcal{B})$, where $V_X \in \mathcal{O}(\mathcal{B}_{pt})$ is determined by the condition

$$\langle V_X,\, y\rangle = b(\deg(X),\, y), \quad \text{for all } y \in U(\mathcal{B}).$$

Here $\langle\, ,\, \rangle : \mathcal{O}(\mathcal{B}_{pt}) \times \widehat{U(\mathcal{B})} \to k^\times$ denotes the canonical pairing (8).

In this situation $\mathcal{E} = \mathcal{B}$, $\mathcal{F}_1 = \mathsf{Vec}$ (so that $\mathcal{F} \subset \mathcal{B}_{pt}$), and $G \subset \mathcal{O}(\mathcal{B}_{pt})$ is the image of the homomorphism $U(\mathcal{B}) \to \mathcal{O}(\mathcal{B}_{pt}) : X \mapsto V_X$.

Conversely, if a datum $(\mathcal{E},\, \mathcal{F},\, G)$ from Theorem 4.1 is such that $\mathcal{E} = \mathcal{B}$ and $\mathcal{F}_1 = \mathsf{Vec}$ (respectively, $\mathcal{F} = \mathcal{B}$ and $\mathcal{E}_1 = \mathsf{Vec}$) then the corresponding braiding is a gauging of the original braiding of $\mathcal{B}$ (respectively, of the reverse braiding).

In the next two examples for a finite group $G$ we denote by $\mathcal{Z}(G)$ the center of $\mathsf{Vec}_G$.

Example 4.9. (This result was independently obtained by Costel-Gabriel Bontea using different techniques). Let $\mathcal{B} := \mathcal{Z}(S_n)$, $n \geq 3$, where $S_n$ denotes the symmetric group on $n$ symbols. Observe that $\mathcal{B}$ has a unique maximal fusion subcategory $\mathcal{B}_{ad}$, which is the subcategory of vector bundles supported on the alternating subgroup $A_n$. Thus, in any presentation $\mathcal{B} = \mathcal{E}\boxtimes_G \mathcal{F}$ either $\mathcal{E} = \mathcal{B}$ or $\mathcal{F} = \mathcal{B}$. Since $U(\mathcal{B}) = \mathbb{Z}_2$ we must have $G = \{1\}$ or $G = \mathbb{Z}_2$. If $\mathcal{E} = \mathcal{B}$ then $\mathcal{F} = \mathsf{Vec}$ or $\mathcal{F} = \mathcal{B}_{pt}$ (note that $\text{FPdim}(\mathcal{B}_{pt}) = 2$). The first possibility gives the standard braiding of $\mathcal{B}$, while the second gives its gauging with respect to the $\mathbb{Z}_2$-grading of $\mathcal{B}$. The situation when $\mathcal{F} = \mathcal{B}$ is completely similar.

Hence, $\mathcal{B}$ has 4 different braidings: the usual braiding of the center, its reverse, and their gaugings with respect to the $\mathbb{Z}_2$-grading of $\mathcal{Z}(S_n)$. The corresponding data are: $(\mathcal{B}, \mathsf{Vec}, 1)$, $(\mathsf{Vec}, \mathcal{B}, 1)$, $(\mathcal{B}, \mathcal{B}_{pt}, \mathbb{Z}_2)$, and $(\mathcal{B}_{pt}, \mathcal{B}, \mathbb{Z}_2)$, respectively.

Example 4.10. Let $G$ be a non-abelian group of order 8, i.e., $G$ is either the dihedral group or the quaternion group. Let $\mathcal{B} = \mathcal{Z}(G)$. We claim that every braiding of $\mathcal{B}$ is a gauging of either its standard braiding or its reverse. The structure of $\mathcal{Z}(G)$ was studied in detail by various authors including [**GMN**, **MN**]. One has $U(\mathcal{B}) = \mathbb{Z}_2^3$ (so in particular, the standard braiding of $\mathcal{B}$ has $2^9 = 512$ different gaugings!) The trivial component of the universal grading is $\mathcal{B}_{pt} = \mathcal{B}_{ad}$, this is a pointed Lagrangian subcategory of the Frobenius-Perron dimension 8. Furthermore, for any non-pointed fusion subcategory $\mathcal{E} \subset \mathcal{B}$ its adjoint subcategory $\mathcal{E}_{ad}$ contains at least 4 invertible objects. In any presentation $\mathcal{B} = \mathcal{E}\boxtimes_G \mathcal{F}$ satisfying the conditions of Theorem 4.1 one of the subcategories $\mathcal{E}$, $\mathcal{F}$ must be non-pointed and and another must be pointed. Indeed, if both are pointed then so is $\mathcal{B}$, a contradiction. If both are non-pointed then $\mathrm{FPdim}(\mathcal{E}_1 \cap \mathcal{F}_1) \geq 2$, a contradiction.

Suppose that $\mathcal{E}$ is non-pointed. Then $\mathcal{F}$ is a pointed fusion subcategory of $\mathcal{B}$ with $\mathrm{FPdim}(\mathcal{B}) = 1, 2, 4$ or $8$.

If $\mathrm{FPdim}(\mathcal{F}) = 1$ then we get the standard braiding of $\mathcal{B}$.

If $\mathrm{FPdim}(\mathcal{F}) = 2$ then either $G = \mathbb{Z}_2$ and the corresponding braiding is a gauging of the standard one, or $G = \{1\}$ and $\mathcal{B} = \mathcal{E} \boxtimes \mathcal{F}$. The latter is impossible since in this case $\mathrm{FPdim}(\mathcal{E}) = 32$ and $\mathcal{E}$ contains $\mathcal{B}_{pt}$ and, hence, $\mathcal{F}$.

If $\mathrm{FPdim}(\mathcal{F}) = 4$ then either $G = \mathbb{Z}_4$ the corresponding braiding is a gauging of the standard one, or $G = \mathbb{Z}_2$ and so $\mathrm{FPdim}(\mathcal{E}) = 32$ and $\mathrm{FPdim}(\mathcal{E}_1 \cap \mathcal{F}_1) = 2$, a contradiction, or $G = \{1\}$ and $\mathcal{B} = \mathcal{E} \boxtimes \mathcal{F}$ which is impossible.

Finally, if $\mathrm{FPdim}(\mathcal{F}) = 8$ then we must have $G = \mathbb{Z}_2^3$ since otherwise we again have $\mathrm{FPdim}(\mathcal{E}_1 \cap \mathcal{F}_1) \geq 2$, which contradicts conditions of Theorem 4.1. So in this case $\mathcal{F}_1 = \mathsf{Vec}$ and the grading of $\mathcal{B}$ is a gauging of the standard one.

Thus, if $\mathcal{E}$ is non-pointed then the corresponding grading is always a gauging of the standard one. Switching $\mathcal{E}$ and $\mathcal{F}$ will give gaugings of the reverse brading.

## 5. Braidings on group-theoretical categories

Let $G$ be a finite group. Let us denote $\mathcal{C}(G) = \mathsf{Vec}_G$ and $\mathcal{Z}(G) := \mathcal{Z}(\mathsf{Vec}_G) = \mathcal{C}(G)^G$.

**5.1. Lagrangian algebras in the center of $\mathsf{Vec}_G$.** It is well known that $\mathcal{Z}(G)$ is identified with the category of $G$-equivariant vector bundles on $G$. The isomorphism classes of simple objects of $\mathcal{Z}(G)$ are parameterized by pairs $(K, \pi)$, where $K \subset G$ is a conjugacy class and $\pi$ is the isomorphism class an irreducible representation of the centralizer $C_G(g_K)$ of $g_K \in K$. The corresponding object $V(K, \pi) = \oplus_{g \in K} V(K, \pi)_g$ is the vector bundle supported on $K$ whose equivariant structure restricted to $C_G(g_K)$ acts by $\pi$ on $V(K, \pi)_{g_K}$.

Recall that equivalence classes of indecomposable $\mathcal{C}(G)$-module categories are parameterized by conjugacy classes of pairs $(H, \mu)$, where $H$ is a subgroup of $G$ and $\mu \in H^2(G, k^\times)$. The module category $\mathcal{M}(H, \mu)$ corresponding to $(H, \mu)$ is the category of modules over the twisted group algebra $k_\mu[H]$ in $\mathcal{C}(G)$. It can be identified with a certain category of $H$-invariant vector bundles on $G$.

Let $\mathcal{Z}(G; H) = \mathsf{Vec}_G^H$ be the category of $H$-equivariant objects in $\mathsf{Vec}_G$. We have $\mathcal{Z}(H) \subset \mathcal{Z}(G; H)$. There is an obvious forgetful functor $F_H : \mathcal{Z}(G) \to \mathcal{Z}(G; H)$. Let $I_H : \mathcal{Z}(G; H) \to \mathcal{Z}(G)$ denote its adjoint.

The following construction was given in [**D2**]. The twisted group algebra $k_\mu[H]$ is a Lagrangian algebra in $\mathcal{Z}(H)$ with the obvious grading and the $H$-equivariant structure given by

(15)
$$k_\mu[H] \to g k_\mu[H] g^{-1} : x \mapsto \varepsilon_g(x) g x g^{-1}, \quad \text{where } \varepsilon_g(x) = \frac{\mu(gxg^{-1}, g)}{\mu(g, x)}, \qquad g, x \in H.$$

Here we abuse notation and identify the cohomology class $\mu$ with a 2-cocycle representing it. Note that

$$\frac{\varepsilon_g(x)\varepsilon_g(y)}{\varepsilon_g(xy)} = \frac{\mu(gxg^{-1}, gyg^{-1})}{\mu(x, y)}, \qquad g, x, y \in H. \tag{16}$$

In particular, $\varepsilon_{g_K}$ restricts to a linear character of $C_G(g_K)$. As an object of $\mathcal{Z}(H)$,

$$k_\mu[H] \cong \bigoplus_K V(K, \varepsilon_{g_K}). \tag{17}$$

Let $A(H, \mu) \in \mathcal{Z}(G)$ be the Lagrangian algebra corresponding to the $\mathcal{C}(G)$-module category $\mathcal{M}(H, \mu)$. It was shown in [**D2**, Section 3.4] that

$$A(H, \mu) \cong I_H(k_\mu[H]). \tag{18}$$

Here $k_\mu[H] \in \mathcal{Z}(H)$ is considered as an algebra in $\mathcal{Z}(G; H)$.

**5.2. Transversality criterion and parameterization of braidings.** Tensor subcategories of $\mathcal{Z}(G)$ were classified in [**NNW**]. They are in bijection with triples $(L, M, B)$, where

(T1) $L$ and $M$ are normal subgroups of $G$ commuting with each other,
(T2) $B : L \times M \to k^\times$ is a $G$-invariant bicharacter.

The corresponding subcategory $\mathcal{S}_G(L, M, B)$ consists of vector bundles supported on $L$ and such that the restriction of their $G$-equivariant structure on $M$ is the scalar multiplication by $B(g, -)$ for all $g \in L$. Equivalently, simple objects of $\mathcal{S}_G(L, M, B)$ are objects $V(K, \pi)$, where $K$ is a conjugacy class contained in $L$ and $\pi$ is contained in the induced representation $\mathsf{Ind}_M^{C(g_K)} B(g_K, -)$. We have

$$\mathrm{FPdim}(\mathcal{S}_G(L, M, B)) = |L|[G : M]. \tag{19}$$

We denote by $\widehat{B} : L \to \widehat{M}$ the group homomorphism associated to $B$.

Let $\mu$ be a 2-cocycle on $G$ with values in $k^\times$. The map $\mathsf{Alt}(\mu) : C_G(M) \times M \to k^\times$ defined by

$$\mathsf{Alt}(\mu)(g, x) = \frac{\mu(x, g)}{\mu(g, x)}, \qquad g \in C_G(M),\ x \in M. \tag{20}$$

is bimultiplicative and $G$-invariant. We have

$$\mathsf{Alt}(\mu)(g, x) = \varepsilon_g(x)$$

for all $g \in C_G(M)$, $x \in M$.

LEMMA 5.1. *The subcategory $\mathcal{S}_G(L, M, B) \subset \mathcal{Z}(G)$ is transversal to the Lagrangian algebra $k_\mu[G]$ if and only if $\frac{B}{\mathsf{Alt}(\mu)} : L \times M \to k^\times$ is non-degenerate in the second argument, i.e., for all $g \in L$, $g \neq 1$, there is $x \in M$ such that*

$$\frac{B}{\mathsf{Alt}(\mu)}(g, x) \neq 1. \tag{21}$$

PROOF. By (17) the transversality is equivalent to the condition

$$\mathsf{Hom}_{C_G(g_K)}(\varepsilon_{g_K}, \mathsf{Ind}_M^{C_G(g_K)} \widehat{B}(g_K)) = 0$$

for all non-identity conjugacy classes $K \subset L$. By the Frobenius reciprocity this is equivalent to

$$\mathsf{Hom}_M(\varepsilon_{g_K}|_M, \widehat{B}(g_K)) = 0, \qquad K \subset L,\ K \neq \{1\},$$

i.e., $\varepsilon_{g_K}|_M \neq \widehat{B}(g_K)$ for all non-identity $K$. This condition means that for each $g_K$ with $K \subset L$ $(K \neq \{1\})$ there is $x \in M$ such that

$$\frac{\mu(x, g_K)}{\mu(g_K, x)} \neq B(g_K, x).$$

Using the $G$-invariance of $B$ and $\mathsf{Alt}(\mu)$ we get the result. □

THEOREM 5.2. *Braidings on* $\mathcal{C}(G)^*_{\mathcal{M}(H,\,\mu)}$ *are in bijection with triples* $(L,\,M,B)$ *satisfying (T1), (T2) and the following conditions:*

(i) $LH = MH = G$,

(ii) *the restriction of* $\frac{B}{\mathsf{Alt}(\mu)}$ *on* $(L\cap H)\times(M\cap H)$ *is non-degenerate.*

PROOF. By Theorem 3.2 braidings on $\mathcal{C}(G)^*_{\mathcal{M}(H,\,\mu)}$ are parameterized by fusion subcategories $\mathcal{S}_G(L,M,B)\subset\mathcal{Z}(G)$ of the Frobenius-Perron dimension $|G|$ transversal to $A(H,\,\mu)$.

The above dimension condition is equivalent to $|L| = |M|$ by (19). Note that this condition follows from (i) and (ii).

In view of (18) we see that a necessary condition for the above transversality is that the restriction of the forgetful functor $F_H : \mathcal{Z}(G)\to\mathcal{Z}(G;\,H)$ to $\mathcal{S}_G(L,M,B)$ is injective. The latter condition is equivalent to transversality of $\mathsf{Rep}(G/M)$ and the function algebra $\mathrm{Fun}(G/H,\,k)$ in $\mathsf{Rep}(G)$, in other words, to $\mathsf{Hom}_G(\mathsf{Ind}_M^G\,k^\times,\,\mathsf{Ind}_H^G\,k^\times) = k$. Here $k^\times$ denotes the trivial module. By the Mackey restriction formula the latter is equivalent to $MH = G$.

If the above condition is satisfied, the transversality of $\mathcal{S}_G(L,M,B)$ and $A(H,\,\mu)$ in $\mathcal{Z}(G)$ is equivalent to the transversality of $F_H(\mathcal{S}_G(L,M,B))$ and $k_\mu[H]$ in $\mathcal{Z}(G;\,H)$.

In this case we have

$$F_H(\mathcal{S}_G(L,M,B))\cap\mathcal{Z}(H) = \mathcal{S}_H(L\cap H,\,M\cap H,\,B|_{(L\cap H)\times(M\cap H)}).$$

Now we can apply Lemma 5.1 (with $G$ replaced by $H$). The transversality of the subcategory $\mathcal{S}_H(L\cap H,\,M\cap H,\,B|_{(L\cap H)\times(M\cap H)})$ and the algebra $k_\mu[H]$ is equivalent to the injectivity of the corresponding homomorphism $L\cap H\to\widehat{M\cap H}$, whence $|L\cap H|\leq|M\cap H|$. This implies

$$|LH| = \frac{|L||H|}{|L\cap H|}\geq\frac{|M||H|}{|M\cap H|} = |MH| = |G|,$$

so that $LH = G$, $|L\cap H| = |M\cap H|$ and, hence, $\frac{B}{\mathsf{Alt}(\mu)}|_{(L\cap H)\times(M\cap H)}$ is non-degenerate. □

EXAMPLE 5.3. Let $G$ be a non-abelian group and let $H\subset G$ be a subgroup such that the only normal subgroup $N$ of $G$ such that $HN = G$ is $G$ itself. Then $\mathcal{C}(G)^*_{\mathcal{M}(H,\,\mu)}$ does not admit a braiding. In particular, if $G$ is simple non-abelian and $H\neq G$ then $\mathcal{C}(G)^*_{\mathcal{M}(H,\,\mu)}$ does not admit a braiding.

REMARK 5.4. (i) Masuoka [**Ma**] showed that that certain self-dual non-commutative and non-cocommutative semisimple Hopf algebras of dimension $p^3$, where $p$ is an odd prime, admit no quasi-triangular structures, thus giving examples of group-theoretical fusion categories that do not admit any braiding. These examples, however, are not of the form considered in Theorem 5.2 (to obtain them one has to generalize Theorem 5.2 by replacing $\mathsf{Vec}_G$ with $\mathsf{Vec}_G^\omega$ for a non-trivial 3-cocycle $\omega$).

(ii) An absence of braidings on certain group-theoretical categories associated to exact factorizations of almost simple groups was established by Natale [**N**].

Example 5.5. The category $\mathcal{C}(G)^*_{\mathcal{M}(G,1)}$ is equivalent to $\mathsf{Rep}(G)$. In this case Theorem 5.2 says that braidings on $\mathsf{Rep}(G)$ are in bijections of triples $(L, M, B)$, where $L$ and $M$ are normal subgroups of $G$ commuting with each other and $B : L \times M \to k^\times$ is a non-degenerate $G$-invariant bilinear form (note that these conditions on $B$ imply that $L$ and $M$ must be Abelian). This classification was obtained by Davydov [**D1**].

Example 5.6. The category $\mathcal{Z}(G)$ is equivalent to $\mathcal{C}(G \times G^{\mathrm{op}})^*_{\mathcal{M}(D,1)}$, where $G^{\mathrm{op}}$ is the group with the opposite multiplication and $D = \{(g, g^{-1}) \mid g \in G\}$. Braidings on this category are parameterized by triples $(L, M, B)$, where $L$, $M$ are normal subgroups of $G \times G^{\mathrm{op}}$ and $B : L \times M \to k^\times$ is a $G \times G^{\mathrm{op}}$-invariant bilinear form such that the following conditions are satisfied:

(i) $L$ and $M$ commute with each other,
(ii) $LD = MD = G \times G^{op}$, and
(iii) the restriction of $B$ on $(L \cap D) \times (M \cap D)$ is non-degenerate.

The standard braiding of $\mathcal{Z}(G)$ corresponds to $L = G \times 1$, $M = 1 \times G^{op}$, and trivial $B$.

This parameterization is an alternative to the description of quasitriangular structures on the Drinfeld double of $G$ given by Keilberg [**K**].

## Acknowledgements

The author thanks Costel-Gabriel Bontea and Alexei Davydov for useful discussions. The author is grateful to the referee for her or his useful comments.

## References


[D1] A. Davydov, *Quasitriangular structures on cocommutative Hopf algebras*, arXiv:q-alg/9706007 (1997).

[D2] Alexei Davydov, *Modular invariants for group-theoretical modular data. I*, J. Algebra **323** (2010), no. 5, 1321–1348, DOI 10.1016/j.jalgebra.2009.11.041. MR2584958

[DGNO] Vladimir Drinfeld, Shlomo Gelaki, Dmitri Nikshych, and Victor Ostrik, *On braided fusion categories. I*, Selecta Math. (N.S.) **16** (2010), no. 1, 1–119, DOI 10.1007/s00029-010-0017-z. MR2609644

[DMNO] Alexei Davydov, Michael Müger, Dmitri Nikshych, and Victor Ostrik, *The Witt group of non-degenerate braided fusion categories*, J. Reine Angew. Math. **677** (2013), 135–177. MR3039775

[EGNO] Pavel Etingof, Shlomo Gelaki, Dmitri Nikshych, and Victor Ostrik, *Tensor categories*, Mathematical Surveys and Monographs, vol. 205, American Mathematical Society, Providence, RI, 2015. MR3242743

[ENO1] Pavel Etingof, Dmitri Nikshych, and Viktor Ostrik, *On fusion categories*, Ann. of Math. (2) **162** (2005), no. 2, 581–642, DOI 10.4007/annals.2005.162.581. MR2183279

[ENO2] Pavel Etingof, Dmitri Nikshych, and Victor Ostrik, *Fusion categories and homotopy theory*, Quantum Topol. **1** (2010), no. 3, 209–273, DOI 10.4171/QT/6. With an appendix by Ehud Meir. MR2677836

[ENO3] Pavel Etingof, Dmitri Nikshych, and Victor Ostrik, *Weakly group-theoretical and solvable fusion categories*, Adv. Math. **226** (2011), no. 1, 176–205, DOI 10.1016/j.aim.2010.06.009. MR2735754

[GNN] Shlomo Gelaki, Deepak Naidu, and Dmitri Nikshych, *Centers of graded fusion categories*, Algebra Number Theory **3** (2009), no. 8, 959–990, DOI 10.2140/ant.2009.3.959. MR2587410

[GN] Shlomo Gelaki and Dmitri Nikshych, *Nilpotent fusion categories*, Adv. Math. **217** (2008), no. 3, 1053–1071, DOI 10.1016/j.aim.2007.08.001. MR2383894

[GMN] Christopher Goff, Geoffrey Mason, and Siu-Hung Ng, *On the gauge equivalence of twisted quantum doubles of elementary abelian and extra-special 2-groups*, J. Algebra **312** (2007), no. 2, 849–875, DOI 10.1016/j.jalgebra.2006.10.022. MR2333187

[K] Marc Keilberg, *Automorphisms of the doubles of purely non-abelian finite groups*, Algebr. Represent. Theory **18** (2015), no. 5, 1267–1297, DOI 10.1007/s10468-015-9540-0. MR3422470

[MN] Ian Marshall and Dmitri Nikshych, *On the Brauer-Picard groups of fusion categories*, Math. Z. **288** (2018), no. 3-4, 689–712, DOI 10.1007/s00209-017-1907-y. MR3778972

[Ma] Akira Masuoka, *Example of almost commutative Hopf algebras which are not coquasitriangular*, Hopf algebras, Lecture Notes in Pure and Appl. Math., vol. 237, Dekker, New York, 2004, pp. 185–191. MR2051739

[Mu] Michael Müger, *On the structure of modular categories*, Proc. London Math. Soc. (3) **87** (2003), no. 2, 291–308, DOI 10.1112/S0024611503014187. MR1990929

[NNW] Deepak Naidu, Dmitri Nikshych, and Sarah Witherspoon, *Fusion subcategories of representation categories of twisted quantum doubles of finite groups*, Int. Math. Res. Not. IMRN **22** (2009), 4183–4219, DOI 10.1093/imrn/rnp084. MR2552301

[N] Sonia Natale, *On quasitriangular structures in Hopf algebras arising from exact group factorizations*, Comm. Algebra **39** (2011), no. 12, 4763–4775, DOI 10.1080/00927872.2011.617625. MR2863465

Department of Mathematics and Statistics, University of New Hampshire, Durham, NH 03824, USA

*Email address*: `dmitri.nikshych@unh.edu`

Contemporary Mathematics
Volume **728**, 2019
https://doi.org/10.1090/conm/728/14661

# Remarks on global dimensions of fusion categories

Victor Ostrik

ABSTRACT. We give a nontrivial lower bound for the global dimension of a spherical fusion category.

## 1. Introduction

**1.1.** This paper is a contribution to the theory of fusion categories over the field of complex numbers. Recall (see [**ENO**, **EGNO**]) that a fusion category is a semisimple rigid tensor category with finitely many simple objects, finite dimensional Hom −spaces, and with simple unit object. The simplest example is the category of finite dimensional vector spaces Vec. A basic invariant of a fusion category $\mathcal{C}$ is its *global dimension* $\dim(\mathcal{C}) \in \mathbb{R}$ introduced by M. Müger in [**Mu1**], see Section 2.1 below. We consider the global dimension as a real valued function on the set of equivalence classes of fusion categories. Here is our first result.

THEOREM 1.1.1 (see Theorem 5.1.1). *The fibers of the map* $\mathcal{C} \mapsto \dim(\mathcal{C})$ *are finite. In other words there are at most finitely many fusion categories (up to a tensor equivalence) of a fixed global dimension.*

It seems natural to ask next about properties of the image of the global dimension, that is of the following subset of $\mathbb{R}$:

$$X := \{\dim(\mathcal{C}) \mid \mathcal{C} \text{ is a fusion category}\} \tag{1.1}$$

It is known that $X$ consists of totally real algebraic integers (and so $X$ is countable), is invariant under the Galois group, and is contained in the interval $[1, \infty)$, see Section 2.1. However other properties of the set $X$ are not well understood. For instance it is an open question (asked in [**ENO**, p. 596]) whether the set $X$ is discrete. A weaker version of this question (also asked in *loc. cit.* and still open) is whether point $1 = \dim(\mathrm{Vec}) \in X$ is isolated. An expected positive answer to this question was called *categorical property T* in *loc. cit.* by analogy with the famous Kazhdan's property T in the group theory. The main result of this paper is a slightly weaker version of the categorical property T. Namely, recall that a fusion category $\mathcal{C}$ is *spherical* if it admits a spherical structure, see Section 2.2. All currently known fusion categories are spherical and it is expected (see e.g. [**ENO**, Conjecture 2.8]) that this is always the case. This conjecture is of great importance for the theory of fusion categories, but it seems to be quite difficult. We consider the following subset of the set $X$:

$$X_s := \{\dim(\mathcal{C}) \mid \mathcal{C} \text{ is a spherical fusion category}\} \tag{1.2}$$

Clearly, the set $X_s$ can be considered as a spherical counterpart of the set $X$ above.

THEOREM 1.1.2 (see Theorem 4.1.1 (i)). *Let $\mathcal{C}$ be a spherical fusion category which is not equivalent to* Vec. *Then*

$$\dim(\mathcal{C}) > \frac{4}{3} \tag{1.3}$$

*In particular point* $1 = \dim(\mathrm{Vec}) \in X_s$ *is isolated.*

The proof of Theorem 1.1.2 is based on the study of the *Drinfeld center* of the category $\mathcal{C}$ and exploiting its properties as a *modular tensor category*, see e.g. [**EGNO**, Chapter 8]. In particular, an important role is played by a result of M. Müger [**Mu2**, Theorem 1.2] computing the *Gauss sum* of the Drinfeld center of a spherical fusion category. This explains partially why we need to assume that $\mathcal{C}$ is spherical in the statement of Theorem 1.1.2.

The constant $\frac{4}{3}$ in Theorem 1.1.2 is not optimal. In the Appendix (joint with P. Etingof) we show that for a spherical fusion category $\mathcal{C} \not\simeq \mathrm{Vec}$ we have

$$\dim(\mathcal{C}) \geq \frac{5-\sqrt{5}}{2} = \dim(YL) \approx 1.381966$$

where $YL$ is the (non-unitary) Yang-Lee category, see e.g. [**EGNO**, Exercise 4.10.7]. Note that the point $\dim(YL) \in X_s$ is isolated by Corollary 4.1.3 or by more precise Proposition A.1.1.

We will also prove Theorem 4.2.1 which extends Theorem 1.1.2 to other numerical invariants of the category $\mathcal{C}$, so called *formal codegrees*, see Section 2.3.

## 1.2. Open questions.

QUESTION 1.2.1. Find a counterpart of Theorem 1.1.2 for a not necessarily spherical fusion categories.

It follows from Corollary 4.1.3 that any points $x \in X_s$ with $x < \sqrt{2}$ is isolated. Thus we have

QUESTION 1.2.2. Prove (or disprove) that $\sqrt{2}$ is not a limit point of $X_s$.

Note that it is clear that $\sqrt{2} \notin X_s$ since $\sqrt{2}$ is not totally positive. Actually we don't know any point $x \in X_s$ which is less than $\sqrt{2}$ but greater than $\dim(YL)$. Thus we ask

QUESTION 1.2.3. Are there any points $x \in X_s$ with $\dim(YL) < x < \sqrt{2}$?

A more ambitious version of this question is to find a next point (or limit point) of $X_s$ after $\dim(YL)$. We suspect that the answer is given by the smallest root of the polynomial $t^3 - 14t^2 + 49t - 49$ (approximately 1.84117) which is the dimension of some fusion category $\mathcal{C}_3$ of rank 3 associated with quantum $\mathfrak{so}_3$ at 7th root of 1, see [**O4**, 4.3].

# 2. Preliminaries

**2.1. Global dimension.** For a fusion category $\mathcal{C}$ over $\mathbb{C}$ we will denote by $\mathcal{O}(\mathcal{C})$ the set of isomorphism classes of simple objects of $\mathcal{C}$. For an object $X \in \mathcal{C}$ let ${}^*X, X^*$ denote the right and left duals of $X$. Recall that for any morphism $f : X \to X^{**}$ one defines its trace $\mathrm{Tr}(f) \in \mathbb{C}$, see e.g. [**EGNO**, 4.7].

For any $X \in \mathcal{O}(\mathcal{C})$ there exists a unique up to scaling isomorphism $a_X : X \to X^{**}$. A choice of $a_X$ gives an isomorphism $(a_X^*)^{-1} : X^* \to X^{***} = (X^*)^{**}$. It is clear that the following scalar introduced by M. Müger [**Mu1**] and called *squared norm* of $X$ in [**ENO**] is independent of the choice of $a_X$:

$$|X|^2 := \mathrm{Tr}(a_X)\,\mathrm{Tr}((a_X^*)^{-1}) \in \mathbb{C}^\times.$$

Now the global dimension of $\mathcal{C}$ is defined as

$$\dim(\mathcal{C}) = \sum_{X \in \mathcal{O}(\mathcal{C})} |X|^2. \tag{2.1}$$

It was proved in [**ENO**, Theorem 2.3 and Remark 2.5] that $|X|^2$ is a totally positive algebraic integer. It follows that $\dim(\mathcal{C})$ is a real algebraic integer $\geq 1$; moreover $\dim(\mathcal{C}) = 1$ if and only if $\mathcal{C} \simeq \mathrm{Vec}$.

For any automorphism $\sigma$ of the field $\mathbb{C}$ let $\mathcal{C}^\sigma$ be the twist of $\mathcal{C}$ by $\sigma$, i.e. $\mathcal{C}^\sigma = \mathcal{C}$ as a category equipped with tensor product but the associativity constraint of $\mathcal{C}^\sigma$ is obtained from the one for $\mathcal{C}$ by applying $\sigma$. It is clear that $\dim(\mathcal{C}^\sigma) = \sigma(\dim(\mathcal{C}))$. In particular the set $X$ in (1.1) is Galois invariant.

**2.2. Spherical fusion categories.** A *pivotal structure* on fusion category $\mathcal{C}$ is an isomorphism of tensor functors $\delta : \mathrm{id} \to ()^{**}$, see e.g. [**EGNO**, 4.7], i.e. functorial tensor isomorphism $\delta_X : X \to X^{**}$ for any $X \in \mathcal{C}$. In presence of such a structure one defines *quantum dimensions* of objects $\dim_\delta(X) = \mathrm{Tr}(\delta_X) \in \mathbb{C}$ which are additive and multiplicative, see e.g. [**EGNO**, Proposition 4.7.12].

A pivotal structure $\delta$ is called *spherical* if $\dim_\delta(X) = \dim_\delta(X^*)$ for any $X \in \mathcal{C}$. A fusion category $\mathcal{C}$ is *sphericalizable* if it has at least one spherical structure. All currently known fusion categories are sphericalizable. A fusion category $\mathcal{C}$ equipped with a choice of spherical structure is called *spherical*. By abusing language we will consider these two notions as synonymous. Namely for each sphericalizable fusion category we will choose and fix one spherical structure $\delta$; we will often omit $\delta$ from notations, say by writing $\dim(X)$ instead of $\dim_\delta(X)$.

It is known that in a spherical fusion category $\dim(X) \in \mathbb{R}$ for any $X \in \mathcal{C}$ and $|X|^2 = \dim(X)^2$ for any $X \in \mathcal{O}(\mathcal{C})$, see [**ENO**, Corollary 2.10]. In particular

$$\dim(\mathcal{C}) = \sum_{X \in \mathcal{O}(\mathcal{C})} \dim(X)^2.$$

**2.3. Formal codegrees.** Let $K(\mathcal{C})$ be the Grothendieck ring of a fusion category $\mathcal{C}$ equipped with a basis $\{b_i\}$ consisting of the classes of simple objects in $\mathcal{C}$. For any $b_i$ representing the isomorphism class of $L \in \mathcal{C}$ let $b_i'$ denote the class of the dual object $L^*$.

Let $E$ be an irreducible $\mathbb{C}-$linear representation of the ring $K(\mathcal{C})$. It was shown in [**L**, Chapter 19] that the following element of $K(\mathcal{C}) \otimes \mathbb{C}$ is central:

$$\alpha_E = \sum_i \mathrm{Tr}(b_i, E) b_i'.$$

Moreover, the element $\alpha_E$ acts by some scalar $f_E$ on the irreducible representation $E$ and by zero on any irreducible representation not isomorphic to $E$. The scalars $f_E$ as $E$ runs over all irreducible representations of $K(\mathcal{C})$ are called *formal codegrees* of

$\mathcal{C}$, see [**O1**]. It is known that the formal codegrees are totally real algebraic integers which are $\geq 1$, see [**O3**, Remark 2.12]. By [**O3**, Proposition 2.10] we have

$$\sum_{E\in \mathrm{Irr}(K(\mathcal{C}))} \frac{\dim(E)}{f_E} = 1. \tag{2.2}$$

A one dimensional representation of $K(\mathcal{C})$ is the same as a ring homomorphism $\phi : K(\mathcal{C}) \to \mathbb{C}$. Such representation is automatically irreducible and the corresponding formal codegree $f_\phi$ is given by

$$f_\phi = \sum_{X\in \mathcal{O}(\mathcal{C})} \phi(X)\phi(X^*) = \sum_{X\in \mathcal{O}(\mathcal{C})} |\phi(X)|^2.$$

An example of the ring homomorphism$K(\mathcal{C}) \to \mathbb{C}$ is the *Frobenius-Perron dimension*homomorphism and its Galois conjugates, see [**ENO**, 8.1]. Thus we see that the *Frobenius-Perron dimension* FPdim$(\mathcal{C})$ (see e.g. [**ENO**, 8.2]) and its Galois conjugates are formal codegrees of $\mathcal{C}$.

If $\mathcal{C}$ is equipped with a spherical structure $\delta$, we have the *dimension homomorphism* $\dim_\delta : K(\mathcal{C}) \to \mathbb{C}$, see Section 2.2. Thus the global dimension $\dim(\mathcal{C})$ and its Galois conjugates are formal codegrees of $\mathcal{C}$. Note that there is no reason for this conclusion to hold for non sphericalizable fusion categories.

## 3. Pseudo-unitary inequality without pseudo-unitarity

**3.1.** Let $f_1, f_2, \ldots, f_r$ be the formal codegrees of $\mathcal{C}$. The following result was established in [**O3**, Theorem 2.21] for a pseudo-unitary $\mathcal{C}$ and was called "pseudo-unitary inequality" there. This is unfortunate as we show now that pseudo-unitarity assumption is not needed.

THEOREM 3.1.1. *For a spherical fusion category $\mathcal{C}$ the formal codegrees satisfy the following inequality:*

$$\sum_{i=1}^r \frac{1}{f_i^2} \leq \frac{1}{2}\left(1 + \frac{1}{\dim(\mathcal{C})}\right). \tag{3.1}$$

PROOF. It is known (see [**O3**, Theorem 2.13]) that the Drinfeld center $\mathcal{Z}(\mathcal{C})$ contains simple objects $Y_1, Y_2, \ldots, Y_r$ of dimensions $\dim(Y_i) = \frac{\dim(\mathcal{C})}{f_i}$ (the objects $Y_i$ are precisely simple objects $Y$ of $\mathcal{Z}(\mathcal{C})$ such that $F(Y)$ contains $\mathbf{1}$ with nonzero multiplicity where $F : \mathcal{Z}(\mathcal{C}) \to \mathcal{C}$ is the forgetful functor). It is also known that the twists of objects $Y_i$ equal 1, see [**O3**, Theorem 2.5].

Let $Z_1, Z_2, \ldots Z_s$ be the simple objects of $\mathcal{Z}(\mathcal{C})$ distinct from $Y_i$; let $\theta_1, \theta_2, \ldots \theta_s$ be their twists (recall that $\theta_i$ are roots of 1).

By the definition of the global dimension of $\mathcal{Z}(\mathcal{C})$ and by [**EGNO**, Proposition 9.3.4] we have the following identity:

$$\dim(\mathcal{C})^2 = \dim(\mathcal{Z}(\mathcal{C})) = \sum_{i=1}^r \dim(Y_i)^2 + \sum_{j=1}^s \dim(Z_j)^2, \tag{3.2}$$

Similarly, by the definition of the *Gauss sum* $G(\mathcal{Z}(\mathcal{C}))$ of $\mathcal{Z}(\mathcal{C})$ (see e.g. [**EGNO**, 8.15]) and by [**Mu2**, Theorem 1.2] (see also [**O3**, Example 2.6]) we have:

$$\dim(\mathcal{C}) = G(\mathcal{Z}(\mathcal{C})) = \sum_{i=1}^{r} \dim(Y_i)^2 + \sum_{j=1}^{s} \theta_j \dim(Z_j)^2. \tag{3.3}$$

It follows from (3.3) that $\sum_{j=1}^{s} \theta_j \dim(Z_j)^2 \in \mathbb{R}$ and

$$\sum_{j=1}^{s} \theta_j \dim(Z_j)^2 \geq -|\sum_{j=1}^{s} \theta_j \dim(Z_j)^2| \geq -\sum_{j=1}^{s} \dim(Z_j)^2. \tag{3.4}$$

Thus adding (3.2) and (3.3) we get

$$\dim(\mathcal{C})^2 + \dim(\mathcal{C}) = 2\sum_{i=1}^{r} \dim(Y_i)^2 + \sum_{j=1}^{s} \dim(Z_j)^2 + \sum_{j=1}^{s} \theta_j \dim(Z_j)^2 \geq \tag{3.5}$$

$$2\sum_{i=1}^{r} \dim(Y_i)^2 = 2\sum_{i=1}^{r} \frac{\dim(\mathcal{C})^2}{f_i^2}$$

which is equivalent to desired inequality. □

REMARK 3.1.2. By applying Galois automorphisms to the category $\mathcal{C}$ we see that we can replace the right hand side of (3.1) by $\frac{1}{2}(1+\frac{1}{f})$ where $f$ is arbitrary Galois conjugate of $\dim(\mathcal{C})$.

It is natural to ask whether the inequality (3.1) can be improved. The following example shows that we cannot hope to improve constant $\frac{1}{2}$ in (3.1) by something better than $\frac{1}{4}$ even if finitely many exceptions are allowed.

EXAMPLE 3.1.3. Let $G$ be a finite group and let $\mathcal{C} = \mathrm{Rep}(G)$ be the category of finite dimensional $G-$modules. Then $K(\mathrm{Rep}(G))$ is commutative, so all representations of $K(\mathrm{Rep}(G))$ are one dimensional. It is well known that all homomorphisms $K(\mathrm{Rep}(G)) \to \mathbb{C}$ are labelled by the conjugacy classes in $G$ and the orthogonality relations for the characters immediately imply that the formal codegrees are $f_i = \frac{|G|}{|C_i|}$ where $C_i$ is a conjugacy class. Thus in this case(3.1) is equivalent to

$$\sum_i |C_i|^2 \leq \frac{1}{2}|G|(|G|+1).$$

Note that for the dihedral group $G = D_p$ with odd prime $p$ one has

$$\sum_i |C_i|^2 = p^2 + \frac{1}{2}(p-1)\cdot 4 + 1 = \frac{1}{4}|G|^2 + |G| - 1.$$

Equivalently,

$$\sum_i \frac{1}{f_i^2} = \frac{1}{4} + \frac{1}{\dim(\mathcal{C})} - \frac{1}{\dim(\mathcal{C})^2}.$$

## 4. Bounds for formal codegrees

**4.1. Global dimensions.** We start with a bound for the global dimension:

THEOREM 4.1.1. *Let $\mathcal{C}$ be a spherical fusion category which is not equivalent to* Vec.

(i) *We have*

$$\dim(\mathcal{C}) > \frac{4}{3}. \tag{4.1}$$

(ii) *Assume that* $\dim(\mathcal{C})$ *has* $k > 1$ *Galois conjugates. Then*

$$\dim(\mathcal{C}) \geq \sqrt{\frac{16k-16}{8k-7}}. \tag{4.2}$$

PROOF. (i) Recall that $\dim(\mathcal{C})$ is an algebraic integer. Thus if $\dim(\mathcal{C}) \in \mathbb{Q}$ then $\dim(\mathcal{C}) \in \mathbb{Z}$, so $\dim(\mathcal{C}) \geq 2$ and the bound holds.

Assume now that $\dim(\mathcal{C}) \notin \mathbb{Q}$. Thus there exists at least one Galois conjugate $f$ of $\dim(\mathcal{C})$ distinct from $\dim(\mathcal{C})$. Then by Remark 3.1.2 we have:

$$\frac{1}{\dim(\mathcal{C})^2} + \frac{1}{f^2} \leq \sum_{i=1}^{r} \frac{1}{f_i^2} \leq \frac{1}{2}(1 + \frac{1}{f}), \tag{4.3}$$

whence

$$\frac{1}{\dim(\mathcal{C})^2} \leq \frac{1}{2} + \frac{1}{2} \cdot \frac{1}{f} - \frac{1}{f^2} = \frac{9}{16} - (\frac{1}{4} - \frac{1}{f})^2 \leq \frac{9}{16}. \tag{4.4}$$

It follows that $\dim(\mathcal{C}) \geq \frac{4}{3}$; however the equality is impossible since $\frac{4}{3}$ is not an algebraic integer.

(ii) Let $f$ be the largest of the conjugates of $\dim(\mathcal{C})$. Then

$$\frac{1}{\dim(\mathcal{C})^2} + (k-1)\frac{1}{f^2} \leq \sum_{i=1}^{r} \frac{1}{f_i^2} \leq \frac{1}{2}(1 + \frac{1}{f}), \tag{4.5}$$

that is $\frac{1}{\dim(\mathcal{C})^2} \leq \frac{1}{2} + \frac{1}{2} \cdot \frac{1}{f} - (k-1)\frac{1}{f^2}$. The maximum value of the right hand side is $\frac{8k-7}{16k-16}$ (achieved when $f = 4(k-1)$), which implies the result. □

EXAMPLE 4.1.2. Let $K_n = \mathbb{Z}[X]/\langle X^2 = nX + 1\rangle$ be based ring of rank 2. There are 2 homomorphisms $K_n \to \mathbb{C}$ with formal codegrees $\frac{n^2+4\pm n\sqrt{n^2+4}}{2}$ which are Galois conjugates if $n \neq 0$. One observes that $\frac{n^2+4-n\sqrt{n^2+4}}{2} < \frac{4}{3}$ for $n > 1$. Thus $K_n$ is not categorifiable by a spherical fusion category for $n > 1$. This falls short from the main result of [**O2**] which states that $K_n$ is not categorifiable for $n > 1$.

COROLLARY 4.1.3. *Let $d$ be a real number with $d < \sqrt{2}$. Then the set $X_s \cap [1, d] \subset \mathbb{R}$ is finite.*

PROOF. Assume $\dim(\mathcal{C}) \leq d$. Then we get from inequality (4.2) $d \geq \sqrt{\frac{16k-16}{8k-7}}$ which gives an upper bound for $k$ as $\lim_{k\to\infty} \sqrt{\frac{16k-16}{8k-7}} = \sqrt{2}$.

Let $f$ be the largest of conjugates of $\dim(C)$. By Remark 3.1.2 we have

$$\frac{1}{d^2} \leq \frac{1}{\dim(\mathcal{C})^2} \leq \sum_{i=1}^{r} \frac{1}{f_i^2} \leq \frac{1}{2}(1 + \frac{1}{f}),$$

that is $\frac{1}{f} \geq \frac{2}{d^2} - 1$ which gives an upper bound for $f$.

Thus the minimal polynomial of algebraic integer $\dim(\mathcal{C})$ has a bounded degree and bounded coefficients (since the coefficients are elementary symmetric functions of the conjugates of $\dim(\mathcal{C})$). The result follows. □

**4.2. More bounds.** It is easy to generalize arguments in the proof of Theorem 4.1.1 to get a bound for arbitrary formal codegree of $\mathcal{C}$.

THEOREM 4.2.1. *Let $\mathcal{C}$ be a spherical fusion category with $r$ simple objects. Then any formal codegree $f_i$ of $\mathcal{C}$ satisfies*

$$f_i \geq \sqrt{\frac{2r}{r+1}}. \tag{4.6}$$

*In particular, if $\mathcal{C} \not\simeq \mathrm{Vec}$ then*

$$f_i > \sqrt{\frac{8}{5}} \approx 1.2649 \tag{4.7}$$

PROOF. We start proving Theorem 4.2.1 with the following

LEMMA 4.2.2. *Let $\mathcal{C}$ be a fusion category with $r$ simple objects. Then there exists a Galois conjugate $f$ of $\dim(\mathcal{C})$ such that $f \geq r$.*

PROOF. For an algebraic integer $a$ let $[a] \in \mathbb{Q}$ be the arithmetic mean of the Galois conjugates of $a$. It is easy to see that $a \mapsto [a]$ is a $\mathbb{Q}-$linear functional on the $\mathbb{Q}-$vector space of algebraic numbers. Also if $a$ is a totally positive algebraic integer then by the arithmetic and geometric mean inequality one has $[a] \geq 1$. This applies to $a = |X|^2$ for $X \in \mathcal{O}(\mathcal{C})$, see Section 2.1. Let $f$ be the largest conjugate of $\dim(\mathcal{C})$. Applying (2.1) we get

$$f \geq [\dim(\mathcal{C})] = \sum_{X \in \mathcal{O}(\mathcal{C})} [|X|^2] \geq \sum_{X \in \mathcal{O}(\mathcal{C})} 1 = r. \tag{4.8}$$

□

REMARK 4.2.3. The equality $f = r$ in Lemma 4.2.2 holds only in the case when $\mathcal{C}$ is a pointed fusion category of dimension $r$. Namely in this case we see from 4.8 that $|X|^2 = 1$ for any $X \in \mathcal{O}(\mathcal{C})$. This implies that the dimension of any simple object of the *pivotalization* $\tilde{\mathcal{C}}$ of $\mathcal{C}$ (see [**EGNO**, Definition 7.21.9]) is $\pm 1$. This in turn implies that the Frobenius-Perron dimensions of simple objects of $\tilde{\mathcal{C}}$ equal 1, see [**EGNO**, Exercise 9.6.2]. Thus $\tilde{\mathcal{C}}$, and hence $\mathcal{C}$, are pointed, see [**EGNO**, Corollary 3.3.10].

Let us prove Theorem 4.2.1. Using Remark 3.1.2 and Lemma 4.2.2 we have

$$\frac{1}{f_i^2} \leq \sum_{i=1}^{r} \frac{1}{f_i^2} \leq \frac{1}{2}(1 + \frac{1}{f}) \leq \frac{1}{2}(1 + \frac{1}{r}) = \frac{r+1}{2r}. \tag{4.9}$$

This implies (4.6). Thus (4.7) holds for $r \geq 4$; in the remaining cases $r = 2$ and $r = 3$ one verifies (4.7) case by case using classifications of fusion categories of rank 2 (see [**O2**] or Example 4.1.2 below) and 3 (see [**O3**]). □

COROLLARY 4.2.4. *Any Galois conjugate of the Frobenius-Perron dimension* $\mathrm{FPdim}(\mathcal{C})$ *of a nontrivial spherical fusion category $\mathcal{C}$ is greater than $\sqrt{\frac{8}{5}}$.*

We note that Theorem 4.2.1 and Corollary 4.2.4 are of interest even for *pseudo-unitary* (see e.g. [**EGNO**, 9.4]) fusion categories while Theorem 1.1.2 is not very interesting for such categories as it follows from [**EGNO**, Proposition 3.3.4 (2)] that $\dim(\mathcal{C}) = \mathrm{FPdim}(\mathcal{C}) \geq 2$ for a pseudo-unitary fusion category $\mathcal{C}$ not equivalent to Vec. We have a stronger version of Corollary 4.2.4 for such categories:

LEMMA 4.2.5. *Let $\mathcal{C}$ be a Galois conjugate of pseudo-unitary fusion category. Then any conjugate $\gamma$ of the Frobenius-Perron dimension* $\mathrm{FPdim}(\mathcal{C})$ *satisfies*

$$\gamma \geq \sqrt{\frac{2\,\mathrm{FPdim}(\mathcal{C})}{\mathrm{FPdim}(\mathcal{C})+1}}. \tag{4.10}$$

PROOF. We can assume that $\mathcal{C}$ is pseudo-unitary. Recall that

$$\dim(\mathcal{C}) = \mathrm{FPdim}(\mathcal{C})$$

(see [**EGNO**, Definition 9.4.4]) and $\mathcal{C}$ is automatically spherical (see [**EGNO**, 9.5]). Thus (3.1) gives

$$\frac{1}{\gamma^2} \leq \sum_{i=1}^{r} \frac{1}{f_i^2} \leq \frac{1}{2}(1+\frac{1}{\dim(\mathcal{C})}) = \frac{1}{2}(1+\frac{1}{\mathrm{FPdim}(\mathcal{C})}) = \frac{\mathrm{FPdim}(\mathcal{C})+1}{2\,\mathrm{FPdim}(\mathcal{C})}.$$

The result follows. □

It would be interesting to improve Lemma 4.2.5. For instance one can ask whether there exists a bound for $\gamma$ which is an unbounded function of $\mathrm{FPdim}(\mathcal{C})$.

## 5. Fusion categories of the same global dimension

**5.1.** Here is the main result of this Section:

THEOREM 5.1.1. *There are at most finitely many fusion categories (up to a tensor equivalence) of a given global dimension.*

PROOF. Let $d$ be a totally positive algebraic integer and let $\mathcal{C}$ be a fusion category with $\dim(\mathcal{C}) = d$. Let $f$ be the largest conjugate of $d$. Then by Lemma 4.2.2 we have that $r = |\mathcal{O}(\mathcal{C})| \leq \lfloor f \rfloor$. The Grothendieck ring $K(\mathcal{C})$ has rank $r$, so it has at most $r$ homomorphisms to $\mathbb{C}$. Hence $\mathrm{FPdim}(\mathcal{C})$ has at most $r$ conjugates.

It is known that for any conjugate $\gamma$ of $\mathrm{FPdim}(\mathcal{C})$ the ratio $\frac{d}{\gamma}$ is an algebraic integer, see [**O3**, Corollary 2.14]. Thus the norm $N$ (i.e. the product of all conjugates) of $\mathrm{FPdim}(\mathcal{C})$ divides $d^k$ where $k$ is the number of the conjugates. Since $k \leq r \leq \lfloor f \rfloor$ we see that $N$ divides $d^{\lfloor f \rfloor}$. Thus $N \leq M$ where $M$ is the largest integer dividing $d^{\lfloor f \rfloor}$. Finally $\mathrm{FPdim}(\mathcal{C}) \leq N$ since all conjugates of $\mathrm{FPdim}(\mathcal{C})$ are $\geq 1$, see Section 2.3. We proved that

$$\mathrm{FPdim}(\mathcal{C}) \leq M = \text{largest integer dividing } d^{\lfloor f \rfloor}.$$

The theorem is proved as there are just finitely many fusion categories of Frobenius-Perron dimension bounded by a given constant (namely it is easy to see that there are just finitely many based rings of bounded Frobenius-Perron dimension and for each such ring there are just finitely many categorifications by [**EGNO**, Theorem 9.1.4]). □

One can try to classify all fusion categories with $\dim(\mathcal{C}) = d$ where $d$ is small number. Here are some examples.

EXAMPLE 5.1.2. (i) Let $d = 2$. Then by Lemma 4.2.2 we have $r \leq 2$; clearly $r = 1$ is impossible and for $r = 2$ we get only pointed categories associated with the group $\mathbb{Z}/2\mathbb{Z}$ by Remark 4.2.3.

(ii) Let $d = 3$. In this case $r \leq 3$; $r = 1$ is impossible and $r = 3$ gives pointed categories. In the case $r = 2$ by the arguments in the proof of Theorem 5.1.1 we see that the norm of the Frobenius-Perron dimension divides $d^r = 9$. However one computes easily that the norm of the Frobenius-Perron dimension of the ring $K_n$ (see Example 4.1.2) is $n^2 + 4$ which never divides 9. Thus this case is impossible.

(iii) Let $d = 4$ and assume that category $\mathcal{C}$ is spherical and not pointed. Thus $r \leq 3$ and we have at most 3 distinct homomorphisms $K(\mathcal{C}) \to \mathbb{C}$. Consider the action of the Galois group on such homomorphisms. If the dimension homomorphism and the Frobenius-Perron homomorphism (see Section 2.3) are in the same orbit then the category is weakly integral (see [**EGNO**, 9.6]), hence *Ising category* (see e.g. [**DGNO**, Appendix B]). Otherwise either orbit of the dimension or of the Frobenius-Perron dimension has exactly one element; this again implies that $\mathcal{C}$ is weakly integral, see [**EGNO**, Exercise 9.6.2].

(iv) Let $d = \frac{5\pm\sqrt{5}}{2}$ and assume that category $\mathcal{C}$ is spherical. Then the Galois conjugate $f = \frac{5\mp\sqrt{5}}{2}$ is also one of formal codegrees of $\mathcal{C}$. Since

$$\frac{1}{\dim(\mathcal{C})} + \frac{1}{f} = 1,$$

equation (2.2) implies that $\mathcal{C}$ has no other formal codegrees. Thus $\mathcal{C}$ is of rank 2, and using Example 4.1.2 we see that $K(\mathcal{C}) = K_1$. It follows that $\mathcal{C} \simeq YL$ or $\mathcal{C} \simeq \overline{YL}$ where $YL$ is the Yang-Lee category and $\overline{YL}$ is its Galois conjugate depending on the value of $\dim(\mathcal{C})$, see e.g. [**O2**, 2.5] and [**EGNO**, Exercise 8.18.7].

(v) Let $d = 5$ and assume that category $\mathcal{C}$ is spherical. Hence $r \leq 5$. In the case $r \leq 3$ the arguments as in (iii) above show that $\mathcal{C}$ is weakly integral, hence pointed, see [**ENO**, Corollary 8.30]. The same conclusion holds if $r = 5$ by Remark 4.2.3.

We assume now that $r = 4$ and $\mathcal{C}$ is not weakly integral. Then there are 4 homomorphisms $K(\mathcal{C}) \to \mathbb{C}$ which split into two orbits of size 2 under the action of the Galois group: orbit of the dimension homomorphism and orbit of the Frobenius-Perron homomorphism. Both formal codegrees for the first orbit equal $\dim(\mathcal{C}) = 5$; let $f_1$ and $f_2$ be the formal codegrees for the second orbit. Then $f_1 f_2$ (= norm of the Frobenius-Perron dimension) is an integer $> 1$ and dividing $d^2 = 25$. Also $\frac{1}{f_1} + \frac{1}{f_2} = 1 - \frac{2}{5} = \frac{3}{5}$ by (2.2). In the case $f_1 f_2 = 5$ we get $f_1 + f_2 = 3$ whence $f_1$ and $f_2$ are not real. Thus $f_1 f_2 = 25$ and $f_1 + f_2 = 15$ whence $\text{FPdim}(\mathcal{C}) = \frac{15+5\sqrt{5}}{2} = \left(\frac{5+\sqrt{5}}{2}\right)^2$. We claim now that $\mathcal{C}$ must be equivalent to $YL \boxtimes \overline{YL}$, cf. [**EGNO**, Exercise 9.4.6]. Indeed, there is a unique expansion of $\text{FPdim}(\mathcal{C}) - 1 = \frac{13+5\sqrt{5}}{2}$ into a sum of three totally positive algebraic integers from the field $\mathbb{Q}(\sqrt{5})$ and maximal among their conjugates:

$$\frac{13 + 5\sqrt{5}}{2} = \frac{3 + \sqrt{5}}{2} + \frac{3 + \sqrt{5}}{2} + \frac{7+3\sqrt{5}}{2} = \left(\frac{1+\sqrt{5}}{2}\right)^2 + \left(\frac{1+\sqrt{5}}{2}\right)^2 + \left(\frac{3+\sqrt{5}}{2}\right)^2.$$

It follows that the simple objects of $\mathcal{C}$ have the Frobenius-Perron dimensions $1, \frac{1+\sqrt{5}}{2}$,

$\frac{1+\sqrt{5}}{2}, \frac{3+\sqrt{5}}{2}$. Let $X$ and $Y$ be the simple objects of $\mathcal{C}$ with $\mathrm{FPdim}(X) = \mathrm{FPdim}(Y) = \frac{1+\sqrt{5}}{2}$. Since $X \otimes X^*$ is a direct sum of $\mathbf{1}$ and an object of dimension $\frac{1+\sqrt{5}}{2}$, we see that at least one of $X$ and $Y$ is self dual. Hence both of them are self dual and $X \otimes Y \simeq Y \otimes X$ is simple. Thus $\mathrm{Hom}(X \otimes X, Y) = \mathrm{Hom}(X, Y \otimes X) = 0$, whence $X \otimes X = \mathbf{1} \oplus X$ and, similarly, $Y \otimes Y \simeq \mathbf{1} \oplus Y$. In other words we have an isomorphism of based rings $K(\mathcal{C}) = K(YL) \otimes K(YL) = K(YL \boxtimes \overline{YL})$.

Let us construct a tensor equivalence $\mathcal{C} \simeq YL \boxtimes \overline{YL}$. We can assume that $\dim(X) = \frac{1-\sqrt{5}}{2}$ and $\dim(Y) = \frac{1+\sqrt{5}}{2}$. Let $YL \subset \mathcal{C}$ be the fusion subcategory consisting of direct sums of $\mathbf{1}$ and $X$ and let $\overline{YL} \subset \mathcal{C}$ be the fusion subcategory consisting of direct sums of $\mathbf{1}$ and $Y$. Consider the Drinfeld center $\mathcal{Z}(\mathcal{C})$ of $\mathcal{C}$, see e.g. [**EGNO**, 7.13]. Using [**EGNO**, Proposition 9.2.2] and [**O3**, Theorem 2.13] we deduce that $\mathcal{Z}(\mathcal{C})$ contains simple objects which map to $\mathbf{1} \oplus X$ and $\mathbf{1} \oplus Y$ under the forgetful functor $\mathcal{Z}(\mathcal{C}) \to \mathcal{C}$; moreover these objects have twists 1. Thus the subcategories $\mathcal{A} \subset \mathcal{Z}(\mathcal{C})$ and $\mathcal{B} \subset \mathcal{Z}(\mathcal{C})$ consisting of objects sent to $YL$ and $\overline{YL}$ under the forgetful functor are non-trivial. We have obvious braided tensor functors $\mathcal{A} \to \mathcal{Z}(YL)$ and $\mathcal{B} \to \mathcal{Z}(\overline{YL})$. Recall that $YL$ has a structure of the modular category (see [**EGNO**, 8.18]) whence $\mathcal{Z}(YL) \simeq YL \boxtimes YL$ as fusion categories. It is easy to see that no proper subcategory of $\mathcal{Z}(YL)$ contains an object with twist 1 and which maps to $\mathbf{1} \oplus X$ under the forgetful functor. We conclude that the functor $\mathcal{A} \to \mathcal{Z}(YL)$ is surjective and, similarly, the functor $\mathcal{B} \to \mathcal{Z}(\overline{YL})$ is surjective. Comparing the Frobenius-Perron dimensions (see e.g. [**DGNO**, Lemma 3.38]) we conclude that both functors are equivalences and $\mathcal{Z}(\mathcal{C}) \simeq \mathcal{A} \boxtimes \mathcal{B} \simeq YL \boxtimes YL \boxtimes \overline{YL} \boxtimes \overline{YL}$. Hence $\mathcal{Z}(\mathcal{C})$ contains 4 fusion subcategories equivalent to $YL \boxtimes \overline{YL}$ and such that the restriction of the forgetful functor yields a desired equivalence $\mathcal{C} \simeq YL \boxtimes \overline{YL}$.

## Appendix A

### By Pavel Etingof and Victor Ostrik

**A.1.** The goal of this appendix is to prove (using a computer) the following proposition.

PROPOSITION A.1.1. *The only number $1 < d < 4\sqrt{3}/5 = 1.38564...$ which can serve as the global dimension of a spherical fusion category over complex numbers is $d = \frac{5-\sqrt{5}}{2} = 1.381966...$*

PROOF. Let $1 < d < \sqrt{2}$ be the global dimension of a spherical fusion category (a totally positive algebraic integer). Let $k$ be the number of algebraic conjugates of $d$. By Theorem 4.1.1 (i) we know that $d > \frac{4}{3}$ and Theorem 4.1.1 (ii) implies that the Proposition holds for $k \geq 4$, so it suffices to prove it for $k = 2, 3$.

Let $d_*$ be the biggest conjugate of $d$. By equation (4.4) we have

$$\frac{1}{d^2} \leq \frac{9}{16} - \left(\frac{1}{4} - \frac{1}{d^*}\right)^2,$$

which yields

$$d^* \leq K, \text{ where } K = K(d) := \left(\frac{1}{4} - \sqrt{\frac{9}{16} - \frac{1}{d^2}}\right)^{-1}. \tag{A.1}$$

Let $k = 2$. Then we have

$$(3d - 4)(3d_* - 4) \geq 1,$$

so by (A.1),

$$(3d-4)(3K(d)-4) \geq 1,$$

which yields

$$d \geq \frac{1}{4}(\sqrt{41}-1) = 1.3507...$$

Now we can use a simple computer program to show that there are no possible values of $d$ in the range

$$\frac{1}{4}(\sqrt{41}-1) = d_- \leq d < d_+ = \frac{4}{5}\sqrt{3}.$$

Namely, let

$$P(x) = x^2 - ax + b$$

be the minimal polynomial of $d$ over $\mathbb{Z}$. Then we have $P(d_-) \geq 0$ and $P(d_+) < 0$, so

$$ad_- - d_-^2 \leq b < ad_+ - d_+^2.$$

Also by [**O1**, Theorem 1.2], $d$ has to be a $d$-number, i.e., must divide all its conjugates, which results in the condition that $a^2$ is divisible by $b$. Finally, we have

$$a = d + d^* \leq \frac{4}{5}\sqrt{3} + K\left(\frac{4}{5}\sqrt{3}\right),$$

which yields $a \leq 23$. Going through all possibilities, we find that there are no solutions except $a = b = 5$, which gives $d = (5-\sqrt{5})/2$. (In fact, there are so few cases that this can be easily checked by hand).

Now suppose that $k = 3$, and $d < d_+$. By Theorem 4.1.1 (ii) we have $d \geq \tilde{d}_- = \sqrt{\frac{32}{17}} \approx 1.37199$. Let

$$P(x) = x^3 - ax^2 + bx - c$$

be the minimal polynomial of $d$. Without loss of generality we may assume that $d$ is the smallest root of $P$. Let $d_1 = d < d_2 < d_3 = d_*$ be the roots of $P$ (they all have to be real and positive). Let us list some more properties of $(a, b, c)$:

(i) $P(\tilde{d}_-) \leq 0$ and $P(d_+) > 0$ ($P(x)$ cannot have 2 roots $< 2$ since it would contradict to (2.2)), so

$$b\tilde{d}_- - a\tilde{d}_-^2 + \tilde{d}_-^3 \leq c < bd_+ - ad_+^2 + d_+^3;$$

(ii) $d_3 = d_* \leq K(\frac{4}{5}\sqrt{3}) \leq 22$, thus $a = d_1 + d_2 + d_3 \leq 45$.

(iii) $b \leq a^2/3$ (by the arithmetic and geometric mean inequality, as all the roots are real).

(iv) $d$ has to be a $d$-number, hence $c$ divides $a^3$ and $c^2$ divides $b^3$.

(v) Inequality (A.1) is satisfied.

Again, going through all possibilities, we find that there are no solutions; the program (in MAGMA) does this for less than a second.

This completes the proof of Proposition A.1.1. $\square$

REMARK A.1.2. Recall that by Example 5.1.2 (iv) the only spherical fusion category with $\dim(\mathcal{C}) = \frac{5-\sqrt{5}}{2}$ is the Yang-Lee category $YL$.

## Acknowledgements

It is my great pleasure to thank Pavel Etingof for his comments which allowed me to significantly improve the results of this note. This work was partially supported by the NSF grant DMS-1702251. The study also has been funded by the Russian Academic Excellence Project '5-100'.


## References


[DGNO] Vladimir Drinfeld, Shlomo Gelaki, Dmitri Nikshych, and Victor Ostrik, *On braided fusion categories. I*, Selecta Math. (N.S.) **16** (2010), no. 1, 1–119, DOI 10.1007/s00029-010-0017-z. MR2609644

[EGNO] Pavel Etingof, Shlomo Gelaki, Dmitri Nikshych, and Victor Ostrik, *Tensor categories*, Mathematical Surveys and Monographs, vol. 205, American Mathematical Society, Providence, RI, 2015. MR3242743

[ENO] Pavel Etingof, Dmitri Nikshych, and Viktor Ostrik, *On fusion categories*, Ann. of Math. (2) **162** (2005), no. 2, 581–642, DOI 10.4007/annals.2005.162.581. MR2183279

[L] G. Lusztig, *Hecke algebras with unequal parameters*, CRM Monograph Series, vol. 18, American Mathematical Society, Providence, RI, 2003. MR1974442

[Mu1] Michael Müger, *From subfactors to categories and topology. I. Frobenius algebras in and Morita equivalence of tensor categories*, J. Pure Appl. Algebra **180** (2003), no. 1-2, 81–157, DOI 10.1016/S0022-4049(02)00247-5. MR1966524

[Mu2] Michael Müger, *From subfactors to categories and topology. II. The quantum double of tensor categories and subfactors*, J. Pure Appl. Algebra **180** (2003), no. 1-2, 159–219, DOI 10.1016/S0022-4049(02)00248-7. MR1966525

[O1] Victor Ostrik, *On formal codegrees of fusion categories*, Math. Res. Lett. **16** (2009), no. 5, 895–901, DOI 10.4310/MRL.2009.v16.n5.a11. MR2576705

[O2] Viktor Ostrik, *Fusion categories of rank 2*, Math. Res. Lett. **10** (2003), no. 2-3, 177–183, DOI 10.4310/MRL.2003.v10.n2.a5. MR1981895

[O3] Victor Ostrik, *Pivotal fusion categories of rank 3* (English, with English and Russian summaries), Mosc. Math. J. **15** (2015), no. 2, 373–396, 405. MR3427429

[O4] Victor Ostrik, *Pre-modular categories of rank 3* (English, with English and Russian summaries), Mosc. Math. J. **8** (2008), no. 1, 111–118, 184. MR2422269



Department of Mathematics, University of Oregon, Eugene, OR 97403, USA

Laboratory of Algebraic Geometry, National Research University Higher School of Economics, Moscow, Russia

*Email address*: vostrik@uoregon.edu

Contemporary Mathematics
Volume **728**, 2019
https://doi.org/10.1090/conm/728/14662

# On Hopf algebras with triangular decomposition

Cristian Vay

Abstract. In this survey, we first review some known results on the representation theory of algebras with triangular decomposition, including the classification of the simple modules. We then discuss a recipe to construct Hopf algebras with triangular decomposition. Finally, we extend to these Hopf algebras the main results of a previous work of the author regarding projective modules over Drinfeld doubles of bosonizations of Nichols algebras and groups.

## 1. Introduction

Over the last years there have been remarkable progress on the problem of classifying finite-dimensional Hopf algebras. New methods to construct families with different properties were introduced and a lot of examples are now at hand. This has motivated an increasing interest in their representation theory. The family of pointed Hopf algebras over abelian groups is deeply related to the universal enveloping of semisimple Lie algebras. It is natural then to draw inspiration from Lie theory to study modules in this context. We would like to generalize, for instance, the following well-known facts in Lie theory:

($*$) The simple $U(\mathfrak{g})$-modules (in a suitable category) are classified by their highest weights.
($**$) The Weyl character formula for the simple modules.

Let us briefly mention some works extending these properties to different families of Hopf algebras. Lusztig **[26]** has demonstrated that ($*$) and ($**$) hold for the *q-analogs* $U_q(\mathfrak{g})$, for any symmetrizable Kac–Moody algebra $\mathfrak{g}$. We can see the $q$-analogs $U_q(\mathfrak{g})$ as examples of a more general family of pointed Hopf algebras introduced by Andruskiewitsch–Schneider. In **[6]**, Andruskiewitsch–Radford–Schneider have extended ($*$) and ($**$) to this family following the strategy of Lusztig. This family also includes the pointed Hopf algebras previously considered in **[36]** (associated to Cartan matrices of finite type) and in **[10, 11, 32]** (multi-parametric version of $U_q(\mathfrak{g})$).

The character formula for the *q-divided power algebra* $\mathcal{U}_q(\mathfrak{g})$ (with $q$ a root of unity of certain order) is a consequence of several results. First, Kashiwara–Tanisaki proved a character formula for affine Lie algebras at a negative level. Then, Kazhdan–Lusztig (and Lusztig) constructed a functor between suitable categories

2010 *Mathematics Subject Classification.* Primary 16T05, 20G05, 16P10.
Partially supported by CONICET, Secyt (UNC), FONCyT PICT 2016-3957 and MathAmSud project GR2HOPF.

of modules over the above algebras which translate the character formulas from one setting to the other. Moreover, Andersen–Jantzen–Soergel [**1**] demonstrated that the character formulas for the small quantum group $\mathfrak{u}_q(\mathfrak{g})$ and the restricted enveloping Lie algebra $U^{[p]}(\mathfrak{g})$ are the same for large primes (and these are independent of the characteristic). Using these formulas and the Steinberg's tensor product theorem, one can deduce the formulas for $\mathcal{U}_q(\mathfrak{g})$, and for the corresponding semisimple connected and simply connected algebraic group in positive characteristic. The reader can find more details and precise references on the various character formulas in the fascinating survey by Jantzen [**20**].

The above Hopf algebras can be constructed via a Drinfeld double procedure, starting from a pointed Hopf algebra over an abelian group. It is natural then to consider other types of Drinfeld doubles. In [**21**], Krop–Radford prove that $(*)$ holds for the Drinfeld double of a finite-dimensional Hopf algebra $H$ such that its coradical $H_0$ is a Hopf subalgebra complemented by a nilpotent Hopf ideal. In this setting, the Drinfeld double of $H_0$ plays the role of the Cartan subalgebra and their simple modules are the corresponding weights (unlike Lie theory, they are not necessarily one-dimensional). Different proofs were given for $H$ being the bosonization of a finite-dimensional Nichols algebra over a finite abelian group by Heckenberger–Yamane [**18**] and by Pogorelsky and the author [**33**] for general finite groups. Andruskiewitsch–Angiono [**2**] recently extended the proof of [**33**] to bosonizations over any finite-dimensional Hopf algebra. In this great generality, a character formula is not expected and we should restrict ourselves, for instance, to certain families of Nichols algebras. We refer the reader to [**3, 13, 22, 33**] to see how the simple modules look like in some explicit examples.

In this survey, we first review a general framework due to Bellamy–Thiel [**9**] and Bonnafé–Rouquier [**12**, §F.2] which ensures that $(*)$ holds. Explicitly, this is guaranteed for any graded algebra with *triangular decomposition* (under some mild extra conditions). Moreover, the simple modules can be obtained as quotients of *standard (Verma) modules* and, under certain assumptions, this leads to the construction of a highest weight category. We then discuss a recipe to construct Hopf algebras with triangular decomposition. We basically rewrite the Drinfeld double construction in order to obtain directly the triangular decomposition. The small quantum groups and the Drinfeld doubles of bosonizations of Nichols algebras can be obtained via this procedure. Finally, we extend to this wider class of Hopf algebras some results of [**39**], regarding projective modules over Drinfeld doubles of bosonizations of finite-dimensional Nichols algebras over groups. These results rely fundamentally on properties of the Nichols algebra.

Roughly speaking, the main idea that we would like to share with the reader is the following: if your favorite Hopf algebra admits a suitable triangular decomposition, then its simple modules are classified by their highest weights and you can construct them as quotients of standard modules. Thus, the major problem to address is to compute the weight decomposition of the simple modules.

The paper is organized as follows. The general framework of algebras with decomposition is summarized in Section 2. We fix the notation and conventions concerning Hopf and Nichols algebras in Section 3. The recipe to construct Hopf algebras with triangular decomposition is explained in Section 4 and the representation theory of them is studied in Section 5.

## 2. A general framework

Throughout this work, $\Bbbk$ is a field and *graded* always means $\mathbb{Z}$-graded. Let $A$ be a graded algebra. We denote by ${}_A\mathcal{M}$ the category of finite-dimensional left $A$-modules and by ${}_A\mathcal{G}$ the category of graded finite-dimensional left $A$-modules with morphisms preserving the grading.

We say that a graded algebra $A = \oplus_{n\in\mathbb{Z}} A_n$ admits a *triangular decomposition* if there exist graded subalgebras $A^-$, $T$ and $A^+$ such that the multiplication

$$A^-\otimes T\otimes A^+ \longrightarrow A \tag{td}$$

gives a linear isomorphism and the following properties hold

(td1) $A^\pm \subseteq \oplus_{n\in\mathbb{Z}_\pm} A_n$ and $T\subseteq A_0$.
(td2) $(A^\pm)_0 = \Bbbk$.
(td3) $B^\pm := A^\pm T = TA^\pm$.

We denote by $\Lambda$ a complete set of non-isomorphic finite-dimensional simple $T$-modules. The elements of $\Lambda$ are called *weights*. If $\lambda\in\Lambda$ is a $T$-submodule of an $A$-module $\mathsf{N}$ such that $(B^+)_{>0}\cdot\lambda = 0$ (resp. $(B^-)_{<0}\cdot\lambda = 0$), we say that $\lambda$ is a *highest weight* (resp. *lowest weight*) of $\mathsf{N}$. In such a case, $\lambda$ is a simple $B^+$-submodule (resp. $B^-$-submodule) isomorphic to $\operatorname{Inf}_T^{B^+}(\lambda)$ (resp. $\operatorname{Inf}_T^{B^-}(\lambda)$), where $\operatorname{Inf}_T^{B^\pm} : {}_T\mathcal{M} \to {}_{B^\pm}\mathcal{M}$ denotes the functor induced by the projection $B^\pm \twoheadrightarrow T$. If $\mathsf{N}$ is also generated by $\lambda$, we say that $\mathsf{N}$ is a *highest weight module* (resp. *lowest weight module*).

The *proper standard module* associated to $\lambda\in\Lambda$ is

$$\overline{\mathsf{M}}(\lambda) = A\otimes_{B^+}\operatorname{Inf}_T^{B^+}(\lambda) \tag{2.1}$$

We denote by $\mathsf{L}(\lambda)$ the head of $\overline{\mathsf{M}}(\lambda)$. Then $\Bbbk\otimes_{B^+}\operatorname{Inf}_T^{B^+}(\lambda)$ is a highest weight of $\overline{\mathsf{M}}(\lambda)$ isomorphic to $\lambda$, and so is its image in $\mathsf{L}(\lambda)$. Hence $\overline{\mathsf{M}}(\lambda)$ and $\mathsf{L}(\lambda)$ are highest weight modules.

### 2.1. The finite-dimensional case.

**2.1. The finite-dimensional case.** In this subsection, we assume that $A$ is a finite-dimensional algebra with triangular decomposition as above. Therefore the simple and projective modules over $A$ admit a grading by [**15**].

Let $P_T(\lambda)$ be the projective cover of $\lambda\in\Lambda$. The *standard module* associated to $\lambda\in\Lambda$ is

$$\mathsf{M}(\lambda) = A\otimes_{B^+}\operatorname{Inf}_T^{B^+}(P_T(\lambda)). \tag{2.2}$$

Notice that

$$\mathsf{M}(\lambda) = \overline{\mathsf{M}}(\lambda) \quad \text{if } T \text{ is semisimple.} \tag{2.3}$$

The *proper costandard module* $\overline{\nabla}(\lambda)$ and *costandard module* $\nabla(\lambda)$ associated to $\lambda$ are defined in a dual way.[1]

THEOREM 2.1 ([**9**, **19**]). *Assume that $T$ is split, i.e.* $\operatorname{End}_T(\lambda) = \Bbbk$ *for all* $\lambda\in\Lambda$. *Then:*

(i) *The head* $\mathsf{L}(\lambda)$ *of* $\overline{\mathsf{M}}(\lambda)$ *is simple for all* $\lambda\in\Lambda$.

[1] In [**9**] the standard modules are denoted by $\Delta$ but, as it is usual in Hopf algebra theory, we keep $\Delta$ to denote the comultiplication.

(ii) *Every simple $A$-module is isomorphic to* $\mathsf{L}(\lambda)$ *for a unique* $\lambda \in \Lambda$.
(iii) *The projective cover* $\mathsf{P}(\lambda)$ *of* $\mathsf{L}(\lambda)$ *admits a standard filtration for all* $\lambda \in \Lambda$, *i.e. there exists a chain of submodules* $0 = \mathsf{N}_0 \subset \mathsf{N}_1 \subset \cdots \subset \mathsf{N}_n = \mathsf{P}(\lambda)$ *such that for each* $i$, $\mathsf{N}_i/\mathsf{N}_{i-1} \simeq \mathsf{M}(\lambda_i)$ *for some* $\lambda_i \in \Lambda$.
(iv) *Brauer Reciprocity* $[\mathsf{P}(\lambda) : \mathsf{M}(\mu)] = [\overline{\nabla}(\mu) : \mathsf{L}(\lambda)]$ *holds for all* $\lambda, \mu \in \Lambda$, *i.e. the number of subquotients isomorphic to* $\mathsf{M}(\mu)$ *in a standard filtration of* $\mathsf{P}(\lambda)$ *is equal to the number of composition factors of type* $\mathsf{L}(\lambda)$ *of* $\overline{\nabla}(\mu)$. *Moreover, this number is independent of the chosen filtration.*

*Dual properties are also satisfied by (proper) costandard modules and the injective hull of* $\mathsf{L}(\lambda)$. □

Thus, the category of modules over a finite-dimensional algebra with triangular decomposition has the flavor of a *highest weight category* [**14**, Definition 3.1]. However, if $A$ is a non-semisimple Hopf algebra[2], ${}_A\mathcal{M}$ is not a highest weight category. In fact, $A$ has infinite global dimension (because it is symmetric and non-semisimple) which contradicts [**31**, Theorem 4.3 (a)].

We also see in the examples (*e.g.* [**39**, §1]) that the axiom of highest weight categories which fails is the existence of a partial order $>$ in $\Lambda$ such that: if $\mathsf{M}(\mu)$ occurs in a standard filtration of $\mathsf{P}(\lambda)$, then $\mu > \lambda$; and $\mathsf{M}(\lambda)$ occurs precisely once. However, Bellamy-Thiel [**9**] realized that this trouble can be fixed by using the grading; that is working inside of the category of graded modules.

We denote by $\mathsf{N}[i]$ the shift of $\mathsf{N} \in {}_A\mathcal{G}$ (or ${}_T\mathcal{G}$) by $i \in \mathbb{Z}$, that is

$$(\mathsf{N}[i])_n = \mathsf{N}_{n-i} \tag{2.4}$$

for all $n \in \mathbb{Z}$. Shifting commutes with taking projective covers and injective hulls [**15**]. Also, it is compatible with the (proper) standard modules and the (proper) costandard modules. In particular,

$$\overline{\mathsf{M}}(\lambda[i]) = \overline{\mathsf{M}}(\lambda)[i], \quad \mathsf{M}(\lambda[i]) = \mathsf{M}(\lambda)[i] \quad \text{and} \quad \mathsf{L}(\lambda[i]) = \mathsf{L}(\lambda)[i]$$

for all $\lambda \in \Lambda$ and $i \in \mathbb{Z}$.

Since $T$ is concentrated in degree 0, the simple objects in ${}_T\mathcal{G}$ are parametrized by $\Lambda \times \mathbb{Z}$. The simple object corresponding to $(\lambda, i)$ is just $\lambda[i]$, the simple $T$-module $\lambda$ concentrated in degree $i$. We can then define a partial order on $\Lambda \times \mathbb{Z}$ by letting $\lambda[i] < \mu[j]$ if and only if $i < j$.

THEOREM 2.2 ([**9**, Theorem 1.1]). *If $T$ is semisimple, then ${}_A\mathcal{G}$ is a highest weight category whose set of weights is $\Lambda \times \mathbb{Z}$ and the standard modules are $\mathsf{M}(\lambda)[d]$ for all $\lambda[d] \in \Lambda \times \mathbb{Z}$. In particular, the properties (i)–(iv) above hold in the category ${}_A\mathcal{G}$ for all $\lambda[d] \in \Lambda \times \mathbb{Z}$.* □

We would like to stress that the simple modules are characterized as follows.

COROLLARY 2.3. *Let $\lambda \in \Lambda$ and $i \in \mathbb{Z}$. Then $\mathsf{L}(\lambda)[i]$ is the unique simple highest weight module with highest weight $\lambda[i]$.* □

COROLLARY 2.4. *Let $\lambda \in \Lambda$ and $i \in \mathbb{Z}$. Then there exist $\overline{\lambda} \in \Lambda$ and $l_\lambda \in \mathbb{Z}_{\leq 0}$ such that $\overline{\lambda}[l_\lambda]$ is a lowest weight of $\mathsf{L}(\lambda)$. Moreover, $\mathsf{L}(\lambda)[i]$ is the unique simple lowest weight module with lowest weight $\overline{\lambda}[l_\lambda + i]$.*

[2]and in other important examples, cf. [**9**, §1].

PROOF. If $\mathsf{N}$ is a graded vector space, we write $\mathsf{N}'$ for the same vector space, but with the reversed grading $(\mathsf{N}')_n = \mathsf{N}_{-n}$ for all $n \in \mathbb{Z}$. Then $A'$ admits a triangular decomposition with subalgebras $T$ and $(A')^{\pm} = A^{\mp}$, and $\mathsf{L}(\lambda)'$ is a simple graded $A'$-module. By the above corollary, there exist $\overline{\lambda} \in \Lambda$ and $l_\lambda \in \mathbb{Z}_{\leq 0}$ such that $\mathsf{L}(\lambda)'$ is the unique simple highest weight $A'$-module with highest weight $\overline{\lambda}[l_\lambda]$. With respect to the grading of $A$, $\overline{\lambda}$ is a lowest weight and hence the corollary follows. □

An object in a highest weight category is *tilting* if it admits both standard and costandard filtrations.

THEOREM 2.5 ([**9**, Theorem 5.1]). *If $T$ is semisimple, then the tilting objects in ${}_A\mathcal{G}$ are precisely the projective-injective objects.* □

2.1.1. *Graded characters.* Recall that the *Grothendieck group* $K(\mathcal{C})$ of an abelian category $\mathcal{C}$ is the abelian group generated by the isomorphism classes $[X]$ of objects in $\mathcal{C}$ subject to the relations $[X] = [Y] + [Z]$ if there exists a short exact sequence $Y \to X \to Z$.

The inclusion $T \to A$ induces a group homomorphism $\mathrm{ch} : K({}_A\mathcal{M}) \to K({}_T\mathcal{M})$. This does not necessarily distinguish simple modules, see for instance [**33**, §1.1]. Once again, we can fix this issue by working inside the graded category.

We consider $K({}_A\mathcal{G})$ and $K({}_T\mathcal{G})$ as $\mathbb{Z}[t, t^{-1}]$-modules where $t$ acts by the shift [1]. As $T$ is concentrated in degree 0, we have that $K({}_T\mathcal{G}) \simeq K({}_T\mathcal{M})[t, t^{-1}]$ and we can define the *graded character* $\mathrm{ch}^\bullet : K({}_A\mathcal{G}) \longrightarrow K({}_T\mathcal{M})[t, t^{-1}]$ by

$$\mathrm{ch}^\bullet \mathsf{N} = \sum_{i \in \mathbb{Z}} \mathrm{ch}\, \mathsf{N}_i\, t^i \tag{2.5}$$

for every $\mathsf{N} \in {}_A\mathcal{G}$. The following is mainly a consequence of the fact that the simple modules $\mathsf{L}(\lambda)$ are distinguished by their highest weights.

PROPOSITION 2.6 ([**9**, Proposition 3.19]). $\mathrm{ch}^\bullet : K({}_A\mathcal{G}) \longrightarrow K({}_T\mathcal{M})[t, t^{-1}]$ *is an injective morphism of $\mathbb{Z}[t, t^{-1}]$-modules.* □

From now on, we identify $K({}_A\mathcal{G})$ with its image via $\mathrm{ch}^\bullet$. The next corollary is a direct consequence of the definition of Grothendieck group. It states that we can read the composition factors of a graded $A$-module from its graded character, see Lemma 5.10.

COROLLARY 2.7. *The set $\{\mathrm{ch}^\bullet \mathsf{L}(\lambda) \mid \lambda \in \Lambda\}$ is a $\mathbb{Z}[t, t^{-1}]$-basis of $K({}_A\mathcal{G})$. More explicitly, for every $\mathsf{N} \in {}_A\mathcal{G}$ there exist unique polynomials $p_{\mathsf{N},\mathsf{L}(\lambda)} \in \mathbb{Z}[t, t^{-1}]$ such that*

$$\mathrm{ch}^\bullet \mathsf{N} = \sum_{\lambda \in \Lambda} p_{\mathsf{N},\mathsf{L}(\lambda)}\, \mathrm{ch}^\bullet \mathsf{L}(\lambda).$$

□

Let ${}_A\mathcal{G}_{proj}$ be the category of finite-dimensional graded projective modules over $A$. The *split Grothendieck group* $K({}_A\mathcal{G}_{proj})$ is the group generated by the isomorphism classes $[X]$ of graded projective modules subject to the relations $[X] = [Y] + [Z]$ if $X \simeq Y \oplus Z$.

Again, we may consider $K({}_A\mathcal{G}_{proj})$ as a $\mathbb{Z}[t, t^{-1}]$-module with the action given by the shift [1]. Clearly, we have a group morphism $K({}_A\mathcal{G}_{proj}) \longrightarrow K({}_A\mathcal{G})$ and we can consider the graded character on projective modules. The next proposition allows us to identify $K({}_A\mathcal{G}_{proj})$ with its image under $\mathrm{ch}^\bullet$.

Proposition 2.8. *If $T$ is semisimple, then* $\mathrm{ch}^\bullet : K({}_A\mathcal{G}_{proj}) \longrightarrow K({}_T\mathcal{M})[t,t^{-1}]$ *is an injective morphism of* $\mathbb{Z}[t,t^{-1}]$*-modules. Moreover, the set* $\{\mathrm{ch}^\bullet \mathsf{M}(\lambda) \mid \lambda \in \Lambda\}$ *is a* $\mathbb{Z}[t,t^{-1}]$*-basis of the image of* $\mathrm{ch}^\bullet$*, i.e. for every* $\mathsf{P} \in {}_A\mathcal{G}_{proj}$ *there exist unique polynomials* $p_{\mathsf{P},\mathsf{M}(\lambda)} \in \mathbb{Z}[t,t^{-1}]$ *such that*

$$\mathrm{ch}^\bullet \mathsf{P} = \sum_{\lambda\in\Lambda} p_{\mathsf{P},\mathsf{M}(\lambda)}\, \mathrm{ch}^\bullet \mathsf{M}(\lambda).$$

Proof. Since projective modules admit standard filtrations in ${}_A\mathcal{G}$ by Theorem 2.2, the graded characters of standard modules generate $\mathrm{ch}^\bullet(K({}_A\mathcal{G}_{proj}))$. Since $T$ is semisimple, $\mathsf{M}(\lambda) = \overline{\mathsf{M}}(\lambda)$ for all $\lambda \in \Lambda$ and hence the standard modules are distinguished by their highest weights. Following the arguments in the proof of [**9**, Proposition 3.19], one can show that $\mathrm{ch}^\bullet$ is injective and the proposition follows. □

**2.2. The infinite-dimensional case.** Following Bonnafé–Rouquier [**12**, §F.2], we can also construct a highest weight category from the category of modules over an algebra $A$ with triangular decomposition without the assumption that neither $A$ is finite-dimensional nor $T$ is semi-simple. This is a generalization of the BGG category $\mathcal{O}$ of a semisimple Lie algebra.

Let $A$, $B^{\pm}$ and $T$ be algebras satisfying (td)–(td3). Let $\Lambda$ be a set of non-isomorphic finite-dimensional simple $T$-modules. We consider now the full subcategory $\mathcal{O}_T$ of $T$-modules whose objects are finite directs sums of modules in $\Lambda$. This is an abelian subcategory. For all $\lambda \in \Lambda$, we assume that:

($\Lambda$1) $\mathrm{End}_T(\lambda) = \Bbbk$,
($\Lambda$2) $B_n^+ \otimes_T \lambda$ and $B_{-n}^- \otimes_T \lambda$ belong to $\mathcal{O}_T$ and
($\Lambda$3) the $A$-module $\overline{\mathsf{M}}(\lambda)$ is noetherian (*e.g.* for $A^-$ noetherian).

We still consider $\Lambda \times \mathbb{Z}$ partially ordered by the degrees and we say that a $B^+$-module $M$ is *locally nilpotent* if $M = \bigcup_{n\in\mathbb{Z}_+}\{m \in M \mid B^+_{>n} m = 0\}$.

Theorem 2.9 ([**12**, Theorem F.2.7 and Proposition F.1.17]). *Let* $\mathcal{O}^{gr}$ *be the category of finitely generated graded $A$-modules, locally nilpotent as $B^+$-modules and whose homogeneous components belong to* $\mathcal{O}_T$*. Then*

(1) $\overline{\mathsf{M}}(\lambda)[d]$ *belongs to* $\mathcal{O}^{gr}$ *and has a unique simple quotient* $\mathsf{L}(\lambda)[d]$ *in* $\mathcal{O}^{gr}$ *for all* $\lambda[d] \in \Lambda \times \mathbb{Z}$.
(2) *Every simple object of* $\mathcal{O}^{gr}$ *is isomorphic to* $\mathsf{L}(\lambda)[d]$ *for a unique* $\lambda[d] \in \Lambda \times \mathbb{Z}$.
(3) $\mathcal{O}^{gr}$ *is a highest weight category whose set of weights is* $\Lambda \times \mathbb{Z}$ *and the standard modules are* $\overline{\mathsf{M}}(\lambda)[d]$ *for all* $\lambda[d] \in \Lambda \times \mathbb{Z}$.

Similar to the finite-dimensional case, we can consider the graded character of $\mathsf{N} \in \mathcal{O}^{gr}$ given by

$$\mathrm{ch}^\bullet \mathsf{N} = \sum_{i\in\mathbb{Z}} \mathrm{ch}\, \mathsf{N}_i\, t^i \in K(\mathcal{O}_T)[[t,t^{-1}]].$$

These are now Laurent formal series.

2.2.1. *Remarks.* In the infinite-dimensional case, there may exist simple quotients of the standard modules that do not admit a grading. For instance, let $A$ be a polynomial algebra $T[x,y]$ for some algebra $T$ with a one-dimensional representation $\varepsilon : T \to \Bbbk$. If we set $-\deg(x) = 1 = \deg(y)$, then $\Bbbk[x] \otimes T \otimes \Bbbk[y] \to A$ is a triangular decomposition. Hence $\mathsf{L}_a = \overline{\mathsf{M}}(\varepsilon)/\langle a-x\rangle$ is a one-dimensional simple $A$-module for all $a \in \Bbbk$ but $\mathsf{L}_a$ does not belong to the category $\mathcal{O}^{gr}$.

Using this algebra, we can also see that not every simple $A$-module is locally nilpotent over $B^+$. Indeed, $\Bbbk_a = T[y]/\langle a-y\rangle$ is not locally nilpotent over $B^+$ and consequently any non-trivial simple quotient of $\mathsf{M}(\Bbbk_a)$ neither.

## 3. Preliminaries on Hopf algebras

We fix here the notation and conventions on Hopf algebras for the rest of the paper.

Let $H$ be a Hopf algebra. We denote by $\Delta_H$, $\varepsilon_H$ and by $\mathcal{S}_H$ the comultiplication, counit and antipode of $H$, respectively. We omit the subscript when it is clear from the context. We always assume that the antipode is invertible. We use Sweedler notation for the comultiplication and coactions, *e.g.* $\Delta(h) = h_{(1)}\otimes h_{(2)}$.

Given a vector space $M$, we set $M^* = \operatorname{Hom}(M, \Bbbk)$ and $\langle f, m\rangle$ denotes the evaluation of $f \in M^*$ in $m \in M$. If $M$ is an $H$-module, then so is $M^*$ via $\langle h\cdot f, m\rangle = \langle f, \mathcal{S}(h)\cdot m\rangle$ for all $h\in H$, $f\in M^*$, $m\in M$.

We recall that the Grothendieck group $K({}_H\mathcal{M})$ is indeed a ring with unit $\varepsilon$ (i.e., the one-dimensional simple module given by the counit) and multiplication induced by the tensor product. Moreover, if $H$ is a graded algebra, then $K({}_H\mathcal{G})$ is a $\mathbb{Z}[t,t^{-1}]$-algebra via the assignment $\varepsilon[\pm 1] \mapsto t^{\pm 1}$. Given two simple $H$-modules $\lambda$ and $\mu$, we denote by $\lambda\mu$ both the $H$-module $\lambda\otimes\mu$ as well as its representative in $K({}_H\mathcal{M})$ (the meaning will be clear from the context).

**3.1. Nichols algebras.** Our main references here are **[4, 7]**. Let ${}^H_H\mathcal{YD}$ be the category of Yetter-Drinfeld modules over $H$. This a braided category with braiding

$$V\otimes W \to W\otimes V, \quad v\otimes w \mapsto v_{(-1)}\cdot w\otimes v_{(0)}$$

for all objects $V, W \in {}^H_H\mathcal{YD}$, $v\in V$ and $w \in W$.

Let $V \in {}^H_H\mathcal{YD}$. The tensor algebra $T(V)$ is a graded braided Hopf algebra with comultiplication induced by $\Delta(v) = v\otimes 1 + 1\otimes v$ for all $v \in V$. The Nichols algebra of $V$ is

$$\mathfrak{B}(V) = T(V)/\mathcal{J}(V)$$

where $\mathcal{J}(V)$ is the largest Hopf ideal of $T(V)$ generated as an ideal by homogeneous elements of degree $\geq 2$.

Let $B$ be a braided Hopf algebra in ${}^H_H\mathcal{YD}$, *e.g.* $T(V)$ or $\mathfrak{B}(V)$. The bosonization $B\#H$ is the Hopf algebra with underlying vector space $B\otimes H$, and multiplication and comultiplication given by

$$\begin{aligned}(x\#h)(\tilde{x}\#\tilde{h}) &= x(h_{(1)}\cdot\tilde{x})\#h_{(2)}\tilde{h} \quad \text{and}\\ \Delta(x\#h) &= x^{(1)}\#(x^{(2)})_{(-1)}h_{(1)}\otimes(x^{(2)})_{(0)}\#h_{(2)}\end{aligned}$$

for all $x, \tilde{x} \in B$ and $h, \tilde{h}\in H$; with $\Delta_B(x) = x^{(1)}\otimes x^{(2)}$.

If $\mathfrak{B}(V)$ is finite-dimensional, we denote by $n_{top}$ its maximum degree. We set

$$\lambda_V = \mathfrak{B}^{n_{top}}(V), \tag{3.1}$$

the homogeneous component of degree $n_{top}$ of $\mathfrak{B}(V)$. It is well-known that $\lambda_V$ is one-dimensional. Therefore $\lambda_V$ is a simple $H$-module and $H$-comodule. We can fix a generator $x_{top} \in \lambda_V$. Thus, there exists a group-like element $g_{top} \in H$ such that the coaction satisfies $(x_{top})_{(-1)}\otimes(x_{top})_{(0)} = g_{top}\otimes x_{top}$.

**3.2. Quasitriangular Hopf algebras.** Here we refer to [**35**]. Let $(H, R)$ be a quasitriangular Hopf algebra, *i.e.* $H$ is a Hopf algebra and $R = R^{(1)} \otimes R^{(2)} \in H \otimes H$ satisfies that

$$\Delta(R^{(1)}) \otimes R^{(2)} = R^{(1)} \otimes r^{(1)} \otimes R^{(2)} r^{(2)}, \quad \varepsilon(R^{(1)}) R^{(2)} = 1,$$

$$R^{(1)} \otimes \Delta(R^{(2)}) = R^{(1)} r^{(1)} \otimes r^{(2)} \otimes R^{(2)}, \quad R^{(1)} \varepsilon(R^{(2)}) = 1 \quad \text{and}$$

$$\Delta^{cop}(h) R = R \Delta(h), \quad \forall h \in H;$$

where $r$ denotes another copy of $R$. The element $R$ is called the *$R$-matrix* of $H$. It turns out that $R$ is invertible and $R^{-1} = \mathcal{S}(R^{(1)}) \otimes R^{(2)} = R^{(1)} \otimes \mathcal{S}^{-1}(R^{(2)})$.

Let $R_{(l)} = \{R^{(1)} f(R^{(2)}) \mid f \in H^*\}$ and $R_{(r)} = \{f(R^{(1)}) R^{(2)} \mid f \in H^*\}$. Then $R_{(l)}$ and $R_{(r)}$ are finite-dimensional Hopf subalgebras of $H$ and $(R_{(l)})^{*cop} \longrightarrow R_{(r)}$, $p \mapsto p(R^{(1)}) R^{(2)}$, is an isomorphism of Hopf algebras [**35**, Proposition 2]. Let $\psi : R_{(r)} \longrightarrow (R_{(l)})^{*cop}$ be its inverse map. We define the pairing

$$R_{(l)} \otimes R_{(r)} \longrightarrow \Bbbk, \quad a \otimes b \mapsto \langle a, b \rangle = \langle a, \psi(b) \rangle \tag{3.2}$$

for all $a \in R_{(l)}$ and $b \in R_{(r)}$. It satisfies that

$$\langle aa', b \rangle = \langle a, b_2 \rangle \langle a', b_1 \rangle, \quad \langle a, bb' \rangle = \langle a_1, b \rangle \langle a_2, b' \rangle, \quad \langle a, \mathcal{S}(b) \rangle = \langle \mathcal{S}^{-1}(a), b \rangle \tag{3.3}$$

for all $a, a' \in R_{(l)}$ and $b, b' \in R_{(r)}$. Moreover, it holds that

$$ab = \langle a_1, b_1 \rangle \langle a_3, \mathcal{S}(b_3) \rangle b_2 a_2 \tag{3.4}$$

for all $a \in R_{(l)}$ and $b \in R_{(r)}$ by [**35**, Theorem 2].

The category of $H$-modules is braided with braiding

$$c = \tau \circ R \quad \text{and} \quad c^{-1} = R^{-1} \circ \tau$$

where $\tau$ is the usual twist map $\tau(a \otimes b) = b \otimes a$.

We can endow any $H$-module $M$ with the coaction $\lambda_c(x) = R^{(2)} \otimes R^{(1)} x$, for all $x \in M$. This gives a braided functor $\eta_c : ({}_H\mathcal{M}, c) \to {}^H_H\mathcal{YD}$. Instead, if we endow $M$ with the coaction $\lambda_{c^{-1}}(x) = \mathcal{S}(R^{(1)}) \otimes R^{(2)} x$, for all $x \in M$, we obtain a braided functor $\eta_{c^{-1}} : ({}_H\mathcal{M}, c^{-1}) \to {}^H_H\mathcal{YD}$.

Nichols algebras can be defined as above for every object in a braided category. Thus, given $M \in {}_H\mathcal{M}$, we can consider the Nichols algebras $\mathfrak{B}(M, c) \in ({}_H\mathcal{M}, c)$ and $\mathfrak{B}(M, c^{-1}) \in ({}_H\mathcal{M}, c^{-1})$. Moreover, these coincide with the Nichols algebras of $\eta_c(M)$ and $\eta_{c^{-1}}(M)$, respectively. This holds because the ideal of relations of a Nichols algebra can be identified with the kernel of the quantum symmetrizer which depends only on the braiding, cf. [**4**, Proposition 3.2.12] or [**7**, Proposition 2.11].

**3.3. Drinfeld double.** Let $K$ be a finite-dimensional Hopf algebra. Following [**27**, Theorem 7.1.1], the Drinfeld double $\mathcal{D}(K)$ of $K$ is a Hopf algebra which is $K \otimes K^*$ as a coalgebra. The multiplication and the antipode are given by

$$\begin{aligned} (a \otimes f)(a' \otimes f') &= \langle f_{(1)}, a'_{(1)} \rangle \langle f_{(3)}, \mathcal{S}_K(a'_{(3)}) \rangle (a a'_{(2)} \otimes f' f_{(2)}), \\ \mathcal{S}(a \otimes f) &= (1 \otimes \mathcal{S}_{K^*}^{-1}(f))(\mathcal{S}_K(a) \otimes \varepsilon), \quad \text{for all } a, a' \in K \text{ and } f, f' \in K^*. \end{aligned} \tag{3.5}$$

It holds that $K$ and $K^{*op}$ are Hopf subalgebras of $\mathcal{D}(K)$.

The Drinfeld double $\mathcal{D}(K)$ is quasitriangular with $R$-matrix given by

$$R = \sum_i f_i \otimes a_i, \tag{3.6}$$

where $\{a_i\}_i$ and $\{f_i\}_i$ are dual basis of $K$ and $K^*$, respectively.

There exists an equivalence of braided categories ${}^K_K\mathcal{YD} \to ({}_{\mathcal{D}(K)}\mathcal{M}, c)$ defined as follows. Given $M \in {}^K_K\mathcal{YD}$, then $M \in {}_{\mathcal{D}(K)}\mathcal{M}$ with

$$(a \otimes f) \cdot m = \langle f, m_{(-1)} \rangle \, (a \cdot m_{(0)}) \tag{3.7}$$

for all $a \in K$, $f \in K^*$ and $m \in M$.

## 4. A method to construct Hopf algebras with triangular decomposition

The Drinfeld double of bosonizations provides examples of Hopf algebras with triangular decomposition. However, we have to work a bit in order to identify the three subalgebras providing such a decomposition. We prefer instead to start at the end, *i.e.* we will consider three suitable algebras which lead to a Hopf algebra with triangular decomposition. These kind of constructions were performed for instance in [**17**, **23**, **28**] in much more generality than we do it here, see §4.2.

From now on, we fix a quasitriangular Hopf algebra $(H, R)$ and a finite-dimensional $H$-module $V$. We consider $V$ in ${}^H_H\mathcal{YD}$ via the braided functor $({}_H\mathcal{M}, c) \to {}^H_H\mathcal{YD}$, *i.e.* the $H$-coaction of $x \in V$ is given by

$$x_{(-1)} \otimes x_{(0)} = R^{(2)} \otimes R^{(1)} x. \tag{4.1}$$

We let $\overline{V} = V^*$ be the Yetter-Drinfeld module over $H$ with action and coaction given by

$$\langle hy, x \rangle = \langle y, \mathcal{S}(h)x \rangle, \tag{4.2}$$

$$y_{(-1)} \otimes y_{(0)} = \mathcal{S}(R^{(1)}) \otimes R^{(2)} y \tag{4.3}$$

for all $y \in \overline{V}$, $x \in V$ and $h \in H$. This corresponds to considering first the dual object of $V$ in ${}_H\mathcal{M}$ and then applying the braided functor $({}_H\mathcal{M}, c^{-1}) \to {}^H_H\mathcal{YD}$. It is immediate that

$$\langle x_{(-1)} \cdot y, x_{(0)} \rangle = \langle y_{(0)}, \mathcal{S}^{-1}(y_{(-1)}) \cdot x \rangle \tag{4.4}$$

for all $x \in V$ and $y \in \overline{V}$.

Given $x \in V$ and $y \in \overline{V}$, we set

$$\begin{aligned} [\![y, x]\!] := {} & yx - (y_{(-1)} \cdot x) y_{(0)} - \langle y, x \rangle \\ & + \langle y_{(-2)}, x_{(-3)} \rangle \langle y_{(0)}, \mathcal{S}(x_{(-1)}) \cdot x_{(0)} \rangle \, y_{(-1)} x_{(-2)}, \end{aligned} \tag{4.5}$$

where $\langle y_{(-2)}, x_{(-3)} \rangle$ denotes the pairing (3.2). The motivation for this formula is Example §4.1.1 below.

Let $\mathbf{J}$ be the ideal of $T(V \oplus \overline{V}) \# H$ generated by the elements $[\![y, x]\!]$ for all $x \in V$ and $y \in \overline{V}$. We define

$$\mathbf{U}_{(H,R)}(V) = {T(V \oplus \overline{V}) \# H}\big/_{\mathbf{J}}. \tag{4.6}$$

PROPOSITION 4.1. $\mathbf{U}_{(H,R)}(V)$ *is a graded Hopf algebra with*

$$\deg V = -1, \quad \deg H = 0, \quad \deg \overline{V} = 1$$

*and* $T(V)\otimes H\otimes T(\overline{V}) \longrightarrow \mathbf{U}_{(H,R)}(V)$ *is a triangular decomposition. Also, the Hopf subalgebra generated by* $V$ *and* $H$*, resp.* $\overline{V}$ *and* $H$*, is isomorphic to*

$$T(V)\#H, \quad \textit{resp. } T(\overline{V})\#H.$$

PROOF. Clearly, $T(V\oplus\overline{V})\#H$ is graded and $\mathbf{J}$ is an homogeneous ideal. Moreover, it is a coideal because

$$\Delta([\![y,x]\!]) = [\![y,x]\!]\otimes 1 + y_{(-1)}x_{(-1)}\otimes[\![y_{(0)},x_{(0)}]\!]$$

for all $x \in V$ and $y \in \overline{V}$. This follows from a straightforward computation using (4.4), (3.4) and the identity $yx_{(-1)}\otimes x_{(0)} = y_{(-3)}x_{(-1)}\mathcal{S}(y_{(-1)})y_{(0)}\otimes y_{(-2)}\cdot x_{(0)}$, which is a consequence of the axioms of the $R$-matrix. Therefore $\mathbf{U}_{(H,R)}(V)$ is a graded Hopf algebra.

The triangular decomposition holds because (4.5) ensures that $yx \in V\overline{V}\oplus H$ in $\mathbf{U}_{(H,R)}(V)$, for all $y \in \overline{V}$ and $x \in V$.

The last part of the statement is clear. □

We are interested in graded quotient Hopf algebras of $\mathbf{U}_{(H,R)}(V)$ admitting a triangular decomposition with $H$ in the middle. The main example is the following. Let $\mathfrak{B}(V)$ and $\mathfrak{B}(\overline{V})$ be the Nichols algebras of $V$ and $\overline{V}$, with defining ideals $\mathcal{J}(V)$ and $\mathcal{J}(\overline{V})$, respectively. We define

$$\mathfrak{u}_{(H,R)}(V) = \mathbf{U}_{(H,R)}(V)\big/\langle\mathcal{J}(V),\mathcal{J}(\overline{V})\rangle. \tag{4.7}$$

PROPOSITION 4.2. $\mathfrak{u}_{(H,R)}(V)$ *is a graded Hopf algebra quotient of* $\mathbf{U}_{(H,R)}(V)$ *and* $\mathfrak{B}(V)\otimes H\otimes\mathfrak{B}(\overline{V}) \longrightarrow \mathfrak{u}_{(H,R)}(V)$ *is a triangular decomposition. Also, the Hopf subalgebra generated by* $V$ *and* $H$*, resp.* $\overline{V}$ *and* $H$*, is isomorphic to*

$$\mathfrak{B}(V)\#H, \quad \textit{resp. } \mathfrak{B}(\overline{V})\#H.$$

PROOF. We only have to prove the triangular decomposition. A direct computation shows that the composition of the braidings $c_{V,\overline{V}} : V\otimes\overline{V} \to \overline{V}\otimes V$ and $c_{\overline{V},V} : \overline{V}\otimes V \to V\otimes\overline{V}$ is the identity. Then, the Nichols algebra $\mathfrak{B}(V\oplus\overline{V})$ is isomorphic to $\mathfrak{B}(V)\otimes\mathfrak{B}(\overline{V})$ as vector spaces by [**16**, Theorem 2.2] and hence $\mathcal{J}(V\oplus\overline{V}) = \mathcal{J}(V)\otimes T(\overline{V}) + T(V)\otimes\mathcal{J}(\overline{V}) = \langle\mathcal{J}(V),\mathcal{J}(\overline{V})\rangle$ as ideals in $T(V\oplus\overline{V})$. Therefore $\langle\mathcal{J}(V),\mathcal{J}(\overline{V})\rangle = \mathcal{J}(V)\otimes H\otimes T(\overline{V}) + T(V)\otimes H\otimes\mathcal{J}(\overline{V})$ as ideals in $T(V\oplus\overline{V})\#H$. Since the generators of $\mathbf{J}$ do not belong to this ideal, we have that

$$\mathfrak{u}_{(H,R)}(V) \simeq \left(T(V\oplus\overline{V})\#H\big/\langle\mathcal{J}(V),\mathcal{J}(\overline{V})\rangle\right)\big/\mathbf{J} \simeq \mathfrak{B}(V)\otimes H\otimes\mathfrak{B}(\overline{V})$$

as vector spaces. □

### 4.1. Examples.

4.1.1. *The Drinfeld double of a bosonization.* Let $K$ be a finite-dimensional Hopf algebra and $R$ the $R$-matrix of $\mathcal{D}(K)$ given by (3.6). Let $V$ be a finite-dimensional Yetter-Drinfeld module over $K$ with $\dim \mathfrak{B}(V) < \infty$. We consider $V$ as a $\mathcal{D}(K)$-module via (3.7).

LEMMA 4.3. *The Drinfeld double of $\mathfrak{B}(V)\#K$ is isomorphic to $\mathfrak{u}_{(\mathcal{D}(K),R)}(V)$. In particular, $\mathfrak{B}(V)\otimes\mathcal{D}(K)\otimes\mathfrak{B}(\overline{V}) \longrightarrow \mathcal{D}(\mathfrak{B}(V)\#K)$ is a triangular decomposition and hence $\mathcal{D}(K)$, $\mathfrak{B}(V)\#\mathcal{D}(K)$ and $\mathfrak{B}(\overline{V})\#\mathcal{D}(K)$ are graded Hopf subalgebras of $\mathcal{D}(\mathfrak{B}(V)\#K)$.*

PROOF. The isomorphism follows as in [**33**, Lemma 7]. We sketch the proof and leave the details for the reader. First, we have to check that $(\mathfrak{B}(V)\#K)^{*op} \simeq \mathfrak{B}(\overline{V})\#K^{*op}$ similar to [**33**, Lemma 5]. Here, we consider $\overline{V} = V^*$ as the Yetter-Drinfeld module over $K^{*op}$ with action and coaction defined by

$$\langle f \cdot y, x\rangle = \langle f, \mathcal{S}^{-1}(x_{(-1)})\rangle\langle y, x_{(0)}\rangle \quad \text{and} \quad \langle y, h \cdot x\rangle = \langle y_{(-1)}, h\rangle\langle y_{(0)}, x\rangle$$

for all $f \in K^{*op}$, $h \in H$, $x \in V$ and $y \in \overline{V}$. This extends [**33**, Definition 4] given for $K = \Bbbk G$. Now, we can consider $\mathfrak{B}(\overline{V})\#K^{*op}$ as a Hopf subalgebra of $\mathcal{D}(\mathfrak{B}(V)\#K)$. We can check that the Hopf algebra generated by $K$ and $K^{*op}$ is isomorphic to $\mathcal{D}(K)$ and that the elements of $V$ and $\overline{V}$ satisfy the relation (4.5). To conclude, we have to see that $V$ and $\overline{V}$ are invariants under the adjoint action of $\mathcal{D}(K)$ and that they are left ideals, *i.e.* left comodules under the comultiplication. Moreover, they are the Yetter-Drinfeld modules over $\mathcal{D}(K)$ defined by (4.1), (4.2) and (4.3). □

4.1.2. *The small quantum groups.* Let $\mathfrak{g}$ be a finite-dimensional semisimple Lie algebra over $\mathbb{C}$ and $q$ a root of unity of odd order. The small quantum group $\mathfrak{u}_q(\mathfrak{g})$ can be constructed following our recipe. Although this is not the most elegant way of doing it, we describe it here for the sake of completeness.

Let $(c_{ij})_{1\le i,j\le\theta}$ be the Cartan matrix of $\mathfrak{g}$ and $(d_i)_{1\le i\le\theta}$ a diagonal matrix such that $(d_i c_{ij})_{1\le i,j\le\theta}$ is symmetric. Let $\Gamma$ be the finite abelian group generated by $g_1,\dots g_\theta$ with $g_i^{\operatorname{ord}(q)} = 1$ for all $1 \le i \le \theta$. Let $\chi_1, \dots \chi_\theta \in \widehat{\Gamma}$ with $\chi_i(g_j) = q_{ij} = q^{d_i c_{ij}}$ for all $1 \le i, j \le \theta$.

If $g = g_1^{n_1}\cdots g_\theta^{n_\theta} \in \Gamma$, we denote $\chi_g = \chi_1^{n_1}\cdots\chi_\theta^{n_\theta} \in \widehat{\Gamma}$. Then $\Bbbk\Gamma$ is quasitriangular with $R$-matrix

$$\overline{R} = \frac{1}{|\Gamma|}\sum_{g,h\in\Gamma} \chi_g(h^{-1})g\otimes h,$$

cf. [**27**, Example 2.1.6]. This is the image of the $R$-matrix of the Drinfeld double $\mathcal{D}(\Bbbk\Gamma)$ under the epimorphism $\mathcal{D}(\Bbbk\Gamma) \twoheadrightarrow \Bbbk\Gamma$ given by $\chi_g = g$ for all $g \in \Gamma$.

Let $V = \Bbbk\{x_1, \dots, x_\theta\}$ be the $\Bbbk\Gamma$-module with action $g \cdot x_i = \chi_i(g)$ for all $1 \le i \le \theta$ and $g \in \Gamma$. If we apply to $V$ and $(\Bbbk\Gamma, \overline{R})$ our construction, we obtain that

$$\mathfrak{u}_q(\mathfrak{g}) \simeq \mathfrak{u}_{(\Bbbk\Gamma,\overline{R})}(V) \simeq \mathfrak{u}_{(\mathcal{D}(\Bbbk\Gamma),R)}(V)\big/\langle\chi_g - g \mid g \in \Gamma\rangle$$

where $R$ is the $R$-matrix of $\mathcal{D}(\Bbbk\Gamma)$, recall (3.6).

**4.2. More general constructions.** We have restricted ourselves to consider quasitriangular Hopf algebras and Nichols algebras because we are interested in the Drinfeld doubles of their bosonizations, and mainly those where $H$ is not an abelian group. However, this recipe is a particular case of more general constructions as we explain below.

4.2.1. Let $(H, R)$ be a quasitriangular Hopf algebra. The double-bosonization construction of Majid [**28**] can be performed for every dually paired braided Hopf algebras $B \in {}_H\mathcal{M}$ and $C \in \mathcal{M}_H$ instead of $T(V)$ and $T(\overline{V})$. Notice that $B$ and $C$ belong to different categories. We prefer in our exposition to consider $T(V)$ and $T(\overline{V})$ in the same category because this fact brings with it some advantages to study modules over the whole algebra $\mathbf{U}_{(H,R)}(V)$ and their quotients. We will implicitly use this fact in §5.

Also, we can then deduce that the action maps

$$V \otimes M \to M \quad \text{and} \quad \overline{V} \otimes M \to M$$

are morphism of $H$-modules for all $M \in {}_H\mathcal{M}$. This allows us to carry out an algorithm in order to compute submodules, cf. [**33**, Remark 18].

4.2.2. In [**23**], Laugwitz analyzes how to endow $T(V)\otimes H \otimes T(V^*)$ (and its quotients) with a structure of Hopf algebra. In particular, he extends the double-bosonization construction to a non-quasitriangular Hopf algebra $H$ [**23**, §3.4]. The data for his construction are two braided Hopf algebras $B, C \in {}^H_H\mathcal{YD}$ *weakly dually paired* [**23**, Definition 6].

4.2.3. There is an important class of Hopf algebras with triangular decomposition which we cannot construct from our recipe. Namely, those with a free abelian group playing the role of $T$. These kind of Hopf algebras were studied for instance by Andruskiewitsch–Radford–Schneider[**6**] and Heckenberger–Yamane [**18**], as we have mentioned in the introduction. The reader can find a systematic way to construct this sort of Drinfeld doubles in [**17**, §4].

**4.3. On the Nichols algebras of $V$, $\overline{V}$ and $V^*$.** Let $V^*$ denote the dual object of $V$ in ${}^H_H\mathcal{YD}$. That is, $V^* = \overline{V}$ as $H$-modules and the coaction of $y \in \overline{V}$ is

$$y_{(-1)} \otimes y_{(0)} = \sum_{i \in I} \mathcal{S}^{-1}(R^{(2)}) \otimes \langle y, R^{(1)} x_i \rangle y_i = \sum_{i \in I} \mathcal{S}^{-1}(R^{(2)}) \otimes \mathcal{S}^{-1}(R^{(1)}) y,$$

cf. [**4**, Proposition 2.1.1]. In the last equality we used (4.2). Notice that, $V^*$ is not necessarily isomorphic to $\overline{V}$ as an $H$-comodule. For instance, if $H$ is the Drinfeld double of a non-abelian group $G$, then $\mathcal{S}^{-1}(R^{(2)}) \in \Bbbk G$ and $R^{(1)} \in (\Bbbk G)^*$, hence there is no isomorphism of comodules between these objects.

We now explain the relation between the Nichols algebras $\mathfrak{B}(V)$, $\mathfrak{B}(\overline{V})$ and $\mathfrak{B}(V^*)$. First, by [**4**, Proposition 3.2.30], we know that

$$\mathfrak{B}(V^*) \simeq \mathfrak{B}(V)^{*bop} \quad \text{as braided Hopf algebras in } {}^H_H\mathcal{YD}. \tag{4.8}$$

This isomorphism also holds in $({}_H\mathcal{M}, c)$ via the forgetful functor ${}^H_H\mathcal{YD} \to ({}_H\mathcal{M}, c)$, since it is a braided functor.

Since the comodule structures do not coincide, $\overline{V}$ and $V^*$ are not equivalent as braided vector spaces. Instead, the braiding of $\overline{V}\otimes\overline{V} \to \overline{V}\otimes\overline{V}$ in ${}^H_H\mathcal{YD}$ coincides with $c^{-1} : V^* \otimes V^* \to V^* \otimes V^*$ in ${}_H\mathcal{M}$ by construction. Thus, we should analyze the Nichols algebra $\mathfrak{B}(\overline{V}, c^{-1})$ in $({}_H\mathcal{M}, c^{-1})$.

By [**5**, Lemma 1.11], we know that $\mathcal{J}(\overline{V}) = \mathcal{J}(V^*)$ and

$$\mathfrak{B}(\overline{V}) = \mathfrak{B}(V^*) \quad \text{as } H\text{-module algebras.} \tag{4.9}$$

Moreover, we also see in the proof of [**5**, Lemma 1.11] that the comultiplication of $\mathfrak{B}(\overline{V})$ satisfies

$$\Delta_{\mathfrak{B}(\overline{V})} = c^{-1} \circ \Delta_{\mathfrak{B}(V^*)}, \tag{4.10}$$

that is the opposite coalgebra in the braided category $({}_H\mathcal{M}, c)$. Therefore, by (4.8) and (4.9), we have that

$$\mathfrak{B}^n(\overline{V}) \simeq \mathfrak{B}^n(V)^* \quad \text{as } H\text{-modules for all } n. \tag{4.11}$$

In particular, if $\dim \mathfrak{B}(V) < \infty$ and $n = n_{top}$, the above isomorphism implies that

$$\lambda_V \lambda_{\overline{V}} = \varepsilon = \lambda_{\overline{V}} \lambda_V \tag{4.12}$$

in the Grothendieck ring $K({}_H\mathcal{M})$, recall (3.1).

## 5. On the representation theory of $\mathfrak{u}_{(H,R)}(V)$

Throughout this section, $(H, R)$ will be a quasitriangular Hopf algebra and $V$ a finite-dimensional $H$-module. We can classify the simple $\mathfrak{u}_{(H,R)}(V)$-modules using the results of §2. Theorems 2.1 and 2.2 imply the following, see also §5.2 below.

THEOREM 5.1. *Assume that $H$ and $\mathfrak{B}(V)$ are finite-dimensional and $H$ is split. Let $\Lambda$ be a complete set of non-isomorphic simple $H$-modules and $\mathsf{L}(\lambda)$ be the head of $\overline{\mathsf{M}}(\lambda)$. Then $\{\mathsf{L}(\lambda)[d]\}_{\Lambda\times\mathbb{Z}}$ is a complete set of non-isomorphic simple graded $\mathfrak{u}_{(H,R)}(V)$-modules.*

*Moreover, if $H$ semisimple, the category of graded $\mathfrak{u}_{(H,R)}(V)$-module is a highest weight category.* □

We would like to stress that the graded character in the Hopf algebra setting is an algebra map. Explicitly, we can restate Proposition 2.6 as follows.

PROPOSITION 5.2. *The functor ${}_{\mathfrak{u}_{(H,R)}(V)}\mathcal{G} \longrightarrow {}_H\mathcal{G}$ given by restriction of scalars is monoidal and hence $\mathrm{ch}^\bullet : K({}_{\mathfrak{u}_{(H,R)}(V)}\mathcal{G}) \longrightarrow K({}_H\mathcal{M})[t, t^{-1}]$ is an injective morphism of $\mathbb{Z}[t, t^{-1}]$-algebras.* □

In the infinite-dimensional case we apply Theorem 2.9.

THEOREM 5.3. *Let $\Lambda$ be a set of non-isomorphic finite-dimensional simple $H$-modules and $\mathcal{O}_H$ the full subcategory of $H$-modules whose objects are finite directs sums of modules in $\Lambda$. Assume for all $\lambda \in \Lambda$ that*

($\Lambda$1) $\mathrm{End}_H(\lambda) = \Bbbk$.
($\Lambda$2) *$\mathfrak{B}^n(V)\otimes_H\lambda$ and $\mathfrak{B}^n(\overline{V})\otimes_H\lambda$ belong to $\mathcal{O}_H$.*
($\Lambda$3) *the $\mathfrak{u}_{(H,R)}(V)$-module $\overline{\mathsf{M}}(\lambda)$ is noetherian (e.g. for $\mathfrak{B}(V)$ noetherian).*

*Let $\mathcal{O}^{gr}$ be the category of finitely generated graded $\mathfrak{u}_{(H,R)}(V)$-modules, locally nilpotent as $\mathfrak{B}(\overline{V})$-modules and whose homogeneous components belong to $\mathcal{O}_H$.*

*Then all the simple objects (up to isomorphism) in $\mathcal{O}^{gr}$ are the heads $\mathsf{L}(\lambda)[d]$ of $\overline{\mathsf{M}}(\lambda)[d]$ for all $\lambda[d] \in \Lambda \times \mathbb{Z}$. Moreover, $\mathcal{O}^{gr}$ is a highest weight category.* □

**5.1. Some problems on the simple modules.** Once we have classified the simple modules, the question regarding their graded characters naturally arises.

QUESTION 5.4. *Describe* $\operatorname{ch}^{\bullet} \mathsf{L}(\lambda)$ *for all* $\lambda \in \Lambda$.

For instance, this was done for the Drinfeld double of the Taft algebra in [**13**]; for the Drinfeld double of the Nichols algebra of unidentified diagonal type $\mathfrak{ufo}(7)$ in [**3**]; for the Drinfeld double of the Fomin-Kirillov algebra over $\mathbb{S}_3$ in [**33**]; and for the small quantum group at a root of unity of large prime-order, a character formula can be deduced from [**1**].

In examples we see that the Hilbert series of $\mathsf{L}(\lambda)$ has symmetric coefficients, *i.e.* $\dim \mathsf{L}(\lambda)_n = \dim \mathsf{L}(\lambda)_{l_\lambda - n}$, where we assume that $\mathsf{L}(\lambda)$ is finite-dimensional with minimum degree $l_\lambda$; recall Corollary 2.4. The Hilbert series of any finite-dimensional Nichols algebra also satisfies this symmetry. The following question for rational Cherednik algebras was posed by Thiel in [**37**, Question 7.7(a)] and [**38**, Problem 6.6]; nevertheless some restrictions are required because he also gives counter-examples in the first paper, see [**37**, Remark 7.8].

QUESTION 5.5. *Are the coefficients of the Hilbert series of* $\mathsf{L}(\lambda)$ *symmetric?*

Assume that $\mathfrak{B}(V)$ and $H$ are finite-dimensional. If $\lambda \in \Lambda$, we denote by $\overline{\lambda}$ the lowest weight of $\mathsf{L}(\lambda)$; recall Corollary 2.4. Then $\overline{\phantom{x}} : \Lambda \to \Lambda$, $\lambda \mapsto \overline{\lambda}$, is a bijection. For the Drinfeld double of the Fomin-Kirillov algebra over $\mathbb{S}_3$, this bijection corresponds to the unique non-trivial braided autoequivalence of the category of $\mathcal{D}(\Bbbk\mathbb{S}_3)$-modules, cf. [**34**, (2)], [**24**, §6.6] and [**29**, §8.1].

QUESTION 5.6. *Does the bijection* $\overline{\phantom{x}} : \Lambda \to \Lambda$ *induce a braided autoequivalence in the category of* $H$*-modules?*

**5.2. On projective modules in the finite-dimensional case.** From now on, we assume that $H$ and $\mathfrak{B}(V)$ are finite-dimensional and $H$ is semisimple.

In this situation Theorem 2.1 gives us information on the projective $\mathfrak{u}_{(H,R)}(V)$-modules:

- Projective modules admit standard filtrations.
- Brauer Reciprocity holds.

We can also extend to $\mathfrak{u}_{(H,R)}(V)$ some of the results in [**39**] where we have considered $H = \mathcal{D}(G)$. Indeed, these results rely on the fact that the homogeneous component of maximum degree of a finite-dimensional Nichols algebra is one-dimensional, recall §3.1. We give the main ideas of the proofs. The reader can complete the details following [**39**].

We set

$$A = \mathfrak{u}_{(H,R)}(V), \quad B^+ = \mathfrak{B}(\overline{V})\#H \quad \text{and} \quad B^- = \mathfrak{B}(V)\#H,$$

and keep the notation of §2. In particular, $\Lambda$ is a complete set of non-isomorphic simple $H$-modules. As $H$ is assumed to be semisimple, the proper standard modules in (2.1) and the standard modules in (2.2) coincide with the Verma modules considered in [**39**], that is

$$\mathsf{M}(\lambda) = A \otimes_{B^+} \operatorname{Inf}_H^{B^+}(\lambda)$$

for all $\lambda \in \Lambda$.

Meanwhile the costandard modules $\nabla(\lambda)$ and the coVerma modules $\mathsf{W}(\lambda)$ of [**39**, (13)] are related by

$$\mathsf{W}(\lambda) := A \otimes_{B^-} \operatorname{Inf}_H^{B^-}(\lambda) \simeq \nabla(\lambda_{\overline{V}}\lambda)$$

for all $\lambda \in \Lambda$.

LEMMA 5.7 ([**39**, Lemma 1]). *For all $\lambda \in \Lambda$, $\mathsf{M}(\lambda)$ is the projective cover of $\operatorname{Inf}_H^{B^-}(\lambda)$ and the injective hull of $\operatorname{Inf}_H^{B^-}(\lambda_V\lambda)$ in ${}_{B^-}\mathcal{G}$.*

PROOF. As $B^-$-module, $\mathsf{M}(\lambda) \simeq B^- \otimes_H \lambda$ is an induced module from a semisimple algebra and hence it is projective and injective (because $B^-$ is Frobenius). It is indecomposable because its socle is simple and isomorphic to $\lambda_V\lambda$ (here we use that $\dim \lambda_V = 1$). $\square$

LEMMA 5.8 ([**39**, (10)]). *For all $\lambda \in \Lambda$, $\mathsf{M}(\lambda)^* \simeq \mathsf{M}((\lambda_V\lambda)^*)$ as $A$-modules.*

PROOF. Using the grading, we see that $(\mathfrak{B}^{n_{top}}(V) \otimes \lambda)^* \simeq (\lambda_V \otimes \lambda)^*$ is a highest weight of $\mathsf{M}(\lambda)^*$. Then there is a morphism $f : \mathsf{M}((\lambda_V\lambda)^*) \longrightarrow \mathsf{M}(\lambda)^*$. If $S$ denotes the socle of $\mathsf{M}((\lambda_V\lambda)^*)$ as $B^-$-module, then $S$ is simple and $f(S) \neq 0$ (cf. the proof of [**39**, (10)]). Hence $f$ is injective and therefore an isomorphism because $\dim \mathsf{M}(\lambda)^* = \dim \mathsf{M}((\lambda_V\lambda)^*)$. $\square$

Given an $H$-module $S$, we set $\mathsf{Ind}(S) := A \otimes_H S$.

LEMMA 5.9 ([**39**, Lemma 4]). *For all $\lambda, \mu \in \Lambda$, $\mathsf{W}(\lambda) \otimes \mathsf{M}(\mu) \simeq \mathsf{Ind}(\lambda\mu)$.*

PROOF. The $H$-submodule $1 \otimes \lambda \otimes 1 \otimes \mu$ of $\mathsf{W}(\lambda) \otimes \mathsf{M}(\mu)$ induces a morphism of $A$-modules $f : \mathsf{Ind}(\lambda\mu) \to \mathsf{W}(\lambda) \otimes \mathsf{M}(\mu)$. Using that $\lambda_V$ is one-dimensional, we can see that $f$ is injective and then it is an isomorphism. $\square$

Let $\mathsf{N} \in {}_A\mathcal{G}$ and $\mathsf{P} \in {}_A\mathcal{G}_{proj}$. By Corollary 2.7 and Proposition 2.8, there exist polynomials

$$p_{\mathsf{N},\mathsf{L}(\lambda)} = \sum_i a_{\mathsf{N},\mathsf{L}(\lambda),i}\, t^i \in \mathbb{Z}[t, t^{-1}] \quad \text{and} \quad p_{\mathsf{P},\mathsf{M}(\lambda)} = \sum_i a_{\mathsf{P},\mathsf{M}(\lambda),i}\, t^i \in \mathbb{Z}[t, t^{-1}]$$

such that

$$\operatorname{ch}^\bullet \mathsf{N} = \sum_{\lambda \in \Lambda} p_{\mathsf{N},\mathsf{L}(\lambda)} \operatorname{ch}^\bullet \mathsf{L}(\lambda) \quad \text{and} \quad \operatorname{ch}^\bullet \mathsf{P} = \sum_{\lambda \in \Lambda} p_{\mathsf{P},\mathsf{M}(\lambda)} \operatorname{ch}^\bullet \mathsf{M}(\lambda).$$

LEMMA 5.10. *The coefficients of $p_{\mathsf{N},\mathsf{L}(\lambda)}$ and $p_{\mathsf{P},\mathsf{M}(\lambda)}$ are given by*

$$a_{\mathsf{N},\mathsf{L}(\lambda),i} = \#\{\textit{composition factors of } \mathsf{N} \textit{ isomorphic to } \mathsf{L}(\lambda)[i]\} \quad \textit{and}$$
$$a_{\mathsf{P},\mathsf{M}(\lambda),i} = \#\{\textit{subquotients isomorphic to } \mathsf{M}(\lambda)[i] \textit{ in a standard filtration of } \mathsf{P}\}.$$

PROOF. The first equality follows by induction on the length of a composition series of $\mathsf{N}$. We use that the right hand side is $\dim \operatorname{Hom}_{{}_A\mathcal{G}}(\mathsf{P}(\lambda)[i], \mathsf{N})$ and the exactness of $\operatorname{Hom}_{{}_A\mathcal{G}}(\mathsf{P}(\lambda)[i], -)$.

For the second one, we note that a standard filtration of $\mathsf{P}$ induces a decomposition of $\mathsf{P}$ (as $B^-$-module) into the direct sum of standard modules. Since the socle of $\mathsf{M}(\lambda)$ (as $B^-$-module) is isomorphic to $\lambda_V\lambda$ by Lemma 5.7, the right hand side is $\dim \operatorname{Hom}_{{}_{B^-}\mathcal{G}}(\lambda_V\lambda[i - n_{top}], \mathsf{P})$. $\square$

The following is a graded version of the well-known BGG Reciprocity. In the general framework of §2.1 this holds if $B^+$ and $B^-$ are related via an isomorphism similar to (4.11), see [**9**, Theorem 1.3], and it is a consequence of Brauer Reciprocity.

Given a polynomial $p(t,t^{-1}) \in \mathbb{Z}[t,t^{-1}]$, we set $\overline{p(t,t^{-1})} = p(t^{-1},t)$. Since $A$ is Frobenius, for $\mu \in \Lambda$ there exist $\widehat{\mu} \in \Lambda$ and $s_{\widehat{\mu}} \in \mathbb{Z}$ such that $\mathsf{L}(\widehat{\mu})[s_{\widehat{\mu}}]$ is the socle of $\mathsf{P}(\mu)$ . Moreover, $\mathsf{P}(\mu)$ is the injective hull of $\mathsf{L}(\widehat{\mu})[s_{\widehat{\mu}}]$.

THEOREM 5.11 ([**39**, Corollary 12]). *Let $\lambda, \mu \in \Lambda$. Then*

$$p_{\mathsf{P}(\mu),\mathsf{M}(\lambda)} = t^{s_{\widehat{\mu}}}\, \overline{p_{\mathsf{M}(\lambda),\mathsf{L}(\widehat{\mu})}}.$$

*In particular, $\widehat{\mu} = \mu$, if $A$ is symmetric, and $s_{\widehat{\mu}} = 0$, if $A$ is graded symmetric. These properties hold if $A = \mathfrak{u}_{(\mathcal{D}(K),R)}(V)$ is a Drinfeld double as in §4.1.1.*

PROOF. Let $\overline{\lambda}[l_\lambda]$ be the lowest weight of $\mathsf{L}(\lambda)$. Then $\mathsf{L}(\lambda)^* \simeq \mathsf{L}(\overline{\lambda}^*)[-l_\lambda]$ as in [**39**, (26)]. We then deduce that $\mathsf{P}(\lambda)^* \simeq \mathsf{P}(\overline{\widehat{\lambda}}^*)[-l_{\widehat{\lambda}} - s_{\widehat{\lambda}}]$ as [**39**, Lemma 8 (vi)]. Notice that [**39**, Theorem 10] also holds in this context because it relies on (4.11). Therefore the equality in the statement follows by combining [**39**, Theorem 10] with the linear isomorphisms

$$\operatorname{Hom}_{{}_A\mathcal{G}}(\mathsf{P}(\mu)[i], \mathsf{M}(\lambda)) \simeq \operatorname{Hom}_{{}_A\mathcal{G}}(\mathsf{M}(\lambda)^*, \mathsf{P}(\mu)[i]^*) \quad \text{and}$$

$$\operatorname{Hom}_{{}_{B^-}\mathcal{G}}(\lambda_V\lambda[i-n_{top}], \mathsf{P}(\mu)) \simeq \operatorname{Hom}_{{}_A\mathcal{G}}(\mathsf{P}(\overline{\widehat{\lambda}}^*)[i-n_{top}-l_{\widehat{\lambda}}-s_{\widehat{\lambda}}], \mathsf{W}(\lambda_V\lambda)^*),$$

see the proof of [**39**, Corollary 11 and 12].

If $A$ is symmetric, then the head and the socle of $\mathsf{P}(\mu)$ are isomorphic as $H$-modules, that is $\widehat{\mu} = \mu$. The definition of graded symmetric is that these modules are isomorphic as graded $H$-modules, that is $s_{\widehat{\mu}} = 0$.

If $A = \mathfrak{u}_{(\mathcal{D}(K),R)}(V)$, then $A$ is symmetric by [**30**, **35**], cf. [**25**, p. 488, (3)]. Moreover, $A$ is graded symmetric by the same proof of [**39**, Lemma 3.1 (i)]. □

If we evaluate the polynomials of the statement at $t=1$, we obtain the BGG Reciprocity:

$$[\mathsf{P}(\mu) : \mathsf{M}(\lambda)] = [\mathsf{M}(\lambda) : \mathsf{L}(\widehat{\mu})]$$

for all $\lambda, \mu \in \Lambda$. The next equivalence is a direct consequence of this equality.

COROLLARY 5.12. *A standard module is projective if and only if it is simple.* □

Proposition 2.8 and Lemmas 5.7 and 5.8 can be restated for $\mathsf{W}(\lambda)$ instead of $\mathsf{M}(\lambda)$. In particular, the projective modules admit costandard filtrations and there exist polynomials $p_{\mathsf{P},\mathsf{W}(\lambda)}$ such that $\operatorname{ch}^\bullet \mathsf{P} = \sum_{\lambda\in\Lambda} p_{\mathsf{P},\mathsf{W}(\lambda)} \operatorname{ch}^\bullet \mathsf{W}(\lambda)$. Thus, the proof of the next result is equal to that of [**39**, Theorem 21] using Lemma 5.9.

THEOREM 5.13 ([**39**, Theorem 21]). *Let $\mathsf{P}$ and $\mathsf{Q}$ be projective in ${}_A\mathcal{G}$. Then*

$$\mathsf{P}\otimes\mathsf{Q} \simeq \oplus_{\lambda,\mu\in\Lambda}\, p_{\mathsf{P},\mathsf{W}(\lambda)}\, p_{\mathsf{Q},\mathsf{M}(\mu)} \operatorname{\mathsf{Ind}}(\lambda\mu).$$

□

**5.3. Rigid modules.** In the general setting of §2, a simple module satisfying $\mathsf{L}(\lambda) \simeq \lambda$ is called *rigid* [**9**, §3.8]. These modules play an important role in the context of rational Cherednik algebras [**8**]. If $A$ is a Drinfeld double, the one-dimensional rigid modules are characterized by [**35**, Proposition 10] since these are the group-like elements of $A^*$.

PROPOSITION 5.14. *Let $\mathfrak{u}_{(\mathcal{D}(K),R)}(V)$ be as in §4.1.1 and $\lambda$ a one-dimensional weight of $\mathcal{D}(K)$. Then $\lambda$ is a rigid $\mathfrak{u}_{(\mathcal{D}(K),R)}(V)$-module if and only if there exist group-like elements $g \in G(K)$ and $\eta \in G(K^*)$ such that $\lambda = \eta \otimes g$ and $g \otimes \eta$ is a central element of $\mathcal{D}(K)$ acting trivially on $V$.*

*In particular, if $K$ is the group algebra of an abelian group, these are all the rigid modules of $\mathfrak{u}_{(\mathcal{D}(K),R)}(V)$.*

PROOF. As we said before, $\lambda$ is rigid if and only if $\lambda$ is a group-like element of $(\mathfrak{u}_{(\mathcal{D}(K),R)}(V))^* \simeq (\mathcal{D}(\mathfrak{B}(V)\#K))^*$. By [**35**, Proposition 10], this is equivalent to the existence of $g \in G(\mathfrak{B}(V)\#K)$ and $\eta \in G((\mathfrak{B}(V)\#K)^*)$ with $\lambda = \eta \otimes g$ and $g \otimes \eta$ in the center of $\mathfrak{u}_{(\mathcal{D}(K),R)}(V)$. We first note that $G(\mathfrak{B}(V)\#K) = G(K)$ and $G((\mathfrak{B}(V)\#K)^*) = G(K^*)$. On the other hand, $g \otimes \eta$ is central if and only if it is central in $\mathcal{D}(K)$ and acts trivially on $V$ and on $\overline{V}$. Since $g \otimes \eta$ is a group-like element in $\mathfrak{u}_{(\mathcal{D}(K),R)}(V)$, $g \otimes \eta$ acts trivially on $V$ if and only if it does on $\overline{V}$. Hence the proposition follows. For the last part, we note that $\mathcal{D}(K)$ is the group algebra of an abelian group if so is $K$, and consequently all its weights are one-dimensional. □

## Acknowedgments

I thank Nicolás Andruskiewitsch and Dmitri Nikshych for the invitation to participate in the special session "Hopf algebras and tensor categories" at the Mathematical Congress of the Americas 2017. This work was carried out during a research stay at the Université Clermont Auvergne supported by CONICET. I thank Simon Riche for his warm hospitality, for the interesting discussions on character formulas and for his willingness to answer all my questions. I also thank Ulrich Thiel for the discussion on [**9**]. I am grateful to the referee and Gastón García for the careful reading of the manuscript and their useful comments.

## References

[1] H. H. Andersen, J. C. Jantzen, and W. Soergel, *Representations of quantum groups at a pth root of unity and of semisimple groups in characteristic p: independence of p* (English, with English and French summaries), Astérisque **220** (1994), 321. MR1272539

[2] N. Andruskiewitsch and I. Angiono. *On Nichols algebras over basic Hopf algebras.* `arXiv:1802.00316v2` (2018).

[3] Nicolás Andruskiewitsch, Iván Angiono, Adriana Mejía, and Carolina Renz, *Simple modules of the quantum double of the Nichols algebra of unidentified diagonal type* $\mathfrak{ufo}(7)$, Comm. Algebra **46** (2018), no. 4, 1770–1798, DOI 10.1080/00927872.2017.1357726. MR3780543

[4] Nicolás Andruskiewitsch and Matías Graña, *Braided Hopf algebras over non-abelian finite groups* (English, with English and Spanish summaries), Bol. Acad. Nac. Cienc. (Córdoba) **63** (1999), 45–78. Colloquium on Operator Algebras and Quantum Groups (Spanish) (Vaquerías, 1997). MR1714540

[5] Nicolás Andruskiewitsch, István Heckenberger, and Hans-Jürgen Schneider, *The Nichols algebra of a semisimple Yetter-Drinfeld module*, Amer. J. Math. **132** (2010), no. 6, 1493–1547. MR2766176

[6] Nicolás Andruskiewitsch, David Radford, and Hans-Jürgen Schneider, *Complete reducibility theorems for modules over pointed Hopf algebras*, J. Algebra **324** (2010), no. 11, 2932–2970, DOI 10.1016/j.jalgebra.2010.06.002. MR2732981

[7] Nicolás Andruskiewitsch and Hans-Jürgen Schneider, *Pointed Hopf algebras*, New directions in Hopf algebras, Math. Sci. Res. Inst. Publ., vol. 43, Cambridge Univ. Press, Cambridge, 2002, pp. 1–68, DOI 10.2977/prims/1199403805. MR1913436

[8] Gwyn Bellamy and Ulrich Thiel, *Cuspidal Calogero-Moser and Lusztig families for Coxeter groups*, J. Algebra **462** (2016), 197–252, DOI 10.1016/j.jalgebra.2016.06.003. MR3519506

[9] Gwyn Bellamy and Ulrich Thiel, *Highest weight theory for finite-dimensional graded algebras with triangular decomposition*, Adv. Math. **330** (2018), 361–419, DOI 10.1016/j.aim.2018.03.011. MR3787549

[10] Georgia Benkart and Sarah Witherspoon, *Representations of two-parameter quantum groups and Schur-Weyl duality*, Hopf algebras, Lecture Notes in Pure and Appl. Math., vol. 237, Dekker, New York, 2004, pp. 65–92. MR2051731

[11] Nantel Bergeron, Yun Gao, and Naihong Hu, *Representations of two-parameter quantum orthogonal and symplectic groups*, Proceedings of the International Conference on Complex Geometry and Related Fields, AMS/IP Stud. Adv. Math., vol. 39, Amer. Math. Soc., Providence, RI, 2007, pp. 1–21. MR2338616

[12] Cédric Bonnafé and Raphaël Rouquier, *Coxeter orbits and modular representations*, Nagoya Math. J. **183** (2006), 1–34, DOI 10.1017/S0027763000009259. MR2253885

[13] Hui-Xiang Chen, *Irreducible representations of a class of quantum doubles*, J. Algebra **225** (2000), no. 1, 391–409, DOI 10.1006/jabr.1999.8135. MR1743667

[14] E. Cline, B. Parshall, and L. Scott, *Finite-dimensional algebras and highest weight categories*, J. Reine Angew. Math. **391** (1988), 85–99. MR961165

[15] Robert Gordon and Edward L. Green, *Graded Artin algebras*, J. Algebra **76** (1982), no. 1, 111–137, DOI 10.1016/0021-8693(82)90240-X. MR659212

[16] Matías Graña, *A freeness theorem for Nichols algebras*, J. Algebra **231** (2000), no. 1, 235–257, DOI 10.1006/jabr.2000.8363. MR1779599

[17] I. Heckenberger, *Lusztig isomorphisms for Drinfel'd doubles of bosonizations of Nichols algebras of diagonal type*, J. Algebra **323** (2010), no. 8, 2130–2182, DOI 10.1016/j.jalgebra.2010.02.013. MR2596372

[18] I. Heckenberger and H. Yamane, *Drinfel'd doubles and Shapovalov determinants*, Rev. Un. Mat. Argentina **51** (2010), no. 2, 107–146. MR2840165

[19] Randall R. Holmes and Daniel K. Nakano, *Brauer-type reciprocity for a class of graded associative algebras*, J. Algebra **144** (1991), no. 1, 117–126, DOI 10.1016/0021-8693(91)90132-R. MR1136900

[20] Jens Carsten Jantzen, *Character formulae from Hermann Weyl to the present*, Groups and analysis, London Math. Soc. Lecture Note Ser., vol. 354, Cambridge Univ. Press, Cambridge, 2008, pp. 232–270, DOI 10.1017/CBO9780511721410.012. MR2528469

[21] Leonid Krop and David E. Radford, *Simple modules for the Drinfel'd double of a class of Hopf algebras*, Groups, rings and algebras, Contemp. Math., vol. 420, Amer. Math. Soc., Providence, RI, 2006, pp. 229–235, DOI 10.1090/conm/420/07978. MR2279242

[22] Leonid Krop and David E. Radford, *Representations of pointed Hopf algebras and their Drinfel'd quantum doubles*, J. Algebra **321** (2009), no. 9, 2567–2603, DOI 10.1016/j.jalgebra.2009.02.011. MR2504492

[23] Robert Laugwitz, *Pointed Hopf algebras with triangular decomposition*, Algebr. Represent. Theory **19** (2016), no. 3, 547–578, DOI 10.1007/s10468-015-9588-x. MR3503231

[24] Simon Lentner and Jan Priel, *A decomposition of the Brauer-Picard group of the representation category of a finite group*, J. Algebra **489** (2017), 264–309, DOI 10.1016/j.jalgebra.2017.05.041. MR3686980

[25] Martin Lorenz, *Representations of finite-dimensional Hopf algebras*, J. Algebra **188** (1997), no. 2, 476–505, DOI 10.1006/jabr.1996.6827. MR1435369

[26] G. Lusztig, *Quantum deformations of certain simple modules over enveloping algebras*, Adv. in Math. **70** (1988), no. 2, 237–249, DOI 10.1016/0001-8708(88)90056-4. MR954661

[27] Shahn Majid, *Foundations of quantum group theory*, Cambridge University Press, Cambridge, 1995. MR1381692

[28] S. Majid, *Double-bosonization of braided groups and the construction of $U_q(\mathfrak{g})$*, Math. Proc. Cambridge Philos. Soc. **125** (1999), no. 1, 151–192, DOI 10.1017/S0305004198002576. MR1645545

[29] Dmitri Nikshych and Brianna Riepel, *Categorical Lagrangian Grassmannians and Brauer-Picard groups of pointed fusion categories*, J. Algebra **411** (2014), 191–214, DOI 10.1016/j.jalgebra.2014.04.013. MR3210925

[30] Ulrich Oberst and Hans-Jürgen Schneider, *Über Untergruppen endlicher algebraischer Gruppen* (German, with English summary), Manuscripta Math. **8** (1973), 217–241, DOI 10.1007/BF01297688. MR0347838

[31] B. Parshall and L. Scott. *Derived categories, quasi-hereditary algebras and algebraic groups.* Lect. at Ottawa-Moosonee Workshop in Alg., 1987.

[32] Yufeng Pei, Naihong Hu, and Marc Rosso, *Multi-parameter quantum groups and quantum shuffles. I*, Quantum affine algebras, extended affine Lie algebras, and their applications, Contemp. Math., vol. 506, Amer. Math. Soc., Providence, RI, 2010, pp. 145–171, DOI 10.1090/conm/506/09939. MR2642565

[33] Barbara Pogorelsky and Cristian Vay, *Verma and simple modules for quantum groups at non-abelian groups*, Adv. Math. **301** (2016), 423–457, DOI 10.1016/j.aim.2016.06.019. MR3539380

[34] B. Pogorelsky and C. Vay. *On the representation theory of the Drinfeld double of the Fomin-Kirillov algebra $\mathcal{FK}_3$.* arXiv:1707.02091 (2017) DOI 10.1007/s10468-018-9826-0.

[35] David E. Radford, *Minimal quasitriangular Hopf algebras*, J. Algebra **157** (1993), no. 2, 285–315, DOI 10.1006/jabr.1993.1102. MR1220770

[36] David E. Radford and Hans-Jürgen Schneider, *On the simple representations of generalized quantum groups and quantum doubles*, J. Algebra **319** (2008), no. 9, 3689–3731, DOI 10.1016/j.jalgebra.2007.11.037. MR2407847

[37] U. Thiel, *Champ: a Cherednik algebra Magma package*, LMS J. Comput. Math. **18** (2015), no. 1, 266–307, DOI 10.1112/S1461157015000054. MR3361642

[38] Ulrich Thiel, *Restricted rational Cherednik algebras*, Representation theory—current trends and perspectives, EMS Ser. Congr. Rep., Eur. Math. Soc., Zürich, 2017, pp. 681–745. MR3644810

[39] C. Vay. *On projective modules over finite quantum groups.* Transf. Groups. (2017) DOI 10.1007/s00031-017-9469-y.

Universidad Nacional de Córdoba, Facultad de Matemática, Astronomía, Física y Computación, CIEM–CONICET, Córdoba, República Argentina

*Email address*: vay@famaf.unc.edu.ar